# Functional Condensation Polymers

# Functional Condensation Polymers

**Edited by**


**Charles E. Carraher, Jr.**

*Florida Atlantic University*
*Boca Raton, Florida and*
*Florida Center for Environmental Studies*
*Palm Beach Gardens, Florida*

**Graham G. Swift**

*G.S.P.C., Inc.*
*Chapel Hill, North Carolina*



Springer Science+Business Media, LLC

Library of Congress Cataloging-in-Publication Data

Functional condensation polymers/edited by Charles E. Carraher, Jr., Graham G. Swift.
p. cm.
Includes bibliographical references and index.
ISBN 978-1-4757-8716-0 ISBN 978-0-306-47563-4 (eBook)
DOI 10.1007/978-0-306-47563-4
1. Polycondensation. 2. Condensation products (Chemistry) I. Carraher, Charles E. II. Swift, Graham, 1939– III. American Chemical Society. Division of Polymeric Materials: Science and Engineering. IV. American Chemical Society. Meeting (221st: 2001: San Diego, Calif.)

QD281.P6 F76 2002
547′.28—dc21

2002019614

ISBN 978-1-4757-8716-0


Originally published by Kluwer Academic/Plenum Publishers, New York in 2002
Softcover reprint of the hardcover 1st edition 2002

http://www.wkap.nl

10 9 8 7 6 5 4 3 2 1

A C.I.P. record for this book is available from the Library of Congress

# Contributors

**Kumudi Abey**, Florida Atlantic University, Boca Raton, Florida

**Stephen Andrasik**, University of Central Florida, Orlando, Florida

**R. Scott Armentrout**, Eastman Chemical Company, Kingsport, Tennessee

**Grant D. Barber**, University of Southern Mississippi, Hattiesburg, Mississippi

**T. Beck**, Pharmacia Corporation, Chesterfield, Missouri

**Kevin D. Belfield**, University of Central Florida, Orlando, Florida

**Carl E. Bonner**, Norfolk State University, Norfolk, Virginia

**K. Botwin**, Pharmacia Corporation, Chesterfield, Missouri

**Timothy L. Boykin**, Bayer Corporation, Pittsburgh, Pennsylvania

**Charles E. Carraher, Jr.**, Florida Atlantic University, Boca Raton, Florida and Florida Center for Environmental Studies, Palm Beach Gardens, Florida

**Shawn M. Carraher**, Texas A&M University, Commerce, Texas

**Donna M. Chamely**, Florida Atlantic University, Boca Raton, Florida

**Victor M. Chapela**, Beremerita Universidad Autonoma de Puebla, Puebla, Mexico

**David M. Collard**, Georgia Institute of Technology, Atlanta, Georgia

**Ann-Marie Francis**, Florida Atlantic University, Boca Raton, Florida

**Holger Frey**, Albert-Ludwigs Universität, Freiburg, Germany

**Sakuntala Chatterjee Ganguly**, Indian Institute of Technology, Kharagpur, India and SAKCHEM, Mowbray, Tasmania, Australia

**Jerome E. Haky**, Florida Atlantic University, Boca Raton, Florida

**Shiro Hamamoto**, Toyobo Research Center Company, Ohtsu, Japan

**Mason K. Harrup**, Idaho National Engineering and Environmental Laboratory, Idaho Falls, Idaho

**James Helmy**, Florida Atlantic University, Boca Raton, Florida

**Samuel J. Huang**, University of Connecticut, Storrs, Connecticut

**R. Jansson**, Pharmacia Corporation, Chesterfield, Missouri

**Michael G. Jones**, Idaho National Engineering and Environmental Laboratory, Idaho Falls, Idaho

**Huaiying Kang**, Virginia Polytechnic Institute and State University, Blacksburg, Virginia

**Kota Kitamura**, Toyobo Research Center Company, Ohtsu, Japan

**D. Kunneman**, Pharmacia Corporation, Chesterfield, Missouri

**G. Lange**, Pharmacia Corporation, Chesterfield, Missouri

**Wesley W. Learned**, Flying L Ranch, Billings, Oklahoma

**Stephen C. Lee**, Pharmacia Corporation, Chesterfield, Missouri and Department of Chemical Engineering and the Biomedical Engineering Center, Ohio State University, Columbus, Ohio

**Timothy E. Long**, Virginia Polytechnic Institute and State University, Blacksburg, Virginia

**Shahin Maaref**, Norfolk State University, Norfolk, Virginia

**Joseph M. Mabry**, University of Southern California, Los Angeles, California

**T. Miller**, Pharmacia Corporation, Chesterfield, Missouri

**Robert B. Moore**, University of Southern Mississippi, Hattiesburg, Mississippi

**Alma R. Morales**, University of Central Florida, Orlando, Florida

**Rolf Mulhaupt**, Albert-Ludwigs Universität, Freiburg, Germany

**David Nagy**, Florida Atlantic University, Boca Raton, Florida

**Junko Nakao**, Toyobo Research Center Company, Ohtsu, Japan

**Rei Nishio**, Teijin Ltd., Iwakuni, Yamaguchi, Japan

**R. Parthasarathy**, Pharmacia Corporation, Chesterfield, Missouri

**Zhonghua Peng**, University of Missouri-Kansas City, Kansas City Missouri

**Judith Percino**, Benemerita Universidad Autonoma de Puebla, Puebla, Mexico

**Fred Pflueger**, Florida Atlantic University, Boca Raton, Florida

**Dirk Poppe**, Albert-Ludwigs Universität, Freiburg, Germany

**Monica Ramos**, University of Connecticut, Storrs, Connecticut

**Alberto Rivalta**, Florida Atlantic University, Boca Raton, Florida

**John R. Ross**, Florida Atlantic University, Boca Raton, Florida

**E. Rowold**, Pharmacia Corporation, Boca Raton, Florida

**Jiro Sadanobu**, Teijin Ltd., Iwakuni, Yamaguchi, Japan

**Yoshimitsu Sakaguchi**, Toyobo Research Center Company, Ohtsu, Japan

**Alicia R. Salamone**, Florida Atlantic University, Boca Raton, Florida

**Katherine J. Schafer**, University of Central Florida, Orlando, Florida

**David A. Schiraldi**, Next Generation Polymer Research, Spartanburg, South Carolina

**Jianmin Shi**, Eastman Kodak, Rochester, New York

**Deborah W. Siegmann-Louda**, Florida Atlantic University, Boca Raton, Florida

**Robin E. Southward**, College of William and Mary, Williamsburg, Virginia

**Herbert Stewart**, Florida Atlantic University, Boca Raton, Florida

**Sam-Shajing Sun**, Norfolk State University, Norfolk, Virginia

**Hiroshi Tachimori**, Toyoba Research Center Company, Ohtsu, Japan

**Satoshi Takase**, Toyoba Research Center Company, Ohtsu, Japan

**D. Scott Thompson**, College of William and Mary, Williamsburg, Virginia

**D. W. Thompson**, College of William and Mary, Williamsburg, Virginia

**C. F. Voliva**, Pharmacia Corporation, Chesterfield, Missouri

**Jianli Wang**, Virginia Polytechnic Institute and State University, Blacksburg, Virginia

**William P. Weber**, University of Southern California, Los Angeles, California

**Alan Wertsching**, Idaho National Engineering and Environmental Laboratory, Idaho Falls, Idaho

**Ozlem Yavuz**, University of Central Florida, Orlando, Florida

**Torsten Zerfaß**, Albert-Ludwigs Universität, Freiburg, Germany

**Shiying Zheng**, Eastman Kodak, Rochester New York

**J. Zobell**, Pharmacia Corporation, Chesterfield, Missouri

# Preface

Most synthetic and natural polymers can be divided according to whether they are condensation or vinyl polymers. While much publicity has focused on funtionalized vinyl polymers, little has been done to bring together material dealing with functionalized condensation polymers. Yet, functionalized condensation polymers form an ever increasingly important, but diverse, group of materials that are important in our search for new materials for the 21st century. They form a major part of the important basis for the new and explosive nanotechnology, drug delivery systems, specific multisite catalysts, communication technology, etc.

For synthetic polymers, on a bulk basis, vinyl polymers are present in about a two to three times basis. By comparison, in nature, the vast majority of polymers are of the condensation variety.

Functionalized or functional condensation polymers are condensation polymers that contain functional groups that are either present prior to polymer formation, introduced during polymerization, or introduced subsequent to the formation of the polymer. The polymers can be linear, branched, hyper-branched, dendritic, etc. They are important reagents in the formation of ordered polymer assemblies and new architectural dendritic-like materials.

Condensation polymers offer advantages not offered by vinyl polymers including offering different kinds of binding sites; the potential for easy biodegradability; offering different reactivities undergoing reaction with different reagents under different reaction conditions; offering better tailoring of end-products; offering different tendencies (such as fiber formation); and offering different physical and chemical properties.

This book is based, in part, on an international symposium given in April 2001 as part of the national American Chemical Society meeting in San Diego, California, which was sponsored by the Division of Polymeric Materials: Science and Engineering. About forty presentations were made at the meeting.

Sample areas emphasized included dendrimers, control release of drugs, nanostructural materials, controlled biomedical recognition, and controllable electrolyte and electrical properties.

Of these presentations, about half were chosen to be included in this volume. Areas chosen for this book are those where functional condensation polymers play an especially critical role. These are nanomaterials, light and energy, bioactivity and biomaterials, and enhanced physical properties.

The book is not comprehensive, but illustrative, with the authors selected to reflect the broadness and wealth of materials that are functional condensation polymers in the areas chosen for emphasis in this book. The authors were encouraged to place their particular contribution in perspective and to make predictions of where their particular area is going.

# Contents

## A. Nano Materials

## C. Bioactivity and Biomaterials

# Functional Condensation Polymers

# A. Nano Materials

Chapter 1

# LANTHANIDE(III) OXIDE NANOCOMPOSITES WITH HEXAFLUOROISOPROPYLIDINE-BASED POLYIMIDES

D. Scott Thompson[1*], D. W. Thompson[1], and Robin E. Southward[2*]
*[1]College of William and Mary, Department of Chemistry, Williamsburg, VA 23197;*
*[2]Structures and Materials Competency, NASA Langley Research Center, Hampton, VA 23681.*
**Corresponding authors.*

## 1. INTRODUCTION

### 1.1 Hexafluoroisopropylidene-containing polyimides

In the mid-1960's Coe (1) and Rogers (2) developed the synthetic route to 2,2-bis(3,4-dicarboxyphenyl)hexafluoropropane dianhydride (6FDA) for use in the preparation of hexafluoroisopropylidene-containing aromatic polyimides. Rogers (2,3) reported the synthesis of 6FDA-based polyimides with diamines including 2,2-bis(4-aminophenyl)hexafluoropropane (4,4´-6F), 4,4´-oxydianiline (ODA), and 1,3-bis(4-aminophen-oxy)benzene (1,3(4)-APB). Early interest in 6F-containing polyimides appears to have centered on the fact that the flexible, non-polarizable, and spatially bulky isopropylidene group lowers the effective symmetry of the dianhydride unit due to the availability

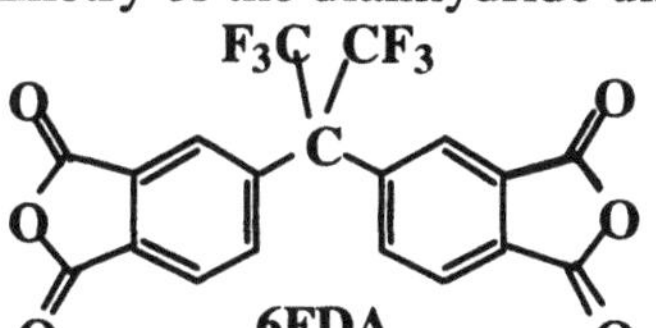
6FDA

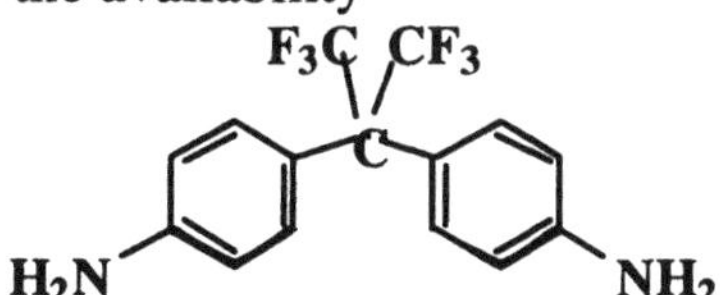
4,4´-6F

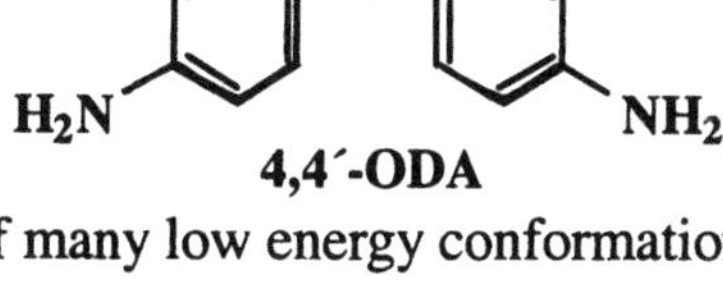
4,4´-ODA

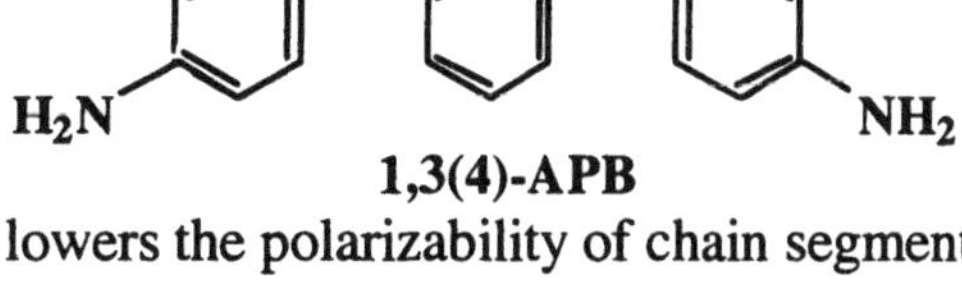
1,3(4)-APB

of many low energy conformations, lowers the polarizability of chain segments, and increases steric constraints between chains. Such properties inhibit non-covalent intermolecular interactions, chain ordering, and crystallinity, and thus yield melt-fusible high-performance polyimides with good solubility and toughness while maintaining the thermal-oxidative stability of traditional aromatic polyimides. It was also noted (2) early that 6FDA-based polyimides were less colored than traditional polyimides such as Kapton (pyromelletic dianhydride - PMDA/ODA). Extending work with 6F-containing monomers, Jones et al. (4-7) in 1975 synthesized 2,2-bis[4-(4-aminophenoxy)-phenyl]hexafluoropropane (4-BDAF) and prepared polyimides of this diamine, including 6FDA/4-BDAF.

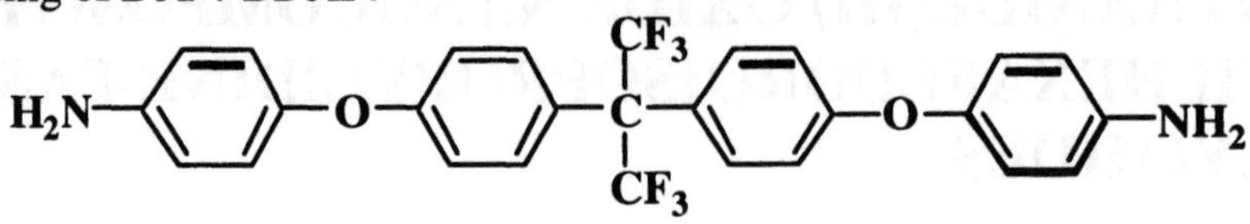

**4-BDAF**

Fluorinated aromatic polyimides with flexible 6F segments have been described by Sasaki and Nishi as "first generation" fluorinated polyimides. (8) The presence of 6F groups, trifluoromethyl groups, and other fluorine-containing entities in polyimide backbone relative to non-fluorinated polyimides such as PMDA/ODA leads to attractive properties including low moisture absorptivity, low dielectric constant, relatively low melt viscosity, resistance to wear and abrasion, low refractive index, and enhanced solubility of the imide form of the polymer. However, uses of first generation fluorinated polyimides have been limited due to a combination of low glass transition temperatures (Tg), high coefficients of thermal expansion (CTE), low adhesive strength, and solvent sensitivity. The synthesis of second generation fluorinated polyimides (8) has focused on developing systems which would be useful in electronic and optoelectronic applications. These new materials would retain the beneficial properties of first generation polyimides but would possess higher Tgs, low CTEs, and tunable low refractive indices.

Extending the earlier patented work of others on polyimides formed from 6FDA, 4-BDAF, and closely related molecules, St. Clair et al. (9-11) reported the synthesis of nine 6F-containing polyimides from purified monomers. Five polyimides were designated as "colorless" with ultraviolet wavelength cutoffs between 310-370 nm at film thicknesses of 5 microns. The motivation for pursuing transparent polyimides came from the need for optically clear thin films which can endure for long periods in space environments. Two of these "colorless" polyimides are prepared from 6FDA with 4-BDAF and 1,3(3)-APB and are prototypical first generation fluorinated polyimides. 6FDA/4-BDAF and 6FDA/1,3(3)-APB have excellent transparency in the visible region of the electromagnetic spectrum, low dielectric constant, low moisture absorptivity, excellent thermal-oxidative stability, resistance to ultraviolet and 1 MeV electron radiation in nitrogen and in vacuum, and reasonable mechanical properties. However, they have been excluded from many applications because of several marginal properties including low Tgs, high CTEs, extreme solvent sensitivity, low tear resistance, and high cost for all but specialty applications.

## 1.2 Potential applications of fluorinated polyimides

There are at least two important areas in which fluorinated polyimides might have a role. First is the area of space materials involving large-area solar collectors, inflatable antennas, solar arrays, and various space optical devices. Secondly, use of aromatic polyimides for electronic applications continues to foster the development of modified polyimides that have appropriate thermal and mechanical properties while meeting the demands of low dielectric constant and low moisture absorptivity. 6F-containing polyimides often offer these properties. (12-16) However, the electronic and steric features of organofluorine groups elevate the CTE. Mismatch of CTEs in the fabrication and application of lamellar and composite electronic devices can lead to cracking, peeling, warping, and the severing of electrical contacts across polymer dielectric layers.

## 1.3 Oxo-metal-polyimide composites

There is substantial interest in the fabrication of composite materials comprised of an organic polymer throughout which nanometer-sized inorganic particles (e.g., silica, two-dimensional montmorillonite silicate sheets, titania, single-wall carbon nanotubes, etc.) are homogeneously dispersed at low weight percents (ca. 2-10%). The most intensely studied inorganic oxide phases are silica and two-dimensional organically modified smectite clays (silicates), particularly montmorillonites. The supposition is that nanometer-based hybrid materials will differ significantly from traditional “filled” polymers, for which the "filler" particle sizes are much larger (>1000 nm), due to the high effective surface area of inorganic oxide nanoparticles and subsequently magnified polymer-inorganic phase interactions leading to enhanced polymer properties at relatively low concentrations of the inorganic oxo-phase.

Currently, the most vigorously pursued oxo-polymer nanocomposites are those containing single (exfoliated) two-dimensional silicate sheets such as the sodium cation type montmorillonite, hectrite, saponite, and synthetic mica. (17-45) Naturally occurring silicate sheet minerals are layered structures with cations in the galleries and are not exfoliated (delaminated) when incorporated into organic polymers due to the intrinsic incompatability of the hydrophilic silicate sheets and the hydrophobic polymers. This exfoliation problem was resolved by the Toyota group in the latter 1980's who found that exchanging the inorganic gallery cations of the layered silicates with large alkyl ammonium cations such as the dodecylammonium ion gave silicate-polymer composites with widely dispersed single silicate sheets. In their seminal work they reported exfoliated montmorillonite-Nylon 6 (17-20) and PMDA/ODA (21,22) nanocomposite materials with ca. 2-5 wt% of the organically modified clay. The Nylon 6 composites exhibited enhanced strength, modulus, and heat distortion temperatures, ca. 100 °C above the parent polyamide. Exfoliated montmorillonite-polyimide composite (2 wt%) films were obtained with increased moduli, decreased CTEs, and markedly decreased gas permeability coefficients. It is generally assumed that both the large surface area and high aspect ratios (ca. 200:1 for montmorillonites) of the silicate sheets are important to the enhancement of polymer properties. (22) Further studies on organically modified montmorillonite-polyimide composites have tended to corroborate the Toyota work. However, more recent work has also revealed that it is more difficult to achieve complete exfoliation of silicate sheets in polyimides than suggested in early work (23,24,25). The extent of cation exchange, the structure of the polyimide, the composition of the organic cation, the order and form of reagent addition, mechanical shearing of the clays, and other considerations play a role in the extent of delamination and dispersion of the silicate sheets. However, even in systems without full exfoliation there are significant property enhancements and modifications with polyimides. Property enhancements include: decreased CTEs (21,22,26-29), decreased gas permeability (5,6,24,30), increased modulus (22,23,26-29), increased resistance to ablative combustion gases (31), decreased solvent uptake and solubility (32), decreased flammibility (33), decreased water absorption (26), decreased imidization temperatures (34), and increased thermal degradation stability. (23,28,29,32,35) For other properties trends are less clear: tensile strengths (23,26,27,29), percent elongation (23,26,27,29), and glass transition temperatures (23,28,29,31,34,35) varied among systems with both increases and decreases of physical properties being observed. Tensile strengths and glass transition temperatures were usually found to increase. Trends similar to those observed with two-dimensional montmorillonites have been observed with three-dimensional silica particles in polyimides formed in situ via the sol-gel hydrolysis of varied silicon alkoxides. (36-45) However, generally the property enhancements observed with silica are significantly less pronounced at low weight percents. In this paper we now report attempts to see if similar property effects can be accomplished through the incorporation of nanometer-sized lanthanum(III) oxide particles.

## 1.4 Research focus of this paper

Traditional polyimides exhibit CTEs in the range of 30-45 ppm/K (46) and have excellent solvent resistance. Typically, metals and inorganic materials such as silicon, quartz, silicon carbide, alumina, and other metal oxides and ceramics have CTEs less than 20 ppm/K. However, polyimides derived from 6FDA have CTEs of 50-60 ppm/K. (13) Since 6FDA/4-BDAF and 6FDA/1,3(3)-APB are easily prepared from readily accessible monomers, herein we report research directed at lowering CTEs of these two colorless polyimides in a controlled manner via the in situ formation of oxo-lanthanide(III)-polyimide nanocomposite materials with low concentrations of the inorganic oxide phase. The oxo-metal(III) phases arise from the hydrolysis and thermal transformation of tris(2,4-pentanedionato)lanthanide(III) complexes which are dissolved initially in a solution of the polyimide. We also report the effects of oxo-metal(III) formation on other selected properties and compare these effects with those seen in montmorillonite-polyimide composites.

# 2. EXPERIMENTAL

## 2.1 Materials

2,2-Bis(3,4-dicarboxypheny)hexafluropropane dianhydride was obtained from Hoechst Celanese and vacuum dried for 17 h at 110 °C prior to use. 1,3-Bis(3-aminophenoxy)benzene (1,3(3)-APB) was purchased from National Starch and 2,2-bis[4-(4-aminophenoxy)phenyl]-hexafluoropropane (4-BDAF)was purchased from Chriskev; both were used as received. 2,4-Pentanedione, lanthanum(III) oxide, and gadolinium(III) oxide were obtained from Fisher, Aldrich, and Alfa-Aesar, respectively. Tris(2,4-pentanedionato)holmium(III) was purchased from REacton as an unspecified hydrate. Thermal gravimetric analysis indicated three water molecules per holmium atom which is consistent with early literature and a recent X-ray crystal structure of tris(2,4-pentanedionato)holmium(III) trihydrate by Kooijman et al. (47) showing the structure to be diaquotris(2,4-pentanedionato)holmium-(III) monohydrate ; we subsequently assumed a trihydrate in the preparation of all films. Holmium(III) acetate tetrahydrate was obtained from Rare Earth Products Limited. All other holmium compounds purchased were at a minimum purity of 99.9%. Other tris(2,4-pentanedionato)-lanthanum(III) complexes were obtained from Alfa-Aesar and used as trihydrates. Dimethylacetamide, DMAc, (HPLC grade) and bis(2-methoxyethyl) ether, diglyme, (anhydrous 99.5 %) were obtained from Aldrich and were used without further purification.

## 2.2 Preparation of diqauotris(2,4-pentanedionato)lanthanum-(III) and diaquotris(2,4-pentadionato)gadolinium(III) monohydrate

Diqauotris(2,4-pentanedionato)lanthanum(III) was made as reported earlier (48) following the recipe of Phillips, Sands, and Wagner (49) who verified the structure by single crystal X-ray analysis. The gadolinium complex was prepared in a manner similar to its lanthanum congener and consistent with the latter procedure of Kooijman et al. (47) who determined the structure to be the same as that for the lanthanum analog but with a molecule of lattice water per gadolinium atom. The resulting crystalline complex was dried at 22 °C in air and used as the trihydrate.

### 2.3 Preparation of the polyimides

Imidized 6FDA/1,3(3)-APB powder was obtained by the addition of 6FDA (0.5% molar excess) to a DMAc solution of 1,3(3)-APB to first prepare the poly(amic acid) at 15% (w/w) solids. The reaction mixture was stirred at the ambient temperature for 7 h. The inherent viscosity of the poly(amic acid) was 1.4 dL/g at 35 °C. This amic acid precursor was chemically imidized at room temperature in an equal molar ratio acetic anhydride-pyridine solution, the pyridine and acetic anhydride each being three times the moles of diamine monomer. The polyimide was then precipitated in water, washed thoroughly with deionized water, and vacuum dried at 200 °C for 20 h after which no odor of any solvent was detectable. The inherent viscosity of the polyimide in DMAc was 0.81 dL/g at 35 °C. $M_n$ and $M_w$ were determined to be 86,000 and 289,000 g/mol by GPC, respectively. Imidized 6FDA/4-BDAF powder was prepared similarly a with a 1 mole percent dianhydride offset. The inherent viscosity of the imide was 1.55 dL/g at 35 °C. GPC gave $M_n$ at 86,000 g/mol and $M_w$ at 268,000 g/mol.

### 2.4 Preparation and characterization of oxo-lanthanum-polyimide composite films

All metal-doped imidized polymer solutions were prepared by first dissolving the metal complex in DMAC and then adding solid imide powder to give a 15% solids (excluding the additives) solution. The solutions were stirred 2-4 h to dissolve all of the polyimide. The clear metal-doped resins were cast as films onto soda lime glass plates using a doctor blade set to give cured films near 25 microns. The films were allowed to sit for 15 h at room temperature in flowing air at 10% humidity. This resulted in a film which was tact free but still had 35% solvent by weight. The films then were cured in a forced air oven for 1 h at 100, 200, and 300 °C. For all cure cycles 30 min was used to move between temperatures at which the samples were held for 1h. The films were removed from the plate by soaking in warm deionized water.

## 3. RESULTS AND DISCUSSION

### 3.1 Film syntheses

6FDA/1,3(3)-APB and 6FDA/4-BDAF films were typically prepared at a molar ratio of polymer repeat unit to Ln(III) of 5:1; concentrations of the Ln complex greater than *ca.* 2.5:1, particularly for 6FDA/1,3(3)-APB films, gave films which fractured on handling. The composite oxo-Ln-polyimide films were prepared by dissolving the tris(2,4-pentanedionato)lanthanide(III) hydrates (i.e., eight coordinate diaquotris(2,4-pentanedionato)lanthanide(III) complexes based on the known crystal structures (47,49) of the La(III) and Gd(III) complexes), or other metal(III) compounds, in DMAc or diglyme followed by addition of the soluble imide form of the polymers. The films were cured to 300 °C. All films were visually clear. TEM data for the 5:1 Ho(III) film of Table 1 indicate oxo-metal particles which are only a few nanometers in diameter. The X-ray diffraction patterns suggest that the oxo-metal(III) phase is not crystalline. The lanthanide-2,4-pentanedionate complexes investigated with 6FDA/1,3(3)-APB and 6FDA/4-BDAF were those of La, Sm, Eu, Gd, Ho, Er, and Tm; additionally, tris(2,4-pentanedionato)aluminum(III) and tetrakis(2,4-pentanedionato) zirconium(IV) were studied to a more limited extent. A series of 6FDA/1,3(3)-APB films was prepared with holmium(III) acetate tetrahydrates and holmium(III) oxide. Holmium(III) acetate tetrahydrate was soluble in DMAc and gave clear films; holmium(III) oxide was not soluble in DMAc and gave opaque heterogeneous films. Tables 1-4 present data for the films that were prepared and characterized.

## 3.2 Film properties: linear coefficients of thermal expansion and thermal and mechanical properties

Table 1 presents CTE data for Ho, Gd, and La films. The CTE of the undoped polyimide film is 49 ppm/K. The CTE decreases regularly from 49 to 33 ppm/K as the concentration of an oxo-holmium(III) phase decreases from a 10:1 (2.6 wt% $Ho_2O_3$) polyimide repeat unit to metal ion ratio to a 2.5:1 (9.4 wt% $Ho_2O_3$) ratio. Figure 1 displays CTE trends for the Ho, La, and Gd-based 6FDA/1,3(3)-APB films. The curves were generated by an exponential fit with $r^2$ values of 0.79, 0.95, and 0.96, respectively.

There has been intense interest in preparing polymer composites containing low weight percentages (<10%) of two-dimensional delaminated nanometer-sized montmorillonite silicate sheets. Such composites have enhanced properties as discussed earlier. Included in Figure 1 is CTE data (exponential fit with $r^2 = 0.96$) for montmorillonite-PMDA/ODA films. (28) The similarity of the data among the four systems suggests that the more spherical nanometer-sized oxo-lanthanide(III) particles may influence physical properties in a manner similar to that of the clay sheets.

One concern is whether any randomly chosen holmium(III) complex, which is soluble in the polyimide-DMAc solution, would give similar CTE lowerings in the cured composite polyimide films. That is, is there anything singular about the 2,4-pentanedionate systems. Thus, 6FDA/1,3(3)APB-holmium(III) acetate tetrahydrate films were prepared and characterized. (Table 2.) Acetate-based transparent films show a minimal decrease in the CTE. Holmium(III) oxide, which is not a soluble additive but is heterogeneously dispersed as micron-sized particles in the resin, gives no lowering of the CTE. Tables 3 and 4 show CTE data for additional tris(2,4-pentanedionato)-lanthanide(III)-6FDA/1,3(3)-APB films. It is apparent that all lanthanide(III)-diketonate complexes lead to significant and similar CTE lowerings at the 5:1 concentrations. This raises the question as to whether non-lanthanide(III) 2,4-pentanedionate metal complexes would give similar film property modifications.

To address this query we prepared 6FDA/1,3(3)-APB films formed with tris(2,4-pentanedionato)aluminum and tetrakis(2,4- pentane-dionato)zirconium. These latter two additives gave minimal CTE lowerings suggesting that there is some unique chemistry attributable to the lanthanide-2,4-pentanedionate complexes. There are no property differences in films cast from DMAc and diglyme.

Consistent with our earlier observations (48), the change in the glass transition temperatures for the 6FDA/1,3(3)-APB samples is minimal at only ± 2°C. For the 6FDA/4-BDAF samples (Table 4) Tg is modestly elevated by 2-8 °C. Since Tg values for the nanocomposite films are similar to those for the parent polyimide, crosslinking interactions must be weak. Such weak interactions would be consistent with the fact that the amide and phenyl ether donors are only weak Lewis bases. David and Scherer (50) found no change in Tg of the polymer up to 20 wt % $SiO_2$, and Leezenberg and Frank (51) found that the in situ precipitation of $SiO_2$ at 20-30 wt% in poly(dimethylsiloxane) "does not affect the Tg." Thus, with the low weight percents of metal(III) used in our work and the minimal changes in Tg found with silicon-oxo phases, it is not surprising that the lanthanide(III)-hybrid films of this work show no dramatic changes in Tg. The essential constancy of Tg values also suggests that there are no metal(III) Lewis acid catalyzed covalent (C-C, C-O, or C-N) crosslinking reactions between chains, which would be expected to increase Tg dramatically as for polystryene, crosslinked with para-divinylbenzene. (52)

Table 1. Concentration dependence of CTE for tris(2,4-pentane-dionato) complexes with Ho(III), La(III), and Gd(III) - 6FDA/1,3(3)-APB films. (The La(III)-containing films were prepared in a mixed DMAc (45%)-diglyme(55%) solvent; Gd and Ho films were prepared in DMAc. All films were cured to a final temperature of 300 °C.)

| *Repeat:Ho Mole Ratio[a]* | *CTE (ppm/K)* | *Repeat:La Mole Ratio[a]* | *CTE (ppm/K)* | *Repeat:Gd Mole Ratio[a]* | *CTE (ppm/K)* |
|---|---|---|---|---|---|
| *Control (0)* | *49* | *Control (0)* | *49* | *Control (0)* | *49* |
| *10:1 (2.6)* | *45* | *16:1 (1.4)* | *45* | *10:1 (2.5)* | *41* |
| *7.5:1 (3.6)* | *37* | *10:1 (2.3)* | *43* | *7.5:1 (3.3)* | *39* |
| *5.0:1 (5.1)* | *33* | *7:1 (3.1)* | *40* | *5:1 (4.9)* | *30* |
| *2.5:1 (9.4)* | *32* | *5:1 (4.4)* | *31* | *2.5:1 (9.7)* | *24* |

a) The values in parentheses are the wt % metal(III) oxide based on only $Ln_2O_3$ remaining in the polyimide after curing.

Table 2. Thermal and mechanical data for a series of holmium(III) acetate tetrahydrate and holmium(III) oxide 6FDA/1,3(3)-3,3′-APB films. Films were prepared in DMAc and cure to a final temperature of 300 °C.

| *Holium Additive* | *Repeat to Ho Ratio* | *CTE (ppm/K)* | *$T_g$ (°C)* | *TGA[c] (°C)* | *Tensile Strength (MPa)* | *% Elongation[b]* | *Modulus (GPa)* |
|---|---|---|---|---|---|---|---|
| *Control* | *zero* | *49* | *202* | *496* | *123* | *5.15* | *3.1* |
| *$Ho(OAc)_3$* | *5:1* | *43* | *202* | *482* | *115* | *4.11* | *3.4* |
| *$Ho(OAc)_3$* | *2.5: 1* | *44* | *202* | *453* | *125* | *4.12* | *4.0* |
| *$Ho(OAc)_3$* | *1.67:1* | *47* | *202* | *448* | *a* | | |
| *$Ho_2O_3$* | *5:1* | *50* | *202* | *495* | *110* | *3.65* | *3.5* |

a) Not able to remove from the casting plate large enough samples for mechanical measurements. Film was brittle.
b) Percent elongation at break.
c) The temperature at which there is ten percent weight loss.

Table 3. Thermal and mechanical data for 6FDA/1,3(3)-APB composite films formed with selected tris(2,4-pentanedionato)-lanthanide(III) complexes, tris(2,4-pentanedionato)-aluminum(III), and tetrakis(2,4-pentanedionato)zirconium(IV). (Films were prepared in DMAc at the 5:1 repeat to metal ratio and cured to a final temperature of 300 °C.)

| *Ln-2,4-pentanedionate Additive* | *CTE (ppm/K)* | *$T_g$ (°C)* | *TGA[a] (°C)* | *Tensile Strength (MPa)* | *% Elongation* | *Modulus (GPa)* |
|---|---|---|---|---|---|---|
| *None* | *49* | *207* | *515* | *123* | *5.1* | *3.1* |
| *$La(acac)_3$* | *31* | *209* | *430* | *113* | *3.4* | *3.8* |
| *$Sm(acac)_3$* | *35* | *212* | *477* | *97* | *3.2* | *3.4* |
| *$Eu(acac)_3$* | *36* | *212* | *461* | *112* | *3.6* | *3.6* |
| *$Gd(acac)_3$* | *30* | *210* | *456* | *104* | *3.3* | *3.5* |
| *$Ho(acac)_3$* | *33* | *210* | *468* | *136* | *4.0* | *4.1* |
| *$Er(acac)_3$* | *32* | *210* | *464* | *119* | *4.2* | *3.4* |
| *$Tm(acac)_3$* | *32* | *210* | *482* | *108* | *3.7* | *3.4* |
| *$Al(acac)_3$* | *44* | *210* | *513* | *117* | *4.5* | *3.2* |
| *$Zr(acac)_4$* | *45* | *210* | *511* | *110* | *3.7* | *3.4* |

a) The temperature at which there is ten percent weight loss.

Table 4. Thermal and mechanical data for 6FDA/4-BDAF composite films formed with selected tris(2,4-pentanedionato)lanthanide(III) complexes and tris(2,4-pentanedionato)aluminum(III). (Films were prepared in DMAc at 5:1 repeat to metal ratio and cured to 300 °C.)

| *Ln- 2,4-pentanedionate Additive* | *CTE (ppm/K)* | *Tg (°C)* | *TGA (°C)* | *Tensile Strength (MPa)* | *% Elongation* | *Modulus (GPa)* |
|---|---|---|---|---|---|---|
| *None* | *51* | *266* | *515* | *93* | *6.1* | *2.3* |
| *La(acac)₃* | *39*[a] | *274* | *561* | *105* | *9.2* | *2.2* |
| *Gd(acac)₃* | *35* | *271* | *473* | *106* | *10.8* | *2.6* |
| *Ho(acac)₃* | *37* | *270* | *478* | *100* | *6.6* | *2.6* |
| *Tm(acac)₃* | *38* | *271* | *489* | *103* | *9.6* | *2.1* |
| *Al(acac)₃* | *45* | *268* | *500* | *104* | *7.7* | *2.4* |

a) CTE of 29 ppm/K is at polymer repeat unit to metal(III) ratio of 3.9:1 or 4.3 wt % $La_2O_3$.

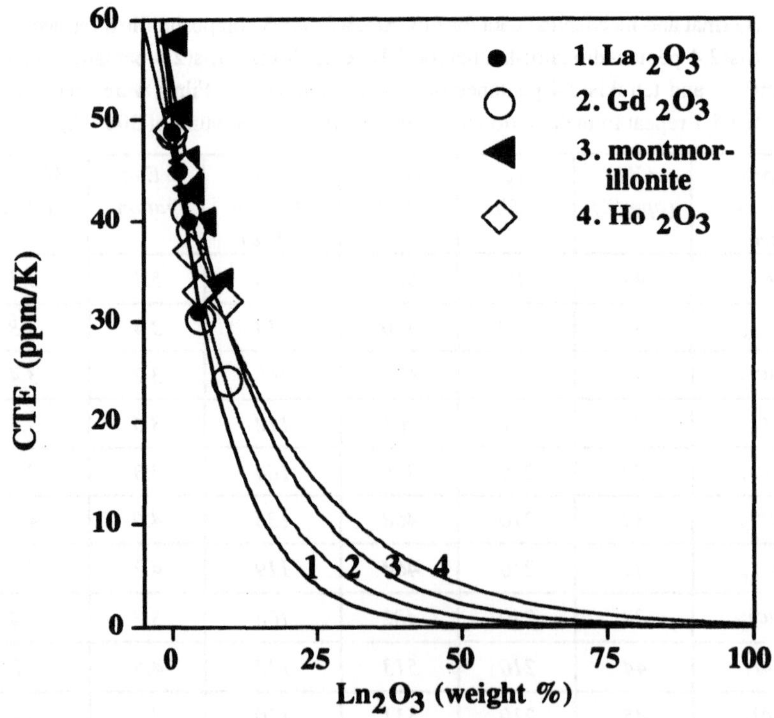


Figure 1. Linear coefficient of thermal expansion versus dopant concentration. The Ln(III) species are in 6FDA/1,3(3)-APB; the montmorillonite is in PMDA/ODA. (Curves are an exponential fit with $r^2$ for curves 1-4 being 0.90, 0.95, 0.79, and 0.96, respectively.)

The temperature at which there is 10% weight loss in air decreases regularly with concentration of the oxo-phase in the composite films. However, at a concentration of 5:1 the polyimide composites still have excellent thermal stability. It is interesting to note that the aluminum(III) and zirconium(IV) β-diketonate complexes give films with minimal CTE lowering and also only modest change in the temperature at which there is 10% weight loss in air.

## 3.3 Rationale for use of lanthanide(III)-based inorganic phases

We chose to investigate lanthanide ions because they exhibit a single stable tervalent oxidation state with crystal radii from 117 to 100 pm, La(III) through Lu(III). The large radii lead to high coordination numbers for lanthanide complexes with eight being most common. Thus, in the lanthanide series one has metal ion additives for polymers which have enlarged coordination spheres and which are hard Lewis acids. These two effects enhance binding of polymer donor atoms, particularly the weakly basic oxygens, as they might occur in imide or ether moieties of 6FDA/1,3(3)-APB and 6FDA/4-BDAF. Polymer-metal coordination during a thermal cure cycle should be of pivotal importance in preventing aggregation of metal(III) species to micron or greater-sized particles within the bulk of the polymer. Such metal-polymer coordination, or "site isolation" as referred to by Sen et al. (53, 54), has been suggested as the basis for the formation of a homogeneous distribution of nanometer-sized oxo-metal clusters throughout a polymer matrix.

## 3.4 Conclusions

The dissolution of the eight coordinate diaquotris(2,4-pentanedionato)lanthanide(III) complex species in solutions of soluble polyimides give thermally cured films with CTEs lowered to a maximum of *ca.* 40%. The CTE lowerings are much greater than those observed in 3-dimensional silica-polyimide hybrids and on the order of those observed with exfoliated 2-dimensional montmorillonite (silicate) sheets incorporated into PMDA/ODA. Also, the increase in modulus for oxo-holmium(III) 6FDA/ODA films parallels that reported for montmorillonite nanocomposites of PMDA/ODA.

**Acknowledgement.** The authors express gratitude to the Petroleum Research Fund administered by the American Chemical Society for partial support of this work.

## 4. REFERENCES

1. Coe, D. G. ; "2,2-Diarylperfluoropropanes," U. S. Patent 3,310,573, E. I. du Pont de Nemours and Co.: U. S., 1967.

2. Rogers, F. E. ; "Polyamide-acids and Polyimides from Hexafluoropropylidene Bridged Diamine," U. S. Patent 3,356,648, E. I. du Pont de Nemours and Co.: U. S., 1967.

3. Rogers, F. E. ; "Melt-Fusible Linear Polyimide of 2,2-Bis(3,4-dicarboxyphenyl)hexafluoropropane dianhydride," U. S. Patent 3,959,350, E. I. du Pont de Nemours and Co.: U. S., 1976.
4. Zakrezewski, G.; Rell, M. K. O.; Vaughan, R. W.; Jones, R. J. "Final Report Contract NAS3-17824, NASA CR-134900," , 1975.
5. Jones, R. J.; O'Rell, M. K.; Hom, J. M. ; "Polyimides Prepared from Perfluoroisopropylidene Diamines," U. S. Patent 4,111,906, TRW Inc.: U. S., 1978.
6. Jones, R. J.; O'Rell, M. K.; Hom, J. M. ; "Fluorinated Aromatic Diamines," U. S. Patent 4,203,922, TRW, Inc.: U. S., 1980.
7. Jones, R. J.; Chang, G. E.; Powell, S. H.; Green, H. E. "Polyimide Protective Coatings for 700 °F Service:" In *Polyimides: Synthesis, Characterization, and Applications*; Mittal, K. L., Ed.; Plenum: New York, 1984.
8. Sasaki, S.; Nishi, S. "Synthesis of Fluorinated Polyimdes:" In *Polyimides: Fundamentals and Applications*; Ghosh, M. K., Mittal, K. L., Eds.; Marcel Dekker: New York, 1996, pp.71-120.
9. Clair, A. K. S.; Clair, T. L. S. ; "Process for Preparing Essentially Colorless Polyimide Film Containing Phenoxy-Linked Diamines," U. S. Patent 4,595,548, NASA: U. S., 1986.
10. Clair, A. K. S.; Clair, T. L. S. ; "Process for Preparing Highly Optically Transparent/Colorless Polyimide Film," U. S. Patent 4,603,061, NASA: U. S., 1986.
11. Clair, A. K. S.; Clair, T. L. S.; Slemp, W. S. "Optically Transparent/Colorless Polyimides:" In *Adv. Polyimide Sci. Tech.*; Weber, W. D., Gupta, M. R., Eds.; Soc. of Plastic Eng., Mid-Hudson Section: Poughkeepsie, 1987, pp.16-34.
12. Auman, B. C. *Mater. Res. Soc. Symp. Proc., Low-Dielectric Constant Materials--Synthesis and Applications in Microelectronics* **1995**, *381*, 12-19.
13. a) Trofimenko, S. "A New Class of Fluorinated Rigid Monomers for Polyimides;" In *Adv. Polyimide Sci. Technol.*, F. Feger, et al., Eds., Society of Plastics Engineers: New York, 1993, pp. 3-14; b) B. C. Auman, "Low Dielectric Constant, Low MoistureAdsorption and Low CTE Polyimides Based on New Rigid Fluorinated Monomers," *ibid.* pp. 15-27.
14. Matsuura, T.; Ando, S.; Sasaki, S.; Yamamoto, F. *Macromolecules* **1994**, *27*, 6665-6670.
15. Ando, S.; Matsuura, T.; Sasaki, S. *Fluoropolymers* **1999**, *2*, 277-303.
16. Lin, S.-H. L., F.; Cheng, S. Z. D.; Harris, F. W. *Macromolecules* **1998**, *31*, 2080-2086.
17. Fukushima, Y.; Inagaki, S. *J. Inclusion Phenom.* **1987**, *5*, 473-82.
18. Kojima, Y.; Usuki, A.; Kawasumi, M.; Okada, A.; Fukushima, Y.; Kurauchi, T.; Kamigaito, O. *J. Mater. Res.* **1993**, *8*, 1185-9.
19. Fukushima, Y.; Okada, A.; Kawasumi, M.; Kurauchi, T.; Kamigaito, O. *Clay Miner.* **1988**, *23*, 27-34.
20. Kojima, Y.; Usuki, A.; Kawasumi, M.; Okada, A.; Kurauchi, T.; Kamigaito, O. *J. Polym. Sci., Part A:Polym. Chem.* **1993**, *31*, 983-6. 21. Yano, K.; Usuki, A.; Okada, A.; Kurachi, T.; Kamigaito, O. *J. Polym. Sci., Part A: Polym. Chem.* **1993**, *31*, 2493-8.
22. Yano, K.; Usuki, A.; Okada, A. *J. Polym. Sci., Part A:Polym. Chem.* **1997**, *35*, 2289-94.
23. Delozier, D. M.; Orwoll, R. A.; Cahoon, J. F.; Johnston, N. J.; Smith, J.G.; Connell, J. W. *Polymer*, in press.
24. Lan, T.; Kaviratna, P. D.; Pinnavaia, T. J. *Chem. Mater.* **1994**, *6*, 573-75.
25. Gu, A.; Kuo, S.-W.; Chang, F.-C. *J. Appl. Polym. Sci.* **2001**, *79*, 1903-10.
26. Gu, A.; Chang, F.-C. *J. Appl. Polym. Sci.* **2001**, *79*, 289-94.

27. Agag, T.; Koga, T.; Takeichi, T. *Polymer* **2001**, *42*, 3399-3408.
28. Tyan, H.-L.; Liu, Y.-C.; Wei, K.-H. *Chem. Mater.* **1999**, *11*, 1942-47.
29. Vaia, R. A.; Price, G.; Ruth, P. N.; Nguyen, H. T.; Lichtenhan, *J. Appl. Clay Sci.* **1999**, *15*, 67-92.
30. Huang, J.-C.; Zhu, Z-K.; Yin, J.; Qian, X.-F.; Sun, Y.-Y. *Polymer*, **2001**, *42*, 873-7.
31. Gilman, J. W. *Appl. Clay Sci.* **1999**, *15*, 31-49.
32. Tyan, H.-L.; Liu, Y.-C.; Wei, K.-H. *Polymer* **1999**, *40*, 4877-86.
33. Morgan, A. B.; Gilman, J. W.; Jackson, C. L. *Macromolecules*, **2001**, *34*, 2735-38.
34. LeBaron, P. C.; Wang, Z.; Pinnavaia, T. J. *Appl. Clay Sci.* **1999**, *15*, 11-29.
35. Hsiao, S.-H.; Liou, G.-S.; Chang, L.-M. *J. Appl. Polym. Sci.* **2001**, *80*, 2067-72. Polyimide/Clay Hybrids
36. Joly, C.; Smaihi, M.; Porcar, L.; Noble, R. D. *Chem. Mater.* **1999**, *11*, 2331-38.
37. Zhu, Z.-K.; Yang, Y.; Yin, J.; Qi, Z.-N. *J. Appl. Polym. Sci.* **1999**, *73*, 2977-84.
38. Huang, J.-C.; Zhu, Z.-K.; Yin, J.; Zhang, D.-M.; Qian, X.-F. *J. Appl. Polym. Sci.* **2001**, *79*, 794-800.
39. Zhu, Z.-K.; Yin, J.; Cao, F.; Shang, X.-Y.; Lu, Q.-H. *Adv. Mater.* **2000**, *12*, 1055-57.
40. Chen, Y.; Iroh, J. O.; *Chem. Mater.* **1999**, *11*, 1218-22.
41. Nunes, S. P.; Peinemann, K. V.; Ohlrogge, K.; Alpers, A.; Keller, M.; Pires, A. T. N. *J. Membr. Sci.* **1999**, *157*, 219-26.
42. Morikawa, A.; Yamaguchi, H.; Kakimoto, M.; Imai, Y. *Chem. Mater.* **1994**, *6*, 913-17.
43. Morikawa, A.; Iyoke, Y.; Kakimoto, M.; Imai, Y. *J. Mater. Chem.* **1992**, *2*, 679-690.
44. Goizet, S.; Schrotter, J.-C.; Smaihi, M.; Deratani, A. *New J. Chem.* **1997**, *21*, 461-8.
45. Ha, C.-S.; Park, H.-D.; Frank, C. W. *Chem. Mater.* **2000**, *12*, 839-844 and references therein.
46. Numata, S.; Fujisaki, K.; Makino, D.; Kinjo N. "Chemical Structures and Properties of Low Thermal Expansion Polyimides:" In *Adv. Polyimide Sci. Tech.*; Weber, W. D., Gupta, M. R., Eds.; Soc. of Plastic Eng., Mid-Hudson Section: New York, 1987, pp. 492-510.
47. Kooijman, H.; Nijsen, F.; Spek, A. L.; Schip, F. v. h. *Acta Cryst.* **2000**, *C56*, 156.
48. Southward, R. E.; Thompson, D. S.; Thompson, D. W.; Clair, A. K. S. *Chem. Mater.* **1998**, *10*, 486-494.
49. Phillips, T.; Sands, D. E.; Wagner, W. F. *Inorg. Chem.* **1968**, *7*, 2295-9.
50. David, I. A.; Scherer, G. W. *Chem. Mater.* **1995**, *7*, 1957.
51. Leezenberg, P. B.; Frank, C. W. *Chem. Mater.* **1995**, *7*, 1784.
52. Glans, J. H.; Turner, D. T. *Polymer* **1981**, *22*, 1540-3.
53. Nandi, M.; Sen, A. *Chem. Mater.* **1989**, *1*, 291.
54. Nandi, M.; Conklin, J. A.; Salvati, L.; Sen, A. *Chem. Mater.* **1990**, *2*, 772-6.

Chapter 2

# FUMARYL CHLORIDE AND MALEIC ANHYDRIDE DERIVED CROSSLINKED FUNCTIONAL POLYMERS AND NANO STRUCTURES

Sam-Shajing Sun[1,2], Shahin Maaref[1] and Carl E. Bonner[1,2]
*[1]Center for Materials Research and [2]Chemistry Department, Norfolk State University, Norfolk, VA 23504*

## 1. INTRODUCTION

### 1.1 The Need for Functional Polymer Nano Structures

Just as the Industrial Revolution was driven by the invention and development of the steam and internal combustion engines over a century ago, the current rapid evolution of an information age has been facilitated by the invention and development of computers and communication devices and systems based on integrated circuits (ICs). The computers and radio wave/microwave based communication systems depend on the development of electronic materials, *i.e.*, conductors, semi-conductors and insulators. The information contained therein is processed by electronic signals. However, due to the explosion of information, particularly since the current widespread use of the Internet, we are rapidly approaching the limit of electronic signal processing systems with regard to speed, capacity, interference or signal/noise ratios. Fortunately, in comparison to electrons, photons (or signals encoded in the form of light) offer tremendous advantages in terms of capacity, speed and signal/noise ratios. For instance, a single fibre optic line has over 10,000 times larger bandwidth compared to a radio frequency

(RF) based TV transmission line (1). It is anticipated that the next generation computers and communication systems will depend upon the development of photonic materials and devices. Metal wires and semiconductor components in IC chips will be replaced by fibers, waveguides and other photonic components. In order to realize variety optical signal processing functions, optical components fabricated from various photonic materials and structures must be developed. In addition, optoelectronic or electro-optical hybrid materials and devices are also needed for systems where essential electronic devices or components are combined with optical components or linked by optic communication channels. Unlike traditional electronic materials where the chemical composition or energy bands are the key materials parameters, many electro-optic or photonic materials require not only specific chemical compositions or molecular structures (this can be called primary structure), but also a specific molecular orientation or domain order (this can be called secondary structure). In addition, a device function typically requires specific material bulk geometry (this can be called tertiary structure). For instance, since light signals can be polarized, any devices affecting the polarization, such as polarizers, polarized light emitting diodes (PLED) must be fabricated from materials with molecules (or atoms) oriented in a specific order. Quadratic nonlinear optical (NLO) polymers (also called EO polymers) (2), which can be used to encode electronic signals into optical signals in an electro-optical modulator device such as a Mach-Zehnder interferometer (Figure 1, bottom), requires not only NLO chromophores of large molecular dipole moment and large first molecular hyperpolarizabilities attached to a polymer matrix (primary structure), but also a bulk dipole non-centrosymmetric order (secondary structure). In addition, the fabricated NLO polymer thin film must be in a specific waveguide pattern (tertiary structure). In addition to the electro-optical application mentioned above, a variety polymer nano structures have also found their actual or potential applications in many other areas including biotechnology, medicine, environment, power systems, etc. (3)

Many inorganic photonic materials and devices have already been developed and commercialised. For instance, most current commercial electro-optical (EO) modulators use lithium niobate ($LiNbO_3$) crystals as NLO waveguide media. These modulators typically require a half-wave switching voltage ($V\pi$) of at least five volts (4). The modulation bandwidth is also relatively small, with the best reported value of 70 GHz (4). In spite of the large bandwidth, the high fabrication cost of these $LiNbO_3$ crystal EO modulators make them too expensive for ordinary homes to afford such a system.

The advantages of using polymer thin films over inorganic crystals in photonic devices include, but are not limited to, more convenient and versatile materials synthesis and device fabrication schemes, lower cost on large scale manufacturing, tunability of materials band gaps and other physical properties via molecular and supra-molecular (multi-level) structure synthesis and processing, small dielectric constants, light weight, flexible shape, ultra-fast signal response, less signal mismatch in RF-light signal modulation, and lower coupling optical loss between the chips and optic fibers (2). As an example, the half-wave switching voltage ($V\pi$) of a recently demonstrated polymer EO modulator was as low as 0.8 volts (4), well below the typical 5 volts in a commercial inorganic crystal based EO modulator. A lower $V\pi$ means reduced power consumption and heat generation, increased signal/noise ratios, and higher device capacity or efficiency. The modulation bandwidth of a prototype polymer modulator has already been demonstrated to reach over 100 GHz (5), and up to 700 GHz is expected (15b). To realize final polymer based EO modulator, all 3 levels of material engineering described above are required, and several additional critical factors also need to be satisfied. These additional critical factors include, good material stability (chemical, thermal, mechanical, orientational, etc.), low optical loss (6) (particularly at the telecommunication wavelength of 1550 nm), and a cost effective materials synthesis, processing, and fabrication scheme.

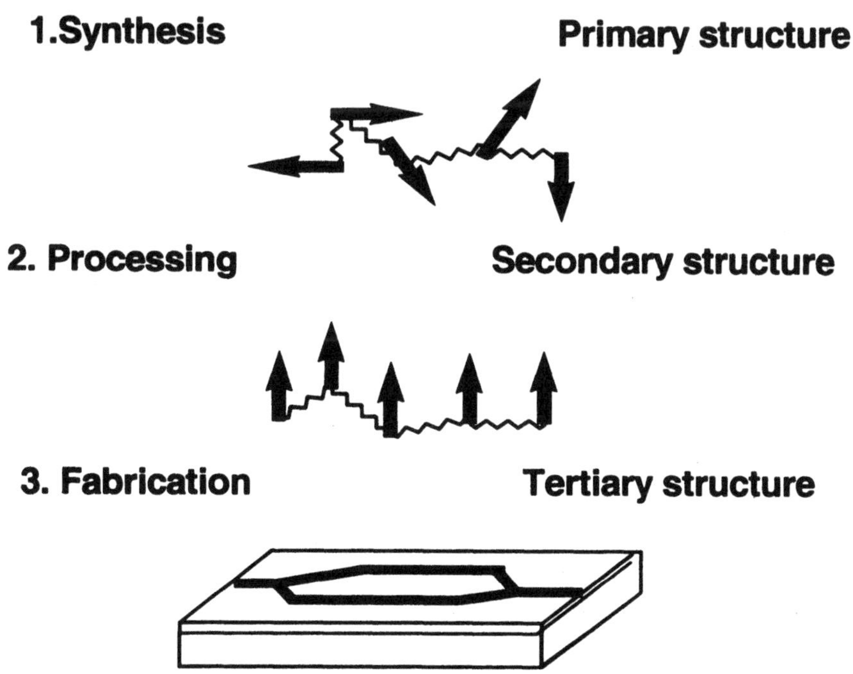


*Figure 1. Three major steps of developing a polymer NLO waveguide*

## 1.2 Polymer NLO Waveguide

To fabricate a basic NLO polymer EO modulator, such as a Mach-Zehnder interferometer waveguide as shown in Figure 1, there are at least three major steps of work involving polymer synthesis or processing.

In the 1st step, a processable functional polymer containing NLO chromophores needs to be designed and synthesized. Processable means the synthesized polymer must be soluble in a solvent and can be easily spin coated to form high optical quality solid thin films on waveguide substrates. For polymeric EO modulator purposes, NLO chromophores are organic molecules that have the form of D-π-A, where D is an electron rich donating unit or Donor, π is a conjugated bridge, and A is an electron withdrawing unit or Acceptor (2). Since materials bulk NLO property $\chi^{(2)}$ can be expressed by

$$\chi^{(2)} = f^2(\omega)\, f(2\omega)\, \langle\cos^3\theta\rangle\, N\, \beta \qquad [1]$$

where N is the NLO chromophore density in polymer matrix, $f^2(\omega)$ and $f(2\omega)$ are Lorenz local field factors at fundamental and second harmonic wavelengths, $\langle\cos^3\theta\rangle$ is an alignment factor reflecting an average non centrosymmetric degree of all NLO chromophore dipolar orientations (2). With EO modulators that are made of $\chi^{(2)}$ materials, the electro-optic coefficient $r_{33}$ is often used instead of $\chi^{(2)}$ to evaluate the materials macroscopic or bulk optical nonlinearity, and $r_{33}$ is linearly proportional to $\chi^{(2)}$. The $V\pi$ of an EO modulator is inversely proportional to $r_{33}$ (4). This is a main reason that a large $r_{33}$ or $\chi^{(2)}$ value is desired. Based on equation [1], in order to achieve a large $\chi^{(2)}$ value, NLO chromophores with large 1st molecular hyperpolarizability (β) values are desired. In addition, NLO chromophores with large dipole moment μ are also desired, since μ not only contributes to the β values (2), it also helps the electric field poling process that will positively contribute to the alignment factor $\langle\cos^3\theta\rangle$. A high chromophore number density N is also desired. However, recent studies have discovered that large μ and N may also negatively affect $\langle\cos^3\theta\rangle$, particularly for those chromophores that have large μβ values (6-7). One explanation for this is that at high chromophore loading density, chromophore dipoles are too close to each other and tend to counter align with each other (therefore decreasing alignment factor $\langle\cos^3\theta\rangle$) in order to minimize the interaction energy. This electro-static interaction becomes even stronger for larger μ chromophores at high chromophore loading density. In addition, due to a low optical loss requirement (at least less then 4 dB/cm) for the polymer waveguide device (6), NLO polymer systems without a charge transfer band tail beyond 1000 nm and without OH and NH functional groups are also desired. This is because OH/NH bonds have strong fundamental vibrational absorptions at around 2800-3200 nm (3600-3200 $cm^{-1}$), and their second harmonic vibrational overtones at around 1400-1600 nm accidentally full into the 1550 nm future telecommunication wavelength (6). From the polymer design point of view, the target NLO polymers should have at least the capability to adjust the NLO chromophore loading density (*N*) in order to achieve an optimum bulk NLO effect for a certain type of polymer architecture.

In the 2nd step, NLO chromophore dipoles in solid polymer thin films need to be oriented noncentrosymmetrically in order to achieve a large alignment factor $<\cos^3\theta>$. While there are a number of ways to do this, the most convenient and commonly used technique is poling in an electric field. In this method, polymer films are first heated to near their glass transition temperatures where the backbones of the polymer become somewhat flexible, then a high voltage DC electric poling field applied to the film to align the chromophores. Unfortunately, once the poling field is withdrawn, aligned NLO chromophore dipoles in the polymer film tend to become randomly oriented again due to the inherent thermal molecular motion and entropy preference. Therefore, stabilization of poling induced chromophore dipole orientation has become a key challenge for polymer EO modulator development. While a number of methods have been investigated in the past decade in order to stabilize or 'lock-in' the poling induced NLO chromophore orientation in polymer thin films, the most convenient, versatile, and widely used method is by crosslinking the polymer in the solid thin film right after the chromophores are poled. Polymer film crosslinking can be initiated either by heat or light. Light initiated thin film crosslinking, like in many photo-resist polymers (8), possess a major advantage for a cost effective photolithographic waveguide fabrication as will be discussed below.

In the 3rd step, an NLO polymer waveguide pattern, also called tertiary structure, is fabricated. For thermally crosslinked NLO polymer thin films, a typical waveguide fabrication protocol is as following (shown in Figure 2a): 1) Poling and crosslinking an NLO polymer layer on a waveguide substrate (containing bottom modulation electrode and an appropriate cladding layer); 2) Spin coating a photo resist polymer layer on top of crosslinked NLO polymer layer; 3) a waveguide pattern, either a negative or positive tune, is created with photo resist polymer layer via photolithography; 4) reactive ion etching (RIE) to etch away the non resist protected NLO polymer region to realize the desired NLO waveguide pattern; 5) remove the resist. However, if the NLO polymer itself is photo-crosslinkable, then the NLO waveguide fabrication becomes much simpler and more convenient. For instance, a waveguide mask can be directly applied to a photo crosslinkable NLO polymer layer (shown in Figure 2b), and after 1) poling and photo crosslinking; 2) mask removal and uncrosslinked NLO polymer dissolution, the NLO polymer waveguide is obtained. There is no need for photo resists patterning and RIE steps.

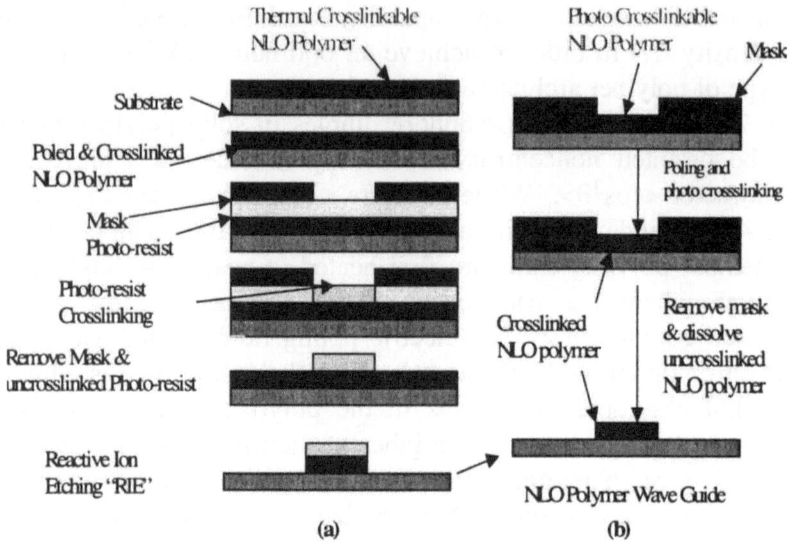


*Figure 2. Simplified fabrication Schemes of an NLO waveguide from (a) Thermally Crosslinkable and (b) Photo Crosslinkable NLO Polymers*

## 2. A BRIEF SURVEY OF CROSSLINKED NLO POLYMERS

Over the last decade, a variety of organic NLO chromophore-polymer systems have been investigated (2, 4-7, 9-11), including guest-host (doped), side chain, main chain, crosslinked and self-assembled polymer systems (Figure 3). In this paper, we will only focus on the crosslinked polymer system since it is by far the most versatile, convenient and relatively stable system. Among the numerous crosslinked NLO polymer systems studied, a few representative systems are summarized below, and are categorized into thermal and photo initiated crosslinking processes.

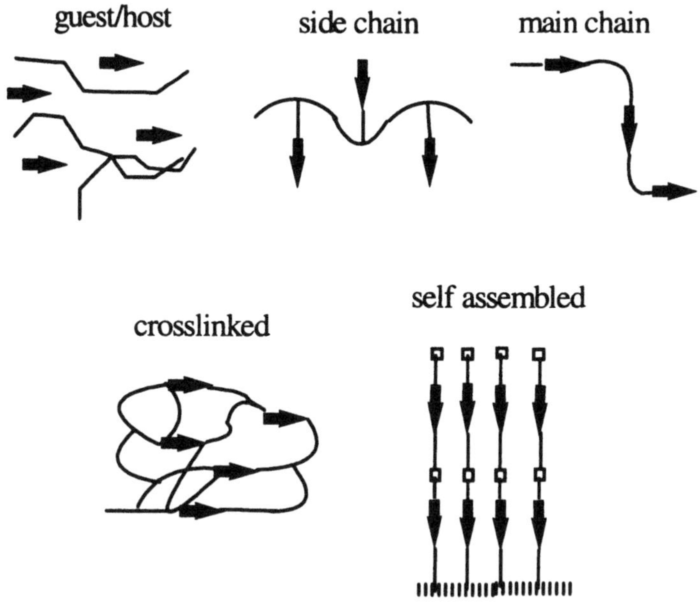


*Figure 3. Main types of NLO polymers*

## 2.1 Thermally crosslinked systems

Thermally crosslinked NLO polymer systems have been widely investigated. This may be due to the large number of solid-state thermal crosslinking reactions available, and the majority are condensation type. Representative systems include: polyurethane NLO oligomers crosslinked by multi-functional alcoholic (10a-b) or epoxy type of crosslinkers (10a, c), polymethylmethacrylate (PMMA) type of polymers crosslinked by covalently attached side chain crosslinkable NLO chromophores (10d), Sol Gel NLO polymers thermal self-crosslinking (10e-f), and interpenetrating NLO polymer network (IPNs) with at least two different polymers and two different crosslinking reactions occur simultaneously (10g). For condensation type reactions, the advantages include the easy control of polymer molecular weight, crosslinking rate and density. The disadvantages include, in the case where small molecules are generated in the crosslinking reaction, thin film morphology and quality may be affected significantly. Also, many of the condensation crosslinking reactions involving OH/NH units, and recent study has found that the OH/NH vibrational overtones contribute significantly to the absorption optical loss at 1550 nm (6). A notable recent progress is using a perflorocyclobutane radical thermal crosslinking scheme (11). One advantage of this system is the potential for low optical loss due to the typical high transparency of CF groups at infrared region.

## 2.2 Photo crosslinked systems

Though photo crosslinked polymer systems offer advantages in the polymer waveguide fabrication step by utilizing a cost effective photolithography protocol (Figure 2b), there are relatively fewer systems developed for NLO purposes so far. The systems that have been studied include acrylic type UV crosslinked system (12a), and most predominantly, cinnamoyl type UV crosslinked polymers (12b-c). One major drawback with UV crosslinking is that certain NLO chromophores (such as azo- type) may be susceptible to radiation damage from UV (<300nm) light. Therefore, photo-crosslinking reactions at chromophore friendly longer visible wavelengths need to be investigated and developed. One report demonstrated a 400 nm light initiated styrene type crosslinking NLO polymer system (12d). Unfortunately, this system also contains NH groups, and the synthetic scheme seems not very convenient and versatile.

# 3. FUMARYL CHLORIDE AND MALEIC ANHYDRIDE DERIVED CROSSLINKED NLO POLYMERS

As discussed above, though many NLO polymer systems have been investigated and developed so far, none of them have satisfied all the required device parameters in a single system. One goal of our research is to develop an NLO polymer system that can simultaneously satisfy most key requirements in one system. Since the magnitude of electro-optical coefficients $r_{33}$ are mainly determined by the NLO chromophores used, and this parameter has been achieved much better then commercial inorganic ones recently (4), the remaining key challenges are better materials stability (chemical, orientational, mechanical, etc), lower optical loss, and more convenience or lower cost of a protocol.

## 3.1 Fumarate type crosslinkable polymers

Both maleic anhydride (MA or its derivatives such as Maleimides) and fumaryl chloride (FC) have been reported for a variety of crosslinked polymer systems (13-14). Specifically, MA is a very widely studied monomer for fabrication of crosslinked polyester products, including coatings, thermoset household objects, etc. (13). As shown in Figure 4, there are several different schemes of using MA to synthesize crosslinked polymers. In the most popular scheme or step [1], maleic anhydride **1** polymerises with a vinyl type of comonomer via radical reactions, and polymer **2** crosslinking can be achieved either through the usage of a multi-vinyl type of comonomer, or through a multi-functional crosslinker that will react with anhydride unit of the MA via condensation reaction. In fact one NLO polymer system has been demonstrated in the later case (10a). An interesting fact is that MA does not polymerise itself. One explanation is that MA is an electron deficient acceptor type of monomer, and it tends to co-polymerise only with an electron rich donor type of vinyl monomers (13d). For instance, donor type of vinyl ethers copolymerises with MA easily even in the absence of initiators and under both thermal and light radiation conditions (13c-d, f). Yet, an initiator is typically needed for other weak donor type of vinyl comonomers such as styrene or methacrylate. In the second scheme or steps [6-7], an amine reacts with MA to form a

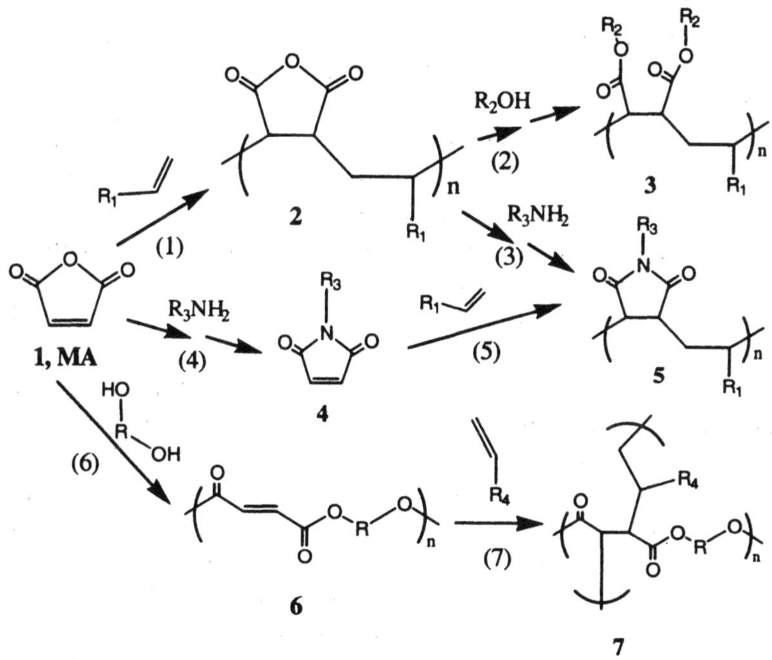


*Figure 4. Polymers Derived from Maleic Anhydride (MA)*

functional maleimide compound **4** first, then polymerise with a vinyl comonomer to form polymer **5**. An uncrosslinked NLO polymer system has been reported using this scheme (13b). As a matter of fact, crosslinking can also be induced either by using a donor type of multi-vinyl ether comonomer, or by using an amine that contains crosslinking sites. In step [9], MA condenses with a diol comonomer to form a linear unsaturated polyester first, and then a vinyl type of crosslinker is used to crosslink the polyester backbones (13a, c-f). A crosslinking scheme using visible light was also demonstrated (13f). For FC, the only scheme of making crosslinked polymers reported so far is to have FC condense with a diol comonomer first to make a unsaturated polyester **6**, then followed by thin film solid state crosslinking using a vinyl type of crosslinker. FC derived crosslinked polyesters have also been investigated for a number of applications (14).

## 3.2 NLO Polymers from Fumarate Type Crosslinked Polyesters

We have recently investigated both Maleic Anhydride (MA) and Fumaryl Chloride (FC) derived crosslinked polyester resins for potential nonlinear optical waveguide applications (15). The synthetic and processing schemes we have investigated are represented in steps (6) and (7) of Figure 4. Either maleic anhydride (MA) or fumaryl chloride (FC) is coupled with an NLO chromophore diol DR-19 and a diol co-monomer (such as Glycol) to form an unsaturated co-polyester **6**. The non-chromophore diol co-monomer is used to fine-tune the NLO chromophore loading and other physical properties of the final NLO polymers. For instance, the co-monomer can be used to minimize the electro-static interactions of the NLO chromophore dipoles that negatively affect chromophore poling efficiency, particularly for large dipole moment $\mu$ NLO chromophore systems (7). Our study shows both FC and MA condense with DR-19 and glycol to yield unsaturated NLO polyesters **6** and final crosslinked NLO polyesters **7** with almost the same chemical, physical, and optical properties at the same chromophore loading. All polymers were characterized by NMR, elemental analysis, IR, DSC, TGA, GPC and UV-VIS (15). For instance, thermal gravimetric analysis (TGA) shows DR-19 functionalized unsaturated polyesters **6** has a much better thermal/chemical stability then the pristine DR-19 (Figure 5). The IR spectrum shows DR-19 exhibited a broad peak at 3400 $cm^{-1}$ due to the fundamental vibrations of hydroxyl (OH) groups, and this peak disappeared after the polyester **6** formation (15a). The disappearance of OH groups and their second harmonic vibrational overtones at 1400-1600 nm is very critical for low optical loss telecommunication applications at 1550 nm (6). There is also a new small shoulder peak appears at 990 $cm^{-1}$ indicating the alkene bonds of the synthesized polyesters **6** are in predominantly trans configuration (16), and this is one reason we call both MA and FC derived polyesters **6** fumarate type.

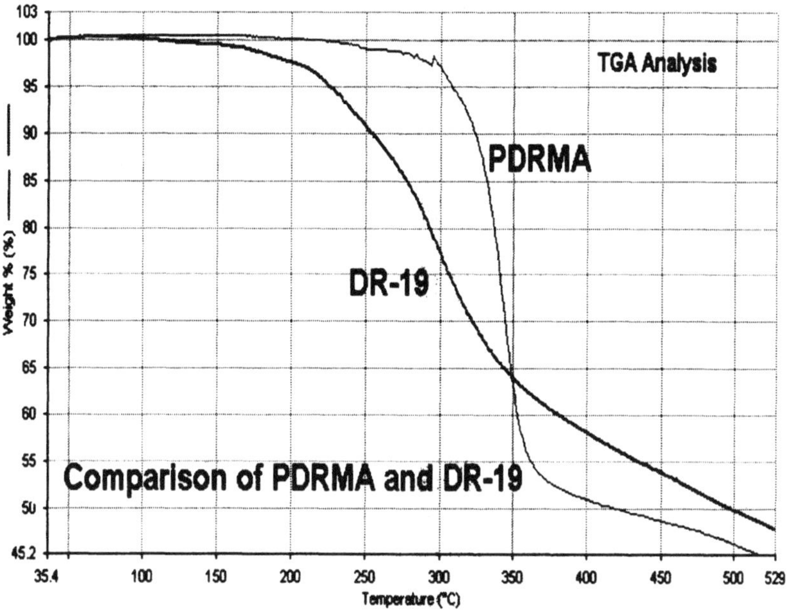


*Figure 5. TGA Analysis of DR-19 vs. Polyester **6**(PDRMA)*

All our crosslinking reactions of polymer thin films were carried out in air. To process polymer thin films, the polyesters **6** were dissolved in dry dioxane or cyclopentanone together with vinyl crosslinkers, and a crosslinking initiator. For thermal or UV crosslinking, initiators such as benzoyl peroxide, and a regular vinyl crosslinker such as methylmethacrylate works well. For visible light crosslinking, a crosslinking initiator such as tetrabromofluoroscine (13f), and a donor type of crosslinker such as a divinyl ether was used. After dissolution, polymer solutions were filtered through a 0.2-μm filter, spin coated on ITO glass slides with thickness between 2-50 μm depending on the purpose of the thin film study. The films were typically dried in a vacuum oven at 50°C for at least 24 hours.

For thermal crosslinking, the dried films were heated to near their glass transition temperatures $T_g$ (typically 120-130°C), and corona poled on a 5-9 kV DC poling stage. The DSC analysis suggested both thermal and photo crosslinking reactions completed in 30 minutes near Tg of polyester **6** in air (15c). FT-IR spectra indicate the reduction or complete disappearance of polyester **6** alkene bond C=C stretching around 1640-1660 $cm^{-1}$ after crosslinking (15a). The crosslinking reaction can also be conveniently characterized by solubility test. The polymer thin films were soft and soluble in dioxane before crosslinking, became rigid and insoluble after crosslinking.

For visible light crosslinking, an Oriel 1 kW QTH Lamp is used as a light source. In crosslinking experiment, the polyester **6** films were heated near the $T_g$ (about 120°C), poled, and then irradiated with a collimated beam (3 inch diameter) visible light of 150 $mW/cm^2$ radiant power on film surface. An insoluble crosslinked film can be obtained within 30 minutes. DSC analysis also indicated the complete Tg disappearance of a 25 minutes visible light crosslinked film (15c).

Second harmonic generation (SHG) measurements were done on the film after corona poling. The set-up employed a Coherent Infinity Nd:YAG laser with 30 mJ/pulse with a 10 Hz repetition rate and 3 ns pulse width at 1064 nm onto the polymer film. Frequency doubled SHG signals were detected at 532 nm.

Normalized SHG signals (reflecting the poling induced NLO chromophore non centrosymmetric nano structure) versus the heating temperature of the polymer films were monitored to over 180°C (15b). As the data demonstrates, the poling induced molecular orientation or nano structure for an uncrosslinked polymer thin film starts to drop at about 50°C. The nano structure of a crosslinked polymer film is maintained up to around 150°C. However, with further optimizations on comonomers, crosslinkers, initiators, poling and crosslinking conditions, more stable nano structures may be achieved.

## 4. SUMMARY AND FUTURE RESEARCH

Functional polymer nano structure fabrication has become an urgent need as well as a major scientific challenge due to a wide variety of emerging applications. Unique molecular orientation need to be controlled and stabilized effectively and precisely at a nano meter scale. All these require the development of more versatile and effective polymer synthesis, processing, and crosslinking schemes. Specific for the FC or MA crosslinked NLO polymer systems we have developed, there are still room to improve the processing and crosslinking efficiency by using different crosslinking schemes, comonomers, crosslinkers and initiators. In combination with certain two photon absorption (TPA) photo initiators, this FC/MA polymer crosslinking system may also be used to fabricate a three dimensional nano structures for other applications such as high capacity optical data storage or 3-D holographic recording devices. This FC/MA crosslinked system may also be useful for fabricating a number of other functional nano structures. For instance, we are currently evaluating a scheme to use MA crosslinked polymer matrix to immobilize enzymes for certain biomedical applications.

In summary, the design, syntheses, processing, and fabrications of functional polymer nano structures has become increasingly critical for variety emerging applications. Polymer crosslinking technique is one of the most versatile and effective methods for fabricating and stabilizing polymer nano structures. Among various polymer crosslinking schemes developed so far, fumaryl chloride (FC) and maleic anhydride (MA) derived crosslinked polyester system seems to be a very versatile, cost effective and convenient protocol. In NLO waveguide fabrications, FC/MA crosslinking system also provides photolithographic fabrication advantage.

## REFERENCES AND NOTES


(1) Hinton, H. S., IEEE Spectrum, February, 42 (1992).

(2) (a) Prasad, P. N.; Williams, D.J.; **Introduction to Nonlinear Optical Effects in Molecules and Polymers**, John Wiley & Sons Inc., New York, 1991. (b) Nalwa, H. S. and Miyata, S.; **Nonlinear Optics of Organic Molecules and Polymers**, CRC Press, Boca Raton, Florida, 1997. (c) Dalton, L. R., *et al.*, Adv. Mater., 7(6), 519 (1995).

(3) (a) Lyshevsky, S. E., **Nano- and Microelectromechanical Systems: Fundamentals of Nano- and Microegnineering**, CRC Press, Boca Raton, FL, 2000. (b) Timp, G., ed., **Nanotechnology**, Springer Verlag, 1999.

(4) Shi, Y.; Zhang, C.; Zhang, H.; Bechtel, J. H.; Dalton, L. R.; Robinson, B. H. and Steier, W. H., Science, *288*, 119 (2000).

(5) Chen, D.; Fetterman, H. R.; Chen, A.; Steier, W. H.; Dalton, L. R.; Wang, W.; Shi, Y., Appl. Phys. Lett., 70, 3335 (1997).

(6) Dalton, L. R., Opt. Eng., 39(3), 589 (2000).

(7) (a) Harper, A. W.; Sun, S.; Dalton, L. R.; Garner, S. M.; Chen, A.; Kalluri, S.;Steier, W. H.; Robinson, B. H., J. Opt. Soc. Am. (B)., 15(1), 329 (1998). (b) Dalton, L. R.; Harper, A. W.; Robinson, B. H., Proc. Natl. Acad. Sci. U.S.A., 94, 4842 (1997). (c) Sun, S.; Zhang, C.; Dalton, L. R.; Garner, S. M.; Chen, A.; Steier, W. H., Chem. Matter., 8(11), 2539 (1996).

(8) Reiser, A., **Photoreactive Polymers: The Science and Technology of Resist**, John Wiley and Sons, New York, 1989.

(9) Dalton, L. R.; *et al.*; Chem. Mater., 7, 1060 (1995). (b) Burland, D. M.; Miller, R. D.; Walsh, C. A., Chem. Rev., 94(1), 31 (1994).

(10) (a) Liang, Z.; Yang, Z.; Sun, S.; Wu, B.; Dalton, L. R.; Garner, S. M.; Kalluri, S.; Chen, A.; Steier, W. H., Chem. Matter., 8, 2681 (1996). (b) Chen, M.; Dalton, L.R.; Yu, L. P.; Shi, Y.Q.; Macromolecules, *25*, 4032 (1992). (c) Wang, X.; Kumar, J.; Tripathy, S. K.; Li, L.; Chen, J.; Marturunkakul, S., Macromolecules, *30*, 219 (1997). (d) Xu, C.; Wu, B.; Todorova, O.; Dalton, L. R.; Shi, Y.; Ranon, R. M. and Steier, W.H.; Macromolecules, 26, 5303 (1993). (e) Yang, Z.; Xu, C.; Wu, B.; Dalton, L. R.; Kalluri, S.; Steier, W. H.; Shi, Y.; Bechtel, J. H., Chem. Mater., 6, 1899 (1994). (f) Hsiue, G.; Lee, R.; Jeng, R., Chem. Mater., 9, 883 (1997). (g) Marturunkakul, S.; Chen, J. I.; Li, L.; Jeng, R. J.; Kumar, J.; Tripathy, S. K., Chem. Mater., 5, 592 (1993).

(11) (a) Ma, H.; Wu, J.; Herguth, P.; Chen, B.; Jen, A. K. Y., Chem. Mater., 12, 1187 (2000). (b) Ma, H.; Chen, B.; Sassa, T.; Dalton, L.R. and Jen, A. K. Y, *J. Am. Chem. Soc.*, **123**, 986 (2001).

(12) (a) Boogers, J. A. F.; Klaase, J. J-V.; Alkema, D. P. W.; Tinnemans, A. H. A., Macromolecules, 27, 197 (1994). (b) Choi, D. H.; Song, S.; Jahng, W.; Kim, N., Mol. Cryst. Liq. Cryst., 280, 17 (1996). (c) Zhu, X.; Chen, Y. M.; Li, L.; Jeng, R. J.; Mandal, B.; Kumar, J.; Tripathy, S. K., Optics Communications., 88, 77 (1992). (d) Godt, A.; Frechet, J. M.; Beecher, J. E.; Willand, C. S., Macromol. Chem. Phys., 196, 133 (1995).

(13) (a) Wicks, Z. W.; Jones, F. N.; Pappas, S. P., **Organic Coatings-Science and Technology**, Wiley Interscience, New York, 1998, pp 290-292. (b) McCulloch, I.; DeMartino, R.; Keosian, R.; Leslie, T., Macromol. Chem. Phys., 197, 687 (1996). (c) Noren, G. K.; in **Photopolymerization Fundamentals and Applications,** ACS Symp. Ser., 673, 121 (1997). (d) Kohli, P.; Scranton, A. B.; Blanchard, G. J., Macromolecules, 31, 5681 (1998). (e) Grobelny, J. and Kotas, A., Polymer, 36(7), 1363 (1995). (f) Jachowicz, J.; Kryszewski, M.; Klosowska-Wolkowicz, Z.; Penczek, P., Die Angewandte Makromolekulare Chemie, 97, 201 (1981).

(14) (a) Jo, S.; Shin, H.; Mikos, A. G., Biomacromolecules, 2, 255 (2001). (b) Sartori, G; Ho; Winston, W. S.; Noone, R. E., US Patent # 5138023, 1991.

(15) (a) Sun, S.; Maaref, S.; Alam, E.; Salter, J.; Wyatt, S., Wang, Y.; Bahoura, M.; Bonner, C. Proc. SPIE, 4106, 177 (2000). (b) Sun, S.; Maaref, S.; Alam, E.; Wang, Y.; Fan, Z.; Bahoura, M.; Higgins, P.; Bonner, C. Proc. SPIE, 4580, 297 (2001). (c) Sun, *et al.*, manuscript in preparation.

(16) White, R.G., **Handbook of Industrial Infrared Analysis**, Plenum Press, N.Y., 1964.

Chapter 3

# HUMERAL IMMUNE RESPONSES TO POLYMERIC NANOMATERIALS

Stephen C. Lee[1, 2], R. Parthasarathy[1, 3], K. Botwin[1], D. Kunneman[1], E. Rowold[1], G. Lange[1], J. Zobell[1], T. Beck[1], T. Miller[1], R. Jansson[1, 4] and C. F. Voliva[1].

*[1]Pharmacia Corporation 700 Chesterfield Village Parkway, Chesterfield, MO USA 63141, [2] Molecular and Cellular Biochemistry, Department of Chemical Engineering and The Biomedical Engineering Center, The Ohio State University, 270 Bevis Hall, 1080 Carmack Road, Columbus, OH 43210, [3]3M Corporation, Minneapolis, MN 92722 [4]Eveready Battery Co., Westlake, OH 43923*

## 1. INTRODUCTION

### 1.1 General

Synthetic polymeric materials are widely used in therapeutic devices and drugs as both structural and active components. Polymers are used not only in macroscale, indwelling devices, such as vascular stents and prosthetic devices, but also in nanoscale bioconjugates to drug molecules, and as formulation excipients that improve the stability, pharmacokinetics, bioavailability or other properties of drug substances.

In applications in which synthetic materials are introduced into the body, host immune responses to the materials must be considered as they can influence therapeutic tolerability, stability and efficacy in the host, and, in the case of circulating bioactive molecules or complexes, half-life and biodistribution. Therapeutic nanodevices may contain synthetic nanomaterials (1-9), and this article focuses primarily on specific humeral immune responses (antibody responses) to these synthetic constituents. A number of such responses have been documented (1-4), and they exhibit certain similarities in terms of the structure of the immunogens triggering them and in the cross-reactivity of the generated antibodies to chemically similar nanostructures. Commonalities between responses to different nanoscale antigens reflect the mechanism by which anti-nanomaterial antibodies are generated, and, ultimately, the structures of nanoscale bioconjugates themselves. Experiences of anti-polymer antibodies are instructive as they illustrate the immune consequences of certain configurations of bioconjugates, and thus can inform later bioconjugate design for nanotherapeutics (5-9).

Antibody responses to drug entities are generally undesirable, so substantial effort is applied in the pharmaceutical industry to avoid or mitigate them. Consequently, as dendrimers and other nanomaterials are incorporated into therapeutics (1-9), their immunogenicity will become an issue. Aside from their therapeutic consequences, antibodies to polymers and synthetic nanostructures are potentially powerful reagents. Antibodies are commonly used in biotechnology to detect and quantitate their cognate antigens, and can be used to follow therapeutic biodistribution and pharmacokinetics, by detecting drugs and their metabolites in biological fluids, and determining the location of therapeutics in tissues by immunohistochemistry. Antibodies are also commonly used in affinity chromatography for the purification of proteins and other macromolecules, and antibodies to synthetic polymers can be used in similar applications for the detection, affinity separation and assembly of nanoscale materials (1-4, 9).

## 1.2 Antigens, immunization and antibodies

Specific immune responses are directed to individual macromolecules or assemblages thereof, and can be divided into cellular immune responses, effected by cytotoxic T-lymphocytes (CTLs), and humeral immune responses, effected by soluble proteins that are secreted by B-cells (antibodies, 10, 11). Regardless of the effector (CTLs or antibodies), specific responses are directed to individual molecular features of antigens called epitopes. In addition to those recognized by immune effectors, some epitopes are recognized by regulatory components of the immune system (helper T-cells). In general, structures or molecules containing epitopes are referred to as antigens or immunogens.

Two distinct classes of epitopes (B-cell epitopes and helper T-cell epitopes), recognized by different cells of the immune system, are involved in the production of specific antibody responses, and potent immunogens contain both of them (10, 11). B-cell epitopes are chemically diverse, and can occur in both biological and non-biological macromolecules. Most such epitopes are polypeptides, though small molecules (with masses under 1,000 AMU) can also act as B-cell epitopes under some circumstances.

Individual B-cells have cell surface receptors that bind antigens containing epitopes recognized by the particular B-cell clone. These receptors are essentially membrane-bound forms of the antibody species that the particular B-cell clone can make, and antigen binding to them stimulates the clone to proliferate and produce secreted, soluble antibody.

Helper T-cell epitopes are a more homogeneous lot than B-cell epitopes,

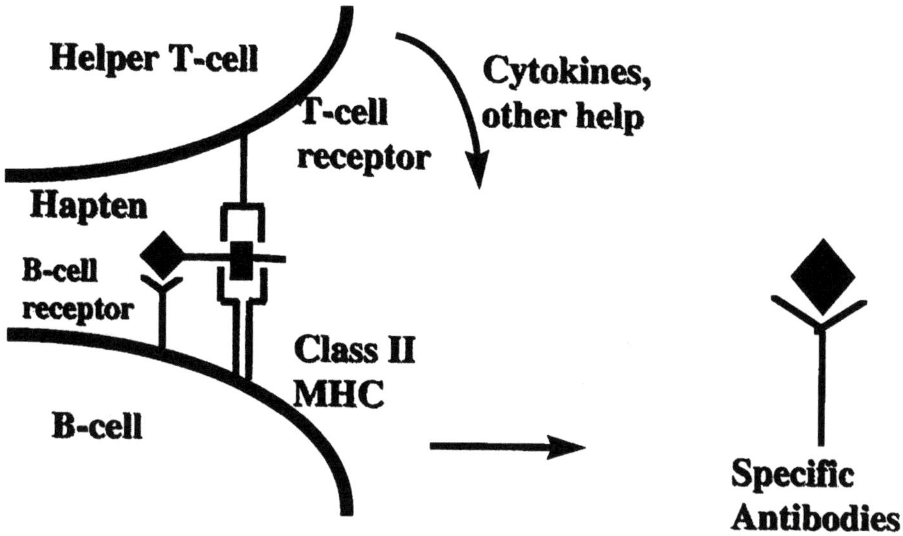


*Figure 1. Antibody responses are most efficiently induced by antigens containing both B-cell epitopes (black diamonds) and helper T-cell epitopes (black rectangle). The B-cell epitope binds a receptor on the B-cell surface that is essentially a membrane-bound form of the antibody species that particular B-cell can produce. The linked, peptidic T-helper epitope engages another receptor complex on the B-cell surface (class II MHC). When presented in the context of the class II MHC complex of the B-cell, the T-helper epitope can be recognized by a cognate T-cell receptor on the helper T-cell. In response to recognition of its appropriately presented, cognate T-helper antigen, the helper T-cell secretes and provides multiple stimulatory substances that cause the B-cell to proliferate, secrete antibody and differentiate. Note that the antibody produced recognizes only the B-cell epitope, and not the T-helper epitope.*

and are typically short peptides derived from larger polypeptides by antigen processing (i.e., the degradation of antigens into smaller fragments for presentation on B-cell surfaces). Helper T-cell epitopes are recognized by helper T-cells only when they are presented on the B-cell surface in the context of the Class II major histocompatibility complex (MHC class II). When specific helper T-cell clones recognize their cognate epitopes. presented on the MHC Class II complex, they provide molecular signals
(cytokines, membrane-bound signaling molecules) to the antigen-presenting B-cell that cause it to proliferate and produce antibodies orders of magnitude more strongly than it can in the presence of its cognate B-cell epitope alone. The process is illustrated in Figure 1.

The number of possible antibodies any individual is theoretically capable

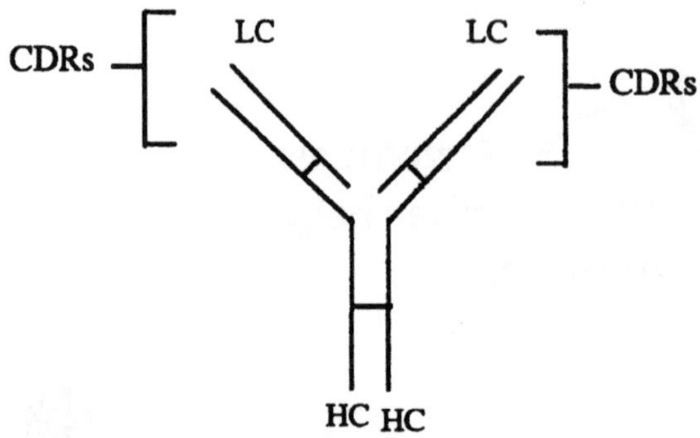


*Figure 2. Antibodies are biological affinity reagents. A typical circulating antibody (an IgG molecule) is comprised of four polypeptide chains (two identical heavy chains or HC and two identical light chains or LC) that are cross-linked by disulfides. Antigen binding specificity is determined by CDRs (complementarity determining regions, to which both HC and LC components contribute). Each IgG molecule contains two sets of identical CDRs, making IgG molecules monospecific but bivalent affinity reagents. IgGs have molecular weights of about 150,000 Daltons and measure approximately 15 nM from arm to arm.*

of producing is vast ($>10^{11}$), but finite (10, 11). Any given epitope is recognized by a limited subset of B-cell or T-cell clones, each clone recognizing a single epitope. The unique collection of immunologically responsive B- and T-cell clones constitutes an individual's immune repertoire. Strong antibody responses can only be mounted in individuals whose immunological repertoires contain T- and B-cell clones that specifically recognize their corresponding epitopes in the immunogen.

Animals initiate antibody responses as the result of exposure to antigens. In the laboratory, animals are immunized for antibody production by injection of cocktails of antigens with adjuvants (such as Fruend's adjuvant or alum), substances known to non-specifically augment specific antibody responses, improving the kinetics and magnitudes of the responses. Typically, animals are subjected to a course of repeated immunogen administration (each repetition of dosing is called a "boost") until the specific antibody titer (a measure of the concentration of antibodies recognizing a specific antigen) plateaus. At this point, antibodies can be harvested from the animal as a polyclonal sera. Alternately, individual B-cells of the immune host can be isolated and immortalized by fusion to tumor cells. The immortalized cells elaborate monoclonal antibodies: single antibody species that recognize only one B-cell epitope. Polyclonal sera are often sufficient for immune detection of antigens (1-4, 9, 10, 11), but the precision of recognition inherent to monoclonal antibodies are likely preferable for some nanotechnological uses of antibodies (2, 3,4).

Though they are required to trigger robust antibody responses, helper T-cell epitopes are neither recognized by the antibodies whose production they stimulate, nor is their presence required in antigens for binding (10,11): antibodies recognize only their cognate B-cell epitopes. Most circulating antibodies (immunoglobulin class G, or IgG molecules, Figure 2) are bivalent species, with two identical antigen-binding domains called CDRs (complementarity determining regions). More polyvalent antibodies can also be made (10, 11): IgA molecules are dimeric (thus having 4 sets of identical CDRs) and IgM molecules are pentameric (10 sets of identical CDRs). Individual antibodies can also be engineered to recognize more than one distinct antigen (bispecific antibodies), and antibody fragments containing individual (monovalent) CDRs can also be made.

## 1.3 Current study

Dendritic polymers are novel nanoscale polymers with fractal architecture (12-14), and are among the most monodisperse synthetic nanomaterials available. They are attractive for applications wherein structural and chemical uniformity are desirable, including pharmaceuticals and nanotechnology. Dendrimers are grown from initiator cores by the addition of successive, discrete layers of homogeneous subunits, and dendrimers are often said to be of a specific "generation," in the case of polyamidoamine (PAMAM) dendrimers, $G_0$ to $G_{10}$. Generations refer to the number of discrete layers added to the initiator core (one layer on the $G_0$ initator core constitutes a $G_1$ dendrimer, two layers a $G_2$, etc.).

Unmodified PAMAM dendrimers are poorly immunogenic (15-17). However, we have recently generated polyclonal murine antibodies that recognize dendrimers using two distinct immunization methods (1-3). The significance of these results is twofold. First, therapeutics containing dendritic polymers are under development by multiple investigators (5-8), and therapeutic constructs configured as were our immunogens may also trigger anti-dendrimer immune responses. Second, the antibodies are intrinsically useful for detection and manipulation of dendrimers.

## 2. EXPERIMENTAL

Generation 0 to generation 10 ($G_0$-$G_{10}$) PAMAM dendrimers were used either as obtained from the manufacturers (Dendritech and Michigan Molecular Institute, both of Midland, MI) or after their surface amines had been "capped" (converted to other functional groups) to support protein conjugation. $G_5$ PAMAM dendrimers were capped (1-3) to have either sulfhydryl (using iminothiolane) or oxiamine surfaces (using boc-aminooxyacetic acid). Proteins (BSA and KLH from Sigma, St. Louis, MO, and a human IL-3 variant, see 1-3) were conjugated to dendrimers using, in

| *Immunogen* | *Anti-dendrimer antibody response* |
|---|---|
| *$G_0$-KLH conjugate* | + |
| *$G_0$-KLH co-administered* | - |
| *$G_0$-BSA conjugate* | + |
| *$G_0$-KLH co-administered* | - |

*Table 1. $G_0$ PAMAM dendrimer-protein bioconjugates are immunogenic. ELISA assays (1-3, 9, 10) indicate that PAMAM dendrimer-protein conjugates can stimulate anti-dendrimer antibody responses. $G_0$-KLH and $G_0$-BSA conjugates were made using carbodiimide conjugation chemistry as described (1, 3). Co-administration refers to immunization by the same schedule and route used for the $G_0$-protein conjugates, but using mixtures of $G_0$ dendrimers and BSA or KLH as non-conjugated, free molecules. Concentrations of $G_0$ dendrimers and BSA or KLH in co-administered doses were matched to those in the $G_0$-BSA and $G_0$-BSA conjugates, respectively. Data summarized from references 1-3 and 9.*

the case of unmodified PAMAM dendrimers, carbodiimide chemistry (17), using a heterobifunctional maleimide linker (Pearce, Rockford, IL) in the case of sulfhydryl-surfaced PAMAMs (17) or using oxiamine orthogonal conjugation (18) with oxiamine-surfaced dendrimers. Poly(triethylenemine) (POPAM) dendrimers were the gift of Dr. Ralph Spindler of Dendritech.

BALB/C and C57BL6 mice (Charles River Laboratory, Wilmington, MA) were immunized by intraperitineal (IP) injection (1). Initial immunization was with 25 μgrams of each immunogen, administered in complete Freund's adjuvant (CFA). At day 14, animals were boosted with another 25 μgram dose in incomplete Freund's adjuvant (IFA) followed by a final dose of 10 μgrams, also in IFA, at day 21. Sera were harvested from mice by intraorbital bleed at 28 days. Anti-dendrimer antibodies were detected in mouse sera using an enzyme linked immunosorbent assay (ELISA), or by dot-blot assays (1-3).

## 3. RESULTS AND DISCUSSION

### 3.1 Immune responses to PAMAM dendrimers

Low immunogenicity of unmodified PAMAM dendrimers notwithstanding, PAMAMs can be immunogenic under some circumstances (Table 1). BALB/c and C57BL6 mice produce anti-dendrimer antibodies (1-3) in response to dendrimer-protein conjugates, though, as previously reported (15-17), PAMAM dendrimers alone do not trigger detectable antibody responses (1-3).

While PAMAM dendrimers themselves are not potent antigens, the murine immune repertoire includes one or more B-cell clones that recognize the dendrimers (i.e., PAMAM dendrimers contain B-cell epitopes). However, the lack of antibody responses to unmodified PAMAMs administered in CFA reflects that the dendrimers lack helper T-cell epitopes. Helper T-cell epitopes are linear peptides which present specific amino acid side chains at specific positions along their lengths (10, 11), features clearly absent from PAMAM dendrimers. However, the lack of T-helper epitopes can be supplemented by proteins or peptides covalently linked to the dendrimer (Table 1).

Antigens that trigger antibody responses independent of helper T-cell epitopes are known (10, 11). These "T-independent antigens" may have some features in common with dendritic polymers as they consist of repetitive polymers (often oligosacharides) wherein each repeating unit constitutes a single B-cell epitope. Such antigens cross-link multiple B-cell receptors to generate a cell proliferation/differentiation signal of sufficient magnitude to drive low affinity, low magnitude antibody responses in the absence of T-cell help. In light of their highly repetitive structure, it is not clear why such responses are not observed to unmodified $G_5$ dendrimers, though it may have to do with the dendritic architecture itself. Perhaps $G_5$ PAMAMs are insufficiently flexible or somehow sterically impeded from making the requisite multiple B-cell receptor contacts to act as T-independent antigens. It remains to be seen whether other generations of PAMAM dendrimers or other dendritic polymers might trigger T-independent responses.

## 3.2 Antibody recognition of PAMAM dendrimers

We investigated the specificity of the antisera using dot blots (1-3). We found that antisera from animals immunized with $G_0$-protein conjugates recognized PAMAM dendrimers of any generation from 0 to 10. This cross-reactivity is likely the result of epitopes shared across generations of PAMAM dendrimers, and reflects the fractal dendrimer architecture. Similar observations have been made with other structurally repetitive nanomaterials and may be a common property of antibodies raised to compositionally homogeneous nanomaterials. For instance, antibodies to C60 fullerenes cross-react with the chemically similar walls of single wall carbon nanotubes (4). That said, antibodies raised to subunits or components of larger

| *B-cell immunogen* | *Antibody detectected to:* | | | |
|---|---|---|---|---|
| | *1* | *2* | *3* | *4* |
| *$G_0$ PAMAM* | + | - | - | - |
| *Sulfhydryl-capped $G_5$ PAMAMs* | - | + | - | - |
| *Oxiamine-capped $G_5$ PAMAMs* | - | - | + | - |
| *POPAMS* | - | - | - | - |

*Table 2. Antibody recognition involves dendrimer end groups and their arrangement. Sera from mice immunized with dendrimer-protein conjugates were assayed by dot-blot (1-3, 9, 10) for specific antibodies (antibody binding indicated by +). B-cell immunogens of the protein conjugates were as shown. Dendritic antigens deployed on membranes are indicated by numbers: 1.Unmodified $G_0$-$G_{10}$ PAMAM dendrimers (each generation of PAMAMs tested were found to bind antibody), 2. Sulfhydryl-capped $G_5$ PAMAMs, 3. Oxiamine-capped $G_5$ PAMAMs and 4. POPAMs. Each anti-sera recognized PAMAM dendrimers with the same end-group as the immunizing antigen, but not differently capped PAMAMs, implicating dendrimer end groups in antibody recognition. End-groups were not the sole determinant of binding: POPAM dendrimers, which have primary amine end groups (as do PAMAMs) but distinct architecture, are not recognized by antibodies raised to $G_0$ PAMAMs. Data summarized from references 1-3 and 9.*

nanostructures can only recognize all of the epitopes of the larger structure if those epitopes are also present in the immunogen. For instance, anti-$G_0$ sera would not be expected to recognize epitopes constituted by structural features that occur only in higher generation PAMAM dendrimers (such as the cavities hypothesized to extend from the surface to the core of the dendrimer 12-14), simply because those structures are not present in the immunogen. Analogously, antibodies raised to fullerenes are unlikely to recognize epitopes unique to the ends of carbon nanotubes because such epitopes do not exist in the topographically closed fullerene used as immunogen.

The end-group of the dendrimer appears to be irrelevant to whether the dendrimers trigger a response (Table 2): there are B-cell clones that recognize either amine-, oxiamine- or sulfhydryl-capped PAMAM dendrimers. We have no reason to believe that the immunogenicity of thedendrimers would significantly differ if they were capped with other functional groups (hydroxyls, carboxylic acids, etc.), though sera raised to other capped dendrimers will likely discriminate between different capped variants of the same dendrimer species. Individual antisera exhibit profound preferences for binding dendrimers with end groups the same as those of the immunogen (Table 2, 1-3): antisera raised to uncapped dendrimers (i.e., $G_0$) recognize neither sulfhydryl- nor oxiamine-capped $G_5$ PAMAMs. Likewise, sera raised to the capped species recognize only their cognate immunogens. Thus, the terminal groups of the PAMAM dendrimers are constituents of the recognized epitopes, but do not fully define the epitope. Antisera raised to $G_0$ PAMAM dendrimers fail to recognize POPAM dendrimers (Table 2), even though POPAMs and PAMAMs exhibit broadly analogous (dendritic) architecture, and both have primary amines on their surfaces.

Our experiments do not fully characterize the immunizing epitopes or the epitope recognized by anti-dendrimer antibodies. For instance, it is uncertain whether PAMAM dendrimers can be processed to smaller fragments, and if not, whether intact dendrimers can act as B-cell epitopes. While technically difficult to address, the issue is relevant to *in vivo* use of dendritic polymers. If PAMAM dendrimers cannot be processed, and if intact dendrimers over some (as yet undetermined) size threshold cannot act as B-cell epitopes, it may be possible to design protein-dendrimer conjugates that would not trigger anti-dendrimer antibodies. Such conjugates might contain extremely pure, high molecular weight dendrimers (i.e., in excess of a hypothetical maximum size for immunogenicity) that are uncontaminated with lower generation dendrimers or dendrimer fragments that might be conjugated to proteins and thereby become immunogenic. Similarly, degradation events (such as reverse Edmunds degradation in PAMAMs) occurring on storage of dendrimer-protein conjugates could be problematic. Non-immunogenic conjugates might become increasingly immunogenic on storage as dendrimer bioconjugates fragment into smaller units that contain dendritic epitopes covalently linked to T-helper epitopes.

## 3.4. Summary and prospects

Despite their low intrinsic immunogenicity, PAMAM dendrimers can stimulate antibody responses when a helper T-cell epitope is covalently linked to the dendrimer. Thus, protein-dendrimer conjugates can be immunogenic, though small molecule-dendrimer conjugates may not be. Since we don't know the specific immunizing and antibody-binding epitopes involved, we can't rule out the possibility that there may be PAMAM dendrimer lots or generations of PAMAM dendrimers that wouldn't trigger an antibody response when conjugated to proteins. We also cannot rule out the possibility that some dendrimers might act as T-independent antigens, and trigger antibody responses in the absence of a linked T-helper epitope. These considerations must be explored in bringing therapeutics forward that contain PAMAM or other dendritic polymers.

The antibody responses triggered are specific to the end groups of the dendrimers and may be influenced by the spatial arrangement of the end groups. The antibodies are amenable to immune detection and quantitation of PAMAM dendrimers in standard immunoassays (ELISA, dot and Western blot assays) and exhibit cross-reactivity across multiple PAMAM dendrimer generations (analogous phenomena have been observed in antibodies directed to other synthetic nanostructures, 4). Cross-reactivity of antibodies between different proteins is rare, but chemical homogeneity of many synthetic nanomaterials may result in cross-reactivity between antibodies directed to chemically similar but topographically or structurally distinct materials. Some uncertainties of the current study arising from use of polyclonal anti-PAMAM dendrimer sera will be resolved by generation of monoclonal antibodies to dendrimers, which we are pursuing. These monoclonal antibodies will be useful for manipulation and processing of dendritic nanomaterials (1-3).

## 5. REFERENCES

1. Lee, S. C.; Parthasarathy, R.; Botwin, K.; Kunneman, D; Rowold, E; Lange, G; Zobel, J; Beck,T.; Miller, T.; Voliva, C., Biomed. Microdev., 1, 53 (2001).

2. Lee, S. C.; Parthasarathy, R.; Botwin, K.; Kunneman, D; Rowold, E; Lange, G; Zobel, J; Beck, T.; Miller, T.; Voliva, C., PMSE, 84, 824 (2001).

3. Lee, S. C.; Parthasarathy, R.; Botwin, K.; Kunneman, D; Rowold, E; Lange, G; Zobel, J; Beck, T.; Miller, T.; Voliva, C., Antibodies to PAMAM dendrimers: Reagents for immune detection, assembly and patterning of dendrimers. *in* **Dendritic polymers-A new macromolecular architecture based on the dendritic state** (Tomalia, D. A.; Frechet, J. Eds.), John Wiley & Co., London, *in press*.

4. Chen, B-X.; Wilson, S. R.; Das, M.; Coughlin, D. J.; Erlanger, B. F., PNAS 95, 10809 (1998).

5. Duncan, R. Chemistry & Industry, 1, 262 (1997).

6. Uhrich, K. TIPS 5, 388 (1997).

7. Baker, J. R., jr. Therapeutic nanodevices. *in* **Biological molecules in nanotechnology: the convergence of biotechnology, polymer chemistry and materials science** (Lee, S. C.; Savage, L., Eds.) , pp 173-183, IBC Press, Southborough, 1998.

8. Baker, J. R., jr.; Quintana, A.; Piehler, L.; Banazak-Hall, M.; Tomalia, D.; Rackza, E. Biomed Microdev 1, 61 (2001).

9. Lee, S. C. Parthasarathy, R. and Botwin, K. Polymer Preprints 40, 449 (1999).
10. Breitling, F.; Dubel, S. **Recombinant antibodies**. John Wiley & Sons, New York. 1999.
11. Janeway, C. A.; Travers, P.; Walport, M.; Capra, J. D. **Immunobiology**, Garland Publishing, New York, 1999.
12. Dvornic, P. R.; Tomalia, D. A. Curr. Opin. Coll. Interfac. Sci. 1, 221 (1996).
13. Frechet, J. M. J., Science 263, 1710 (1994).
14. Tomalia, D. A.; Baker, H.; Dewald, J.; Hall, M.; Kallos, G. ; Roeck, J.; Ryder, J.; Smith., P. B., Polymer J. 17, 117 (1985).
15. Barth, R. F.; Adams, D.; Soloway, A. H.; Alam, F. Darby, M. V. Bioconjugate Chem. 5: 58 (1994).
16. Roberts, J.; Bhalgat, M.; Zera, R. T. J. Biomedical. Materials Res. 30, 53 (1996).
17. Toyokuni, T.; Singhal, A. K. Chem. Soc. Rev. 22: 231 (1995).
18. Aslam, M; Dent, A. **Bioconjugation**, Grove Press, New York, 1998.
19. Lemieux, G. A.; Bertozzi, C. R. TIBTech. 16: 506 (1998).

Chapter 4

# PREPARATION AND CHARACTERIZATION OF NOVEL POLYMER/SILICATE NANOCOMPOSITES

Mason K. Harrup, Alan K. Wertsching, and Michael G. Jones
*Energy and Environmental Sciences, Idaho National Engineering and Environmental Laboratory, P. O. Box 1625, Idaho Falls, ID 83415-2208*

## 1. INTRODUCTION

### 1.1 Nanocomposite Classification System

Nanocomposite materials with an inorganic glass and an organic polymer constitute a relatively new and unique area in material science. The term "ormocers", "ormosils" and "ceramers" are often utilized to describe this class of nanocomposite (1, 2). By combining at the molecular level inorganic and organic polymeric material a blending of unique physical properties can be achieved. The value in these materials is apparent, from fiber optics to paints these materials may provide the requisite physical properties to achieve the next technological advance.

There are several different ways of synthesizing this class of nanocomposite; therefore a means of classification is necessary. Most developed nomenclature is based on synthetic techniques; Wilkes has a relatively recent and exhaustive categorization (3). However we chose to classify these materials upon a simpler system first suggested by Novak (4). Five categories cover the majority of composites synthesized with more recent techniques being modifications or combinations from this list.

Type I: Organic polymer embedded in an inorganic matrix without covalent bonding between the components.

Type II: Organic polymer embedded in an inorganic matrix with sites of covalent bonding between the components.

Type III: Co-formed interpenetrating networks of inorganic and organic polymers without covalent bonds between phases.

Type IV: Co-formed interpenetrating networks of inorganic and organic polymers with covalent bonds between phases.

Type V: Non-shrinking simultaneous polymerization of inorganic and organic polymers.

For our purposes here, we will limit discussion to only Type I, II and V because of their prevalence in the literature over the other two Types.

Before discussing different kinds of nanocomposites it is appropriate to discuss the sol-gel process. Although zirconium, titanium, aluminum and boron oxides have been utilized as the inorganic component (5), the great majority of nanocomposites incorporate silica from tetraethoxysilane (TEOS). The formation of the inorganic component involves two steps, hydrolysis and condensation as seen in Scheme 1. The important insight in the formation of this part of the composite is based upon the relative kinetic rates of each step. For example, if the rate of hydrolysis is fast compared to condensation, then simple particles or highly branched silicate matrices are formed. Conversely, if the condensation step is quicker than hydrolysis, then string-like filaments are formed. These changes in morphology of the silicate matrix can be manipulated by the choice of sol-gel catalyst with dramatic effect on the physical properties of the nanocomposite (6).

**Hydrolysis**

$$Si(OR)_4 + nH_2O \longrightarrow (RO)_{4-n}Si(OH)_n + nROH$$

**Condensation**

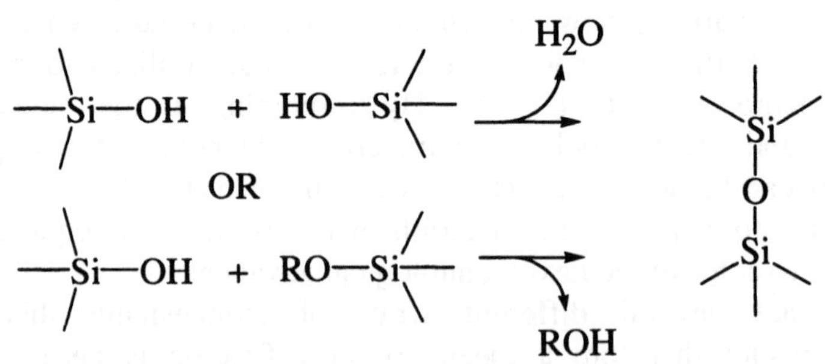


**Overall**

$$Si(OR)_4 + H_2O \xrightarrow[\text{Catalyst}]{\text{Acid, base or salt}} SiO_2 \text{ glass}$$

*Scheme 1.* Generalized scheme for the hydrolysis and condensation of ceramic species.

Perhaps the most common and straightforward nanocomposites found in the literature are the Type I composites. Typically, TEOS is used to form an inorganic network around an organic polymer component. The goal in the process is to form a completely interpenetrating network (IPN) of both inorganic and organic phases. Homogeneous nanocomposites with good IPNs are often stronger, more resilient, and optically transparent, whereas heterogeneous composites are often mechanically weaker and opaque. Choice of co-solvents is critical throughout the formation of the inorganic component. As the hydrolysis and/or condensation reaction occurs, changes in the polarity of the solvent mixture can result in undesired phase separation of the inorganic and organic components. Phase separations can produce a range of results from small zones of one component dominance to total inorganic/organic separations. However, the result is that the same sites of instability are created. The timing and physical conditions in which phase separations occur can produce isolated silicate formations within composite formulations as seen in Figure 1.

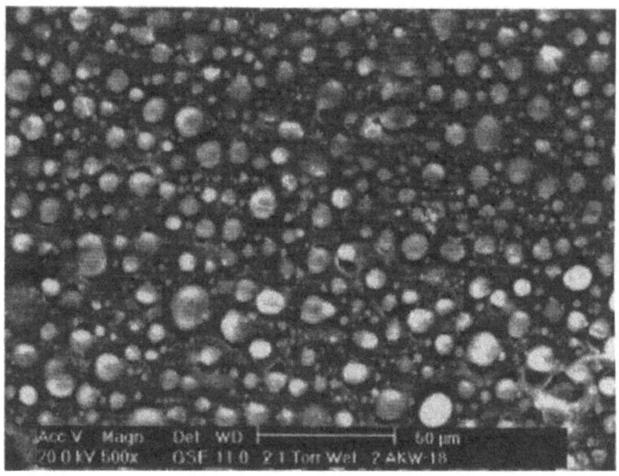

*Figure 1.* Formation of isolated ceramic forms in a phase separated composite.

Type II composites utilize modified organic polymers with sites capable of covalently bonding directly to the forming inorganic phase of the nanocomposite. Most often an organosilane with a reactive pendant group is used to form the covalent bond to the organic polymer component. These pendant groups can be isocyanates, which easily react with polymers containing alcohols or amines (4). Another option that has been explored is hydrosilation reactions with polymers containing a terminal alkene (7). Once bound, the silane's remaining pendant ethoxide or chloride groups will be available to bond with such sites on other silane species, forming the inorganic component.

Type V composites are unique examples of these materials because of the physical conditions in which they are formed. Novak and Grubbs developed a method of simultaneously polymerizing both inorganic and organic components without shrinkage of the material (8). The insight was that by replacing the ethoxide found in TEOS with an organic monomer oxide, solvent loss created during the hydrolysis step is eliminated. It is the loss of ethanol (or other solvents) in the sol-gel process, which creates material shrinkage and leads to cracking from capillary pressures. Although extremely innovative, unfortunately a limited number of composites have been formed successfully with this process (4). This is presumably because of the difficulties in precisely matching the rates of polymerization of the organic component with both the rates of hydrolysis and condensation of the ceramic component to prevent disastrous phase separations.

## 2. EXPERIMENTAL

### 2.1 Syntheses

A general synthesis for a base, acid, or salt catalyzed polyphosphazene, polyethylene oxide (PEO), and polyethylene oxide/polypropylene oxide (PPO/PEO) block nanocomposite is as follows: 300 mg of polymer is dissolved into 10 mL of a 50/50 by volume tetrahydrofuran (THF)/ethanol mixed solvent in a capped vial. To this solution is added TEOS (336 mg). A catalyst is then introduced as an aqueous solution (150 μl) and the mixture is capped and sonicated at 50 °C for 30 minutes. The solution is aged from hours to days depending upon the catalyst used in a sealed vial and poured into a Teflon mold and loosely covered at room temperature. The nanocomposite self assembles as the volatile solvent slowly escapes during the condensation process.

The synthesis of polyvinyl acetate (PVAc)/silicate nanocomposites requires a different approach from the other nanocomposites. PVAc (300 mg) is dissolved into an 50/50 by volume acetic acid/methanol (10 mL) mixed solvent in a capped vial. To this solution is added TEOS (373 mg). The solution is then sonicated for 5 minutes in a sealed vial at room temperature and poured into a Teflon mould and loosely covered at room temperature. The nanocomposite self assembles during the curing process, which typically lasts up to 24 hours. Additional heating at 100 °C for 30 minutes aids in removing lingering acetic acid from the nanocomposite.

### 2.2 Mechanical Analyses

Various membrane samples were tested on a Dynamic Mechanical Analyzer (TA Instruments, Model 2980, New Castle, DE) fitted with a Tension (film) or Penetration (compression) clamp. A simple linear ramp-force protocol was developed to evaluate the membrane strength in tension and compression. Both tension and compression protocols consist of the establishment of 0.03 Newton (N) static force and equilibration to 25 °C. A force is then applied to the specimen at a specific rate, 0.5 N per minute for tension, and 1.0 N per minute for compression (18 N maximum) or until the maximum limit of travel is reached (24.7 mm). The membrane specimens are cut with a laboratory fixture consisting of two razor blades mounted on a parallel plate yielding a 6.20 mm by 39.0 mm sample. The specimen's thickness is measured with a caliper (Mitutoyo, Model ID-C112EBS, Japan) in several places along the length to obtain an average membrane thickness.

## 2.3 ESEM Measurements

Various thin films were imaged as an unstressed, swelled, or stretched nanocomposite with a Philips XL30ESEM. A 10 to 20 kV electron beam was utilized. Wet mode analyses employed a 500 μm pressure-limiting aperture (PLA). To image the topography either a gaseous secondary electron detector (wet mode) or secondary electron detector (HiVac) was used.

# 3. RESULTS AND DISCUSSION

## 3.1 Polyphosphazene nanocomposites

The polyphosphazene family of polymers was chosen for our initial investigations into nanocomposites due to their diverse array of physical and solubility properties. We were specifically interested in some heretofore inaccessible phosphazenes that lacked the mechanical stability to be useful for membrane and barrier applications. The polymer poly[*bis*-(2-(2-methoxyethoxy)ethoxy)phosphazene] (MEEP) is such a polymer - one where the chemical properties, such as selective transport of ions are extremely interesting, yet the polymer alone lacks the requisite mechanical integrity to form a stable membrane for evaluation and subsequent use. This water-soluble polyphosphazene is a viscous liquid with two phosphorus bound polyether ligands on an alternating nitrogen phosphorus backbone. In all of the nanocomposites TEOS was employed as the ceramic precursor. The condensation of the ceramic component was catalyzed in a solvent composed of various ratios of water, ethanol and THF. Depending upon the nature of the catalyst employed, differing properties were observed in the resulting nanocomposites. These catalysts are best separated into three categories; acid, base or ionic salt; and each will be discussed separately below.

Acid catalyzed nanocomposites that were examined in this study were found to have the highest tensile strength of any of the catalyst types studied; yet they were also found to be glassy and brittle when compared to salt or base catalyzed analogues. While providing a maximum of physical stabilization for the liquid-like MEEP polymer, this translates directly into extremely poor durability when wetted, as exhibited in aqueous swelling tests. Figure 2 is an ESEM image of an aqueous swelled/deswelled membrane, which has fractured from the internal hydrostatic pressure. The dominant reason for this lack of durability when wet is the nature of the ceramic morphology of these acid catalyzed nanocomposites. It is known from the ceramic literature that the general morphology of acid catalyzed sol-gel condensations forms long filament structures instead of star-like or particulate structures (9, 10). This more rigid morphology imparts a high tensile strength to these nanocomposites, yet leaves them less elastic and highly susceptible to damage upon contact with liquids.

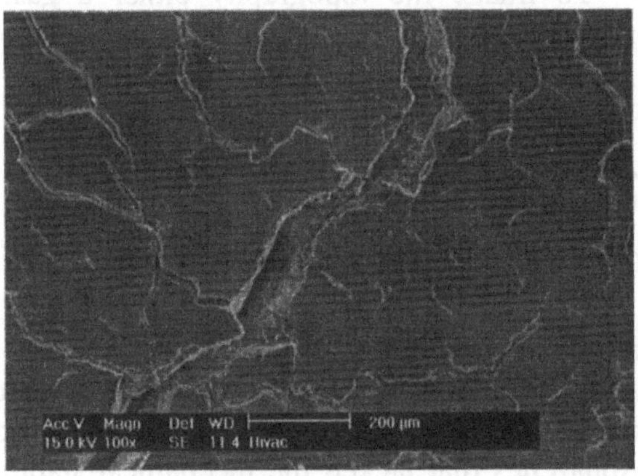


*Figure 2.* ESEM of an acid catalyzed nanocomposite after wetting.

Base catalyzed nanocomposites have the best elastic and swelling characteristics of all the nanocomposites studied. As seen in the acid catalyzed nanocomposites, the morphology of the condensing ceramic component dominates the physical properties of the material. Figure 3 is an ESEM image of a MEEP nanocomposite after being soaked to equilibrium in water. Before swelling, the nanocomposite has a featureless uniform appearance, as it does after drying. During wetting, the hydrophilic polymer component expands as it absorbs the water yet the ceramic is not capable of expansion. This behaviour produces the creases seen in the micrograph. The villae-like features reveal the underlying structure of the ceramic component of the nanocomposite, i.e. areas of higher silicate composition. This is consistent with the morphology expected for a base catalyzed nanocomposite – one that contains interlocking star-like structures (9). In regions between the ceramic star structures, the polymer is capable of absorbing water and expanding, while regions within the ceramic star structures maintain their structural rigidity. Because a single material contains both of these types of different regions, this renders base catalyzed nanocomposites simultaneously both tough and flexible.

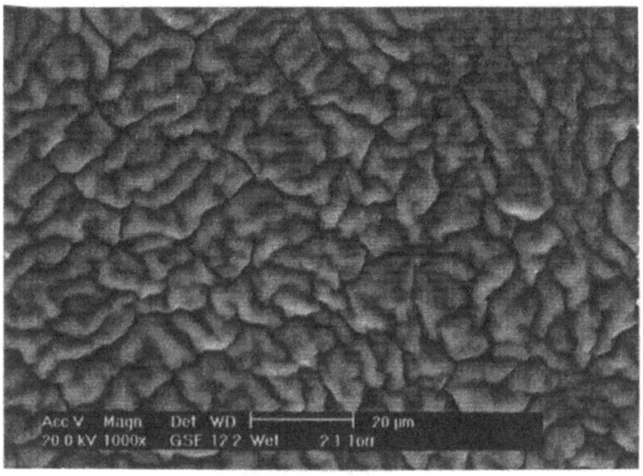


*Figure 3.* ESEM of a base catalyzed nanocomposite taken while still wet.

Salt catalyzed nanocomposites exhibit properties that are between those found for acid and base catalyzed nanocomposites. Again, the nature of the ceramic dominates the physical properties of the nanocomposite. There is considerable variation in the properties of this class of nanocomposites as there are a variety of simple salts that have proven to be effective catalysts. To investigate the nature of ionic salt catalysts in MEEP nanocomposites, a course of fluoride salts with various alkali counter-cations was examined, as it is known that fluoride anion salts are superior ceramic condensation catalysts over other halide salts (11). As was expected, the mechanical properties of fluoride catalyzed nanocomposites more closely resembled their base catalyzed analogues because the condensation was driven by an anionic catalyst, similar to $OH^-$.

The initial results indicated salt catalysts that contain large cations like cesium performed better than small ones such as lithium. Intuitively, hard ions such as lithium, the closest of the alkali metal ions to the proton (an extremely effective catalyst) should outperform the softer ionic catalysts, yet exactly the opposite trend is observed. The reason behind the observed trend becomes apparent when one considers that the strongly electronegative fluoride ion has a poorer affinity towards large polarizable cations (like cesium) than harder cations (like lithium). Catalyst availability in these polar, protic organic solvent systems becomes the dominant factor in the total catalytic effectiveness of these systems. As shown in Figure 4, there is a linear correlation between the lattice energy of the alkali fluoride salt employed and the final strength of the resulting nanocomposite, with the lowest energy salt yielding the strongest material. This is due to higher availability of the "naked" fluoride anion in systems possessing the lower lattice energies. In these systems, the catalyst ions spend less time during synthesis associated as intimate ion pairs in solution and are consequently more effective in driving the ceramic condensation.

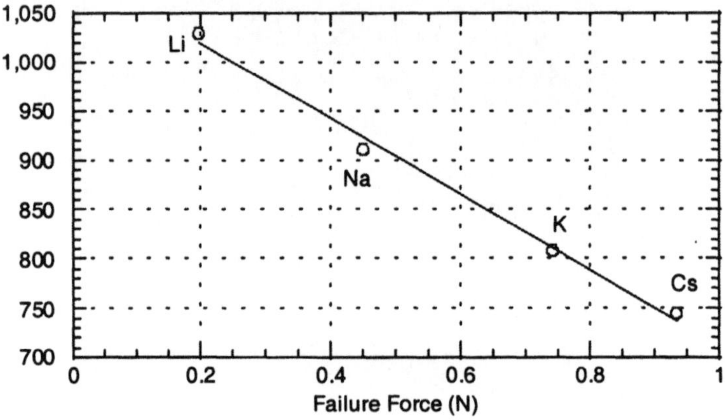


*Figure 4.* Correlation between catalyst lattice energy and resulting nanocomposite strength.

## 3.2 Organic Polymer Nanocomposites

A better understanding of the formation, properties, and morphology of Type I nanocomposites was gained during our initial studies with polyphosphazene polymers. We used this information to try to better understand the properties of nanocomposites employing more conventional organic polymers. Commercially available polymers were used in the preparation of these polymer/silicate materials. A block polymer consisting of PEO/PPO ($M_w$ = 2K), PEO (2K), polyvinylalcohol (PVA, 30–70K), PVAc (83K and 167K) and poly(acrylonitrile) (PAN, 150K) were the primary organic polymers of interest. Drawing upon previous experience, a 50/50 (w/v) polymer to silicate ratio was used to form these nanocomposites. Acid catalyzed composites were the focus of most of these studies.

Numerous attempts were made with low molecular weight PEO and PEO/PPO block polymers to prepare a nanocomposite with acceptable mechanical integrity. Unfortunately, all attempts resulted in materials that were too brittle to be lifted as a single unit suitable for mechanical analyses. This was surprising in light of the fact that acid catalyzed composites were previously found to yield the strongest nanocomposites. Our explanation for this behaviour lies in the low molecular weight of the polymer component. Typical molecular weights for the MEEP polymer were in the $10^6$-$10^7$ range, providing a directing effect for the condensing ceramic. The PEO polymer was sufficiently light that the ceramic exhibits a directing effect on the orientation of the polymer leading to a significantly different ceramic morphology, more base-like in nature. (10)

Polyvinyl alcohol (PVA) and polyvinyl acetate (PVAc) were found to readily form tractable nanocomposites with good mechanical integrity, good flexibility and interesting wetting characteristics. In general, the PVA nanocomposites yielded a significantly higher storage modulus then the PVAc nanocomposites. However, results of a simple destructive force ramp test, at the individual material's glass transition temperature (PVA 85 °C, PVAc 30 °C), suggests that the PVAc nanocomposite actually is mechanically stronger. Another interesting feature of these two nanocomposites that was noted is that they tend to behave somewhat as a thermoplastic. That is, once heated above the polymer component's glass transition temperature, the material can be easily deformed or bent, and the material will retain its shape once cooled. This is a rare example of where the polymer dictates the physical properties of the entire nanocomposite.

The behaviour of PAN nanocomposites was also explored. However, PAN has limited solubility in solvents commonly used to form this type of nanocomposite due to polarity differences between the polymer (lower polarity) and the solvents or TEOS (higher polarity). Attempts to form PAN nanocomposites through our usual synthetic protocols resulted in an unusual material unlike the other homogeneous nanocomposites previously produced. The polymer and ceramic components retreated into separate phase domains, forming an inhomogeneous composite material. As was expected, the mechanical properties of this material were poor due the poor interfacial characteristics between the polymer and ceramic phases. Additional attempts were made at producing a true homogeneous nanocomposite by using N-methylpyrrolidone as the solvent for the condensation. Though a homogeneous nanocomposite was formed with this method, difficulties in the condensation/curing process resulted in a highly fractured material such that a mechanically acceptable material was never obtained.

## 4. APPLICATIONS

There are a host of applications for the kind of nanocomposites described above. One of the most interesting of these applications is as solid polymer electrolytes (SPE) for lithium batteries. The polyphosphazene MEEP is a well-known SPE with very high room temperature conductivity, however it lacks the mechanical stability to be used in a practical device (12). Traditional stabilization methods, such as deep UV or electron beam crosslinking methods do improve the physical stability of SPEs, however this crosslinking lowers ionic conductivity – tests performed in our laboratory revealed this to be a factor of 30-45 for MEEP-like phosphazene polymers. This reduction is due to the additional covalent linkages formed during the crosslinking process that inhibit chain segmental motion and ion transfer. Since the nanocomposites formed by the ceramic condensation process do not form bonds to the polymer component, (Type I nanocomposites) mechanical stabilization is achieved without a great loss of ionic conductivity (13). As illustrated in Figure 5, the ceramic component forms a "scaffolding" around the polymer providing the needed stability, while not interfering with the polymer's ability to easily transport ions. It is hoped that these nanocomposites will afford the development of all-solid lithium batteries that will be better, safer, and more environmentally friendly than those available today.

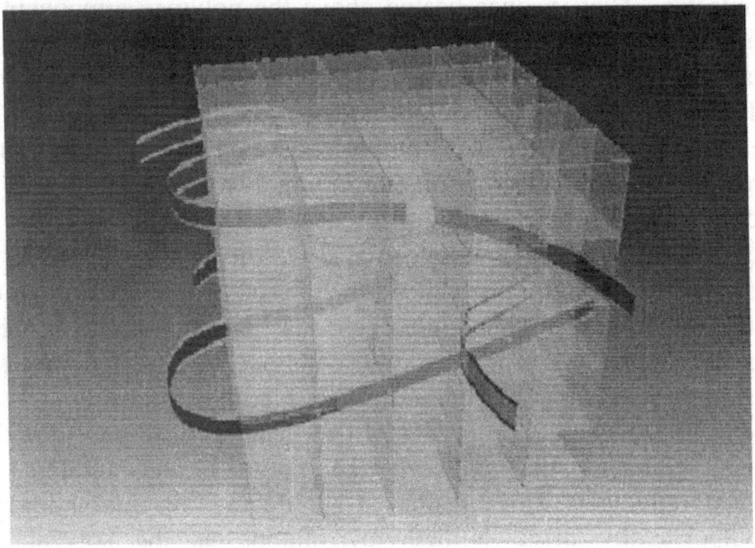

*Figure 5.* An illustration of a nanocomposite SPE. The cubes represent the ceramic component, while the ribbons represent the polymer.

Another set of applications for the nanocomposites is in separations membranes and other selective barriers. Polymers that are used in selective gas separations (14), and liquid pervaporation separations (15) must usually be mechanically stabilized, typically by crosslinking, before they become practically usable. Since the nanocomposite is able to physically stabilize the polymer component in a manner very different than these traditional methods, a real advantage may be gained through nanocompositing rather than crosslinking. Work in our laboratories is ongoing to describe the separations behaviour of nanocomposites formed from well-known separations polymers and to compare their performance to crosslinked polymer-only analogues.

One final separation application that is being explored is using nanocomposites as subsurface permeable reactive barriers. The nanocomposite matrix is being used to provide a platform for selective chelating agents that remove contaminants from subsurface water. The design employs reproducibly swellable hydrophilic materials (such as the base catalyzed materials discussed above) that allow subsurface water to flow through the barrier, allowing the entrained chelators to selectively sequester the contaminants of concern. This type of *in-situ* remediation technology is of particular interest to the U. S. Department of Energy, but is also applicable to other contaminated sites.

## 5. ACKNOWLEDGEMENTS

The authors thank Dr. Thomas A. Luther for helpful discussions. This work was supported by the United States Department of Energy through contract DE-AC07-99ID13727.

## 6. REFERENCES

1 Schmidt, H., J. Non-Cryst. Solids, 73, 681 (1985).
2 Wilkes, G. L.; Orler, B.; Huang, H., Polym. Prepr., 26, 300 (1985).
3 Wen, J.; Wilkes, G. L., Chem. Mater., 8, 1667 (1996).
4 Novak, B. Advanced Materials, 5, 422 (1993).
5 Brinker, C. J.; Scherer, G. W., Sol-Gel Science, New York, Academic (1990).
6 Harrup, M. K.; Wertsching, A. K.; Stewart, F. F. Amer. Chem. Soc. PMSE Preprints, 82, 307 (2000).
7 Speier, J. L., Adv. Organomet. Chem., 17, 407 (1979).
8 Ellsworth, M. W.; Novak, B. M. J. Am. Chem. Soc., 113, 2756 (1991).
9 Pope, J. A.; Mackenzie, J. D., Journal of Non-Crystalline Solids, *87*, 185 (1986).
10 Jonkinen, M.; Gyorvary, E.; Rosenholm, J., Coll. and Surf. 141, 205 (1998).
11 Schaefer, D. W.; Keefer, K. D., Mater. Res. Soc. Symp. Proc., 32, 1 (1984).
12 Blonsky, P. M.; Shriver, D. F.; Austin, P.; Allcock, H. R., J. Am. Chem. Soc. 106, 6854 (1984).
13 Harrup, M. K.; Stewart, F. F.; Wertsching, A. K., "Self-Doped Molecular Nanocomposite Battery Electrolytes" *U. S. Patent Pending*.
14 Orme, C. J.; Harrup, M. K.; Luther, T. A.; Lash, R. L.; Houston, K. S.; Weinkauf, D. H.; Stewart, F. F., J. Mem. Sci. 186, 249 (2001).
15 Stewart, F. F. and Harrup, M. K., Polymer News, 26, 78 (2001).

Chapter 5

# METALLOCENE HEMATOPORPHYRINS AS ANALYTICAL REAGENTS-NICKEL (II) METAL ADSORPTION STUDIES OF GROUP IVB METALLOCENE POLYMERS DERIVED FROM HEMATOPORPHYRIN IX

Charles E. Carraher, Jr., Jerome E. Haky, and Alberto Rivalta
*Florida Atlantic University, Department of Chemistry and Biochemistry, Boca Raton, FL 33431, and Florida Center for Environmental Studies, Palm Beach Gardens, FL 33410*

## 1. INTRODUCTION

There is an increasing interest in understanding and accomplishing selective separations on a molecular level, etc. Some of this effort is aimed at looking at natural and natural-like materials such as the porphyrins that selectively uptake metal ions that eventually perform selective functions. It should be possible to synthesize such porphyrin-containing assemblies that in themselves selectively bind ions that subsequently perform specific functions that may include selective catalytic, electrical, binding, and photon activities. Through combining porphyrins with other moieties that can also exert specific behaviors it may be possible to construct materials that contain several important sites, with each site performing a separate or related activity.

The cyclopentadienyl derivatives of titanium, zirconium, and hafnium were first described by Wilkinson and co-workers in the early 1950's (1). These compounds have a tetrahedral geometry, possess a high degree of covalent character within their composite structure and behave as organic acid chlorides in condensation polymerization reactions. Group IV B metallocene containing materials are know to be active catalytic sites and to offer potentially useful photonic and other properties (for instance 2-6).

While Group IVB metallocene derivatives are know to be active stereoregular catalysts and some exhibit potentially useful biological activities, the current report focuses on their potential use as selective adsorptive materials for metal ions.

As noted above, in typical Lewis acid-base reactions the metallocene dichlorides act similarily to typical organic acid dichlorides allowing chain extension through reaction at both chlorides giving products such as the following for reaction with a diol where Cp represents a cyclopentadiene group.

$$Cp_2MCl_2 \quad + \quad \text{HO-R-OH -----> -(-O-R-O-}\underset{Cp}{\overset{Cp}{\text{M}}}\text{-)}$$

Porphyrins are heterocyclic compounds containing pyrrole units. They constitute major components in the human body and within plants and are essential agents for life. For instance, the iron-hematoporphyrin complex is the active component of hemoglobin and serves as the site for oxygen transfer. Another important porphyrin is the cobalt-porphyrin complex which is know as vitamin B12. Chlorophyll is a magnesium-porphyrin.

Hematoporphyrin IX, HPIX, was the first porphyrin isolated from natural materials and was obtained by the sulfuric acid treatment of blood. HPIX, and some derivatives, are commercially available and have been reported to offer a wide variety of uses. They are used as sensitizes in cancer physiotherapy (7). These compounds are selectively concentrated by neoplastic tissues, and accordingly the tumors can be distinguished by selective fluorescence under UV light.

HPIX is a macrocyclic "tetrapyrrole" structure with four methyl, two propionic acid, and two ethanolic side chains. Because of the presence of the two propionic acid and two ethanolic chain chains, the Lewis acid-base reaction with Group IVB metallocene dichlorides can, and does, give crosslinked products as shown below.

```
               |            |
            Cp-M-Cp      Cp-M-Cp
               |            |
               O            O
               |            |
            Me-C-H       Me-C-H
               \            /
                  Porphyrin
                    Core
Cp               /           \               Cp
- M-O-C-CH2 -CH2              CH2 -CH2 -C-O-M-
Cp   ||                                 ||   Cp
     O                                  O
```

HPIX is one of the most unstable of the natural porphyrins commonly employed in the laboratory because of the presence of labile hydroxyethyl side chains (8). Thus the employed synthesis of polymers from it must be accomplished rapidly and under mild conditions.

In the present study the adsorptive species (the substance being adsorbed) is a nickel ion and the adsorbent species (solid phase) are the three Group IVB products obtained from the interfacial polycondensation reaction between HPIX and the metallocene dichlorides and two commercially available ion exchange resins.

Adsorption is a term which simply means to be attached to the surface, but it can be attached either physically or chemically bonded.
The synthesis of the products from M = Ti and Zr was previously reported (9,10). Latter we will report the synthesis of the analogous Hf product.

## 2. EXPERIMENTAL

The following chemicals were used as obtained without further purification. Hematoporphyrin IX ({21H,23H-porphine-2,18-dipropionic acid-7,12-bis-(1-hydrosyethyl)-3,8,13,17-tetramethyl]; Aldrich Chemical Co.); titanocene dichloride (bis(cyclopentadienyl)titanium dichloride; 98%, Eastman Kodak Co.), Zirconcene dichloride (bis(cyclopentadienyl)zirconium dichloride; 98%; Aldrich), hafnocene dichloride (bis(cyclopentadienyl)hafnium dichloride;

Thiokol/Ventron Division, Alfa Products), nickel acetate (Delray Chem. Co.), Dowex 50X8-200 Ion Exchange Resin (Aldrich), and Amberlite MB1 Ion Exchange Resin (Mallinckrodt Chemical Works).

The synthesis of all three products is similar. The synthesis of the titanocene product is briefly described. An aqueous solution (50 ml) of hematoporphyrin IX (0.5987 gr., 1.00 mmole), to which four equivalents of sodium hydroxide (0.1600 gr., 4.00 mmole) was added, was prepared. This was added to a one quart Kimax emulsifying jar which was placed on top of a Waring Blender (Model 1120; no load stirring rate of about 18,000 rpm). A chloroform solution (50 ml) containing titanocene dichloride (0.4980 gm., 2.00 mmole) was added to the aqueous solution. After 30 seconds, a dark brownish red gelatinous precipitate was recovered employing vacuum filtration.

Atomic Absorption, AA, studies were carried out employing a Thermal Jerrol Ash Atomic Absorption Spectrophotometer. Samples were prepared in a 1:1 nitric acid solution and analyzed utilizing an air/acetylene flame.

The surface area was determined using a microscope. The diameter was measured by observing the particle size of each product. The products were approximately spherical and so the surface area was calculated using the relationship relating the surface area of a sphere. The surface area for the commercial resins was determined by water retention. The resin was treated with water, the surface dried and the weight of the wet resin was obtained. The resin was dried and the difference between the two weights related to the volume of water retained in the resin.

## 3. RESULTS AND DISCUSSION

Product yields, colors, and decomposition temperatures are given in Table 1.

*Table 1. Product yield, color, and decomposition temperature.*

| *Metallocene* | *Average Yield (Grams)* | *%-Yield* | *Decomposition Temperature, C* | *Color* |
|---|---|---|---|---|
| *Ti* | *0.203* | *20* | *200-250* | *Dk. Brown* |
| *Zr* | *0.88* | *79* | *230-260* | *Dk. Brown* |
| *Hf* | *1.1* | *97* | *250-285* | *Dk. Brown-Black* |

Elemental analysis results are given in Table 2. Two theoretical calculations are included. One is the minimum percentage of metal necessary to form a 1:1 polymer between the metallocene and HPIX. The second calculation, column 4, is the percentage metal required to correspond to the metal/HPIX ratio given in column 5. The numbers given in column 5 were determined from calculations based on the observed percentage metal.

*Table 2. Elemental analysis results and theoretical calculations related to the number of metallocenes per HPIX moiety.*

| *Metallocene* | *%-Metal* | *Calculated Minimum* | *Metal % Corresp.* | *Ratio Metal/HPIX* |
|---|---|---|---|---|
| *Ti* | *15.9* | *6.2* | *14.5* | *4* |
| *Zr* | *19.3* | *11.0* | *17.3* | 2 |
| *Hf* | *28.0* | *19.8* | *29.6* | 2 |

The zirconium and hafnium products are probably less crosslinked while the titanium product is probably more crosslinked with a greater number of titanocene end-groups.

Possible "end-groups" for the titanocene product include Ti-OH and Ti-Cl. The chlorine content was low (0.06 %) consistent with the absence of significant amounts of Ti-Cl being present. The IR shows a large band centering about 3400 1/cm consistent with the presence of hydroxyl groups that may be derived from Ti-OH end-groups and/or unreacted HPIX hydroxyl groups.
Porphyrin-containing materials are known to retain selected metal ions through chelation and through adsorption involving the pi electron cloud. The present polymeric materials may offer advantages in comparison to the monomeric porphyrin with respect to selectivity because of the added structural features within the polymeric structures. Three major ions known to be retained by porphyrins, in general, are zinc, copper, and nickel.

Metal ion incorporation into porphyrin molecules is believed to occur through an intermediate, called the sitting-atop-complex, SAT (11). Here, the metal ion rests on top of the porphyrin before it is completely incorporated. This mechanism has been verified by UV and IR spectroscopy. The active species involved in metal incorporation are the metal cation and the neutral porphyrin (12). For chelation to occur, the porphyrin must be in the liquid state. Since the current materials are insoluble, chelation studies were not performed, rather liquid-solid interface metal adsorption studies were carried out.

Adsorption processes are dependent on the chemical and physical properties of the phases, interfaces, and adsorptives. The solid or bulk phase is a critical factor that typically controls the adsorption process. The physical composition of the solid surface also plays a role in adsorption. Spaces available on the surface may be of several kinds. Generally, the edges, corners, and surface defects are more exposed giving them a higher capability to interact than less exposed portions. Thus, surface roughness is important because this roughness allows more exposure of the solid to the metal ion.

Coulombic and polar factors can also influence an absorption process. for instance, if the surface of the adsorbent molecules are polarizable and the molecules of the adsorptive contain permanent dipoles, then an interaction between the two can occur favoring adsorption.

The surface area and specific surface area for the polymers and two commercially available resins was determined. Results are given in Table 3. The surface area of the commercially available resins is much greater than the surface areas of the polymers. This is mainly due to a designed high degree of porosity found for the two resins.

*Table 3. Surface area of materials employed in the present study.*

| *Material* | *Surface Area (cm2)* | *Specific Surface Area (cm2/gram)* |
|---|---|---|
| *Ti* | $6.61X10^{-8}$ | $6.61X10^{-7}$ |
| *Zr* | $8.76X10^{-8}$ | $8.76X10^{-7}$ |
| *Hf* | $1.42X10^{-8}$ | $1.42X10^{-7}$ |
| *Dowex 50X8-200* | *2.99* | *0.299* |
| *Amberlite MB1* | *1.91* | *0.191* |

Metal adsorption studies were performed by treating the polymeric materials with solutions of nickel acetate at known concentrations followed by a determination of the concentration of the nickel solution after equilibrium. The exchange capacity of two commercially available resins was also determined. Results appear in Table 4.

*Table 4. Adsorption of nickel plus two ion.*

| *Material* | *Ni+2 Adsorbed* | | |
|---|---|---|---|
| | *mmole* | *mmole/gram Resin used* | *mmole/cm2 spec. surface area* |
| *Ti* | *$2.02X10^{-5}$* | *$2.02X10^{-4}$* | *306* |
| *Zr* | *0.00* | *0.00* | *0.00* |
| *Hf* | *0.00* | *0.00* | *0.00* |
| *Dowex 50X8-200** | *$2.37X10^{-5}$* | *$2.37X10^{-4}$* | *$1.03X10^{-3}$* |
| *Amberlite MB1*** | *$2.31X10^{-5}$* | *$2.31X10^{-4}$* | *$1.21X10^{-3}$* |

**Initial nickel ion concentration = $2.77X10^{-5}$ mmole*
***Initial nickel ion concentration = $2.59X10^{-5}$ mmole*

The results are consistent with the titanocene product having a greater exchange capacity efficiency than either of the two commercial resins. It must be remembered that these studies were performed at low concentrations of the adsorptive, the nickel ion which may correspond to the lower region of equilibrium concentration of an isotherm. It is possible that if the study was performed at saturated equilibrium concentrations, the exchange capacity for the commercial resins would be higher than the values found for the titanocene product. Even so, under the employed conditions the titanocene product out performed the commercial resins.

The reason(s) why the titanium derived material absorbed the nickel ion are currently unknown. As noted before, the titanocene product differs from the hafnocene and zirconcene products in that the titanocene product has a much greater percentage of the titanocene moiety in the product.
Preliminary results indicate that further study is merited. Further study should involve a better understanding of the influence of the surface and nature of the metalocene on the adsorption process. Also, the materials should be studied under more varied concentrations of a number of metal ions including copper, zinc, and iron as well as nickel.

## 4. REFERENCES

1. Wilkinson, G.; Pauson, P. L.; Birmingham, J. M.; Cotton, F. A. J. Amer. Chem. Soc. 75, 1011 (1953).
2. Carraher, C. **Polymer Chemistry**, Dekker, NY, pages 299 and 306, 2000.
3. Chien, J. C. **Homogeneous Polymerization Catalysis**, Prentice-Hall, Englewood Cliffs, New Jersery, 1993.

4. Kaminsky, W.; Sinn, H., **Transition Metals and Organometallics as Catalysts for Olefin Polymerization**, Springer-Verlag, NY, 1988.
5. Carraher, C. ; Linville, R.; Manek, T.; Blaxall, H.; Taylor, R.; and. Torre, L. **Conductive Polymers** (R. Seymour, Ed.), Plenum, NY, 1984.
6. Carraher, C.; Foster, V.; Linville, R.; Stevison, D.; and. Venkatachalam, R. **Adhesives, Sealants, and Coatings for Space and Harsh Environments**, Plenum, NY, 1988..
7. Dougherty, T. J.; Kaufman, J. E.; Goldfarb, A.; Weishaupt, K. R.; Boyle, D.; Mittleman, A. Cancer Research 38, 2628 (1987).
8. Saitoh, K.; Sugiyama, J.; Suzuki, N. J. Chromatography 358, 307 (1986).
9. Carraher, C.; Haky, J.; Rivalta, A.; Sterling, D. PMSE 70, 329 (1993).
10. Carraher, C.; Rivalta, A.; Haky, J. PMSE 74, 149 (1996).
11. Fleischer, E. B.; Wang, J. H., J. Amer. Chem. Soc. 82, 3498 (1960).
12. Dolphin, D., **The Porphyrins**, Volume V, Physical Chemistry Part C, page 479, Academic Press, NY, 1978.

Chapter 6

# POLYESTER IONOMERS AS FUNCTIONAL COMPATIBILIZERS FOR BLENDS WITH CONDENSATION POLYMERS AND NANOCOMPOSITES

Robert B. Moore*, Timothy L. Boykin[1] and Grant D. Barber
*Department of Polymer Science, University of Southern Mississippi, P.O. Box 10076, Hattiesburg, Mississippi 39406-0076 and [1]Plastics Division, Bayer Corporation, 100 Bayer Road, Pittsburgh, PA 15205-9741*

## 1. INTRODUCTION

Immiscible blends can be made compatible, and in some cases miscible, by introducing mutually interacting functional groups to one or both of the blend components. For example, ionomers are capable of forming strong, specific interactions with various polar groups and even at relatively low ion contents, these specific interactions tend to decrease the overall heat of mixing for immiscible blends resulting in enhanced miscibility (1). The degree of control that can be obtained over the compatibilization process is very high due to the ability to systematically vary the strength of ionic interactions by changing ion content, counterion type, and degree of neutralization. In the first section of this chapter, polyester ionomers are evaluated as compatibilizers for blends of commercially important condensation polymers.

The commercial importance of polyesters and polyamides has driven recent research into the compatibility and subsequent properties of polyester/polyamide blends (2-5). Although both homopolymers exhibit ductile behavior, non-oriented PET/N6,6 blends exhibit brittle behavior in both tensile and impact properties, under the same test conditions (3). Kamal and coworkers (3) have suggested two possible mechanisms for the decreased mechanical behavior of these blends: (i) poor compatibility leading to weak interphase behavior, and (ii) enhanced crystallinity of the polyester due to increased nucleation by polyamide crystals. The poor interphase has been attributed to phase separation upon crystallization, which leads to decreased interfacial strength between the weakly interacting polar groups of the dissimilar polymers. Consequently, methods to increase the interfacial bonding between the two phases have been of recent interest.

Compatibilization of polyester/polyamide blends has been accomplished by *in-situ* copolymer formation via ester-amide interchange reactions during melt processing (4). However, without an acid catalyst the exchange reaction must be conducted at elevated temperatures for several hours; whereas, with a catalyst, conversion of up to twenty-four percent of the base polymers was reported with reaction times of only 2-3 minutes. Consequently, reactive extrusion can be employed in the formation of *in-situ* ester-amide copolymers for compatibilization of polyester/polyamide blends. As an alternative, we have recently demonstrated the application of polyester ionomers as compatibilizers for polyester/polyamide blends (6). In this study, the effects of specific interactions and reactions between the ionomers and other commercially important polymers are evaluated.

The second section of this chapter is focused on the application of polyester ionomers as compatibilizers (or exfoliants) in clay nanocomposite systems (7). Polymer layered silicate nanocomposites often have superior physical and mechanical properties over their microcomposite counterparts, including improved modulus (7-10), reduced gas permeability (8,10), and flame retardantcy (10,11). These enhanced properties are achieved by adequately dispersing the layered silicate particles (e.g., montmorillonite clay) within the polymer matrix. The silicate polymer mixture is termed *intercalated* when at least one extended polymer chain is absorbed between the host platelets; the result is a well ordered multi-layer system with alternating polymer/inorganic host layers and a repeat distance of a few nanometers. When the silicate platelets are isotropically dispersed in a continuous polymer matrix, the material is termed *exfoliated*. Particularly in the exfoliated state, considerable enhancements in physical properties have been observed (8,12).

Silicate nanocomposites may be formed using several different dispersion techniques such as in-situ polymerization (8, 13-15), melt mixing (7-9, 16,17), and solution mixing (7,18). However, the silicate and host polymer must be chemically compatible for uniform dispersion to occur. Several compatibililization techniques have been developed to increase the favorable interactions between the ionic silicate with polar and nonpolar polymers. Generally, the chemical nature of clay particle surfaces are modified by exchanging the inorganic cations naturally associated with the clay particles with organic cations such as alkyl ammonium or alkyl phosphonium ions (19). The presence of these organic cations alters the original ionic, hydrophilic silicate surface to a more hydrophobic, organophilic surface and increases the interfacial adhesion between the inorganic silicate and the polymer. Further modifications have been made to this technique such as the use of organic cations with reactive functionalities. These reactive cations have been utilized for reaction in *in-situ* polymerizations between the platelets, which resulted in increased dispersion and increased decomposition temperature of the resulting nanocomposite (14,15).

Other compatibilization techniques have involved the utilization of oligomers and/or polymers modified with polar functionalities to increase interactions between the silicate layers and the host polymer (20). Maleic anhydride modified polypropylene has also been used as a compatibilizer for polypropylene silicate nanocomposites (16,21,22). With the addition of the maleic anhydride modified polypropylene, strong interactions between the maleic anhydride groups and inorganic clay surfaces resulted in a more uniform dispersion of the clay platelets (21). As the dispersibility of the clay was increased, the mechanical reinforcement effect of the clays was increased. Kyu and coworkers (13), formed a hybrid via *in-situ* polymerization of nylon 6 in the presence of an inorganic filler (kaolin). The hybrid material exhibited increased glass transition temperature with increased kaolin loading suggesting strong filler/matrix interactions. This material was then mixed with nylon 6,6 to form the composite. The nylon 6, chemically bound to the surface of the inorganic particles afforded increased miscibility with nylon 6,6 and resulted in stronger and stiffer samples relative to the conventional melt-mixed samples prepared without the nylon 6 compatibilizer. Fischer and coworkers (9), utilized block copolymers containing one block that was miscible with the organic polymer and the other miscible with the clay as compatibilizers various systems. Intercalation/exfoliation was achieved by variation and adjustment of the block lengths of the block copolymers.

We have recently found that ion-containing polymers (i.e., ionomers) are effective polymeric compatibilizers that increase the favorable interactions between the homopolymer host and organically-modified montmorillonite clays. In this study, the effect of ionic content in polyester and polyamide ionomers on the dispersion of clay platelets in the host polymer is evaluated.

# 2. EXPERIMENTAL

## 2.1 MATERIALS

The sodium form of a sulfonated polyester ionomer (AQ® 55; EW ≅ 1500 grams of polymer per mole of sulfonate groups; IV = 0.42) was obtained from Eastman Chemical Company. A non-sulfonated form of the AQ polymer (ns-AQ; IV = 0.33), which has an analogous chemical composition to AQ® 55 minus the addition of the sulfonated comonomer (sulfoisophthalic acid), was also supplied by Eastman Chemical Company. Poly(ethylene terephthalate) (Eastar® 7352; $M_n$ = 25,000 g/mol) was supplied by Eastman Chemical Company, poly(butylene terephthalate) (Valox® 315; $M_n$ = 110,000 g/mol) was supplied by General Electric Company, and nylon 6,6 (Vydine® 21; $M_n$ = 18,000 g/mol) was supplied by Solutia, Inc. To study the effect of counterion type on blend behavior, the "as-received" form of the amorphous polyester ionomer (NaAQ) was converted to both the $Zn^{2+}$ and $Mn^{2+}$ forms by a solution-state ion-exchange process (6). The sulfonated PET (NaSPET) ionomer was used as received in the sodium ion form.

Organically-modified montmorillonite clay (Cloisite 30A) was received from Southern Clay Products containing the methyl, tallow, bis-2-hydroxyethyl ,quaternary ammonium cation at a loading of ca. 95 meq/100g. The sodium form of sulfonated poly(ethylene terephthalate) [Na-SPET] containing ca. 2, 6, and 10 mol% sulfoisophthalic acid respectively was also obtained from Eastman Chemical Company. Polyamide 6,6 (Vydine 21® $M_n$ = 18,000 g/mol) [PA] was supplied by Solutia Inc. The sodium form of sulfonated polyamide 6,6 [Na-SPA] containing ca. 2.3 mol% sulfoisophthalic acid was obtained from Solutia, Inc. All materials were dried under vacuum at 80 °C for 12 hr and stored in a vacuum desiccator. The 2,2,2,-trifluoroethanol was obtained from Aldrich. The 1,1,1,3,3,3-hexafluoro-2-propanol (99.5%) was obtained from Acros Organics. All other materials were obtained from Aldrich and used as received without further purification.

## 2.2 PREPARATION OF BLEND SAMPLES

Due to limited amounts of the different ion forms of the amorphous polyester ionomer, a small melt-mixer with a mechanical stirrer was developed to prepare melt blends for comparison with solution blends. Blend compositions ranged from 0 to 50 wt% AQ polymer with both PET and N66.

Phenol/tetrachloroethane (Ph/TCE), 1:1 (v/v) mixture, was used as the common solvent for the binary blends. The blend components were separately dissolved to obtain a 10% (w/v) polymer solution. The solution blends were prepared by combining the appropriate polymer solutions (by volume) to the desired blend composition. The same compositions as the melt blends were used for the solution blends. After precipitation, the blends were then dried under vacuum (80 °C) for 2 days to remove any residual solvent.

Ternary PBT/N66 blends containing the NaSPET compatibilizer were prepared by a two-step compounding method using a Haake counter-rotating twin-screw extruder equipped with a Rheocord 90 controller. In order to promote ester-ester interchange between PBT and NaSPET-6, the polyester ionomer was pre-extruded with PBT at a screw speed of 15 rpm using the following temperature profile: $Z_1$ = 260 °C; $Z_2$ = $Z_3$ = 270 °C; die = 250 °C. These pre-extruded blends were then blended at the appropriate composition with N66 in order to obtain ternary blends containing 2 and 5 wt% NaSPET-6. Each of the N66/PBT blends was extruded at a screw speed of 15 rpm using the following temperature profile: $Z_1$ = 260 °C; $Z_2$ = $Z_3$ = 280 °C; die = 265 °C.

Samples for tensile testing were prepared by injection molding using a BOY 15S injection molding machine. The N66/PBT blends and blend components were injection molded under the following conditions: barrel temperature; rear zone = front zone = 275 °C; mold temperature ca. 25 °C (i.e., water-cooled); screw speed: 175 rpm; injection pressure: 100 bar; cycle time: injection = 30 s, cure = 30 s.

## 2.3 BLEND CHARACTERIZATION

Thermal analysis of the pure polymers and blends was performed using a Perkin Elmer DSC 7. The samples ranged in weight from 10-15mg, and were scanned from 45 °C to 300 °C under a nitrogen atmosphere at a heating rate of 20 °C/min. Prior to data collection, each sample was heated to ca. 300 °C then quenched at a rate of -200 °C/min to the starting temperature. The melting points of the blends and pure polymers were used to determine the Flory-Huggins interaction parameter ($\chi_{12}$) for each blend system using the Nishi-Wang approach (23).

*Dynamic Mechanical Analysis.* Samples were evaluated using a Seiko DMS 210 dynamic mechanical analyzer (DMA) in the tensile mode at a frequency of 1 Hz with the amplitude of the dynamic deformation set at 10 μm. The glass transition temperature ($T_g$) was taken as the peak maximum of the α-relaxation from the tan δ versus temperature plot. Thin films (ca. 0.1 mm) for the DMA analysis were prepared by compression molding samples between polyimide sheets (Theramalimide®) at 270 °C, followed by rapidly cooling to room temperature. The sample dimensions were ca. 30 mm x 8 mm x 0.1 mm.

*Static Mechanical Analysis.* The tensile properties of the N66/PBT blends and blend components were determined by stress-strain analysis of tensile samples at room temperature using a Mechanical Testing System 810 following the D 638 ASTM standard procedure. The stress-strain behavior of conditioned (50% relative humidity for 3 months) samples were evaluated for each of the N66/PBT blends and blend components.

*TEM Analysis.* The phase morphology of the N66/PBT blends was evaluated by transmission electron microscopy (TEM) using a Zeiss EM 10C Electron Microscope. Thin sections were cut at room temperature using a Reichert-Jung Ultracut E ultramicrotome equipped with a glass knife. Thin sections of the blends were collected and stained using an osmium tetraoxide vapor for an exposure time of ca. 30 minutes. Staining allowed for increased contrast between the aromatic polyester (PBT) and the aliphatic polyamide (N66).

## 2.4 PREPARATION OF NANOCOMPOSITE SAMPLES

For polyester nanocomposites, 1,1,1,3,3,3-hexafluoro-2-propanol and chloroform, 1:1 (v/v) mixture were used as the common solvent. The nanocomposite components were separately dissolved to obtain a 10% (w/v) polymer solution. The ionomer and organically-modified montmorillonite were stirred in solution for 10 min then subjected to ultrasonic agitation for 20 min. The homopolymer solution was then added to the ionomer-clay solution, stirred in solution for 10 min and then subjected to ultrasonic agitation for 20 min. The films were cast in a crystallization dish overnight at room temperature. These materials were then dried for 12 hr under vacuum at 80 °C to remove residual solvent and moisture.

For the polyamide nanocomposites, 2,2,2-trifluoroethanol was used as the common solvent. The nanocomposite components were separately dissolved to obtain a 10% (w/v) polymer solution. The ionomer and organically-modified montmorillonite were mixed in solution for 30 min followed by the addition of the homopolymer and mixed for another 30 min. The films were cast in a crystallization dish overnight at room temperature.

## 2.5 NANOCOMPOSITE CHARACTERIZATION

X-ray diffraction data were obtained in the transmission mode using a Siemens XPD-700P polymer diffraction system equipped with a two-dimensional, position-sensitive area detector. The sample-to-detector distance was 30 cm, yielding an angular scan from 1 to 10 degrees (2θ). The two-dimensional scattering patterns were analyzed using the GADDS™ software package, and the x-ray diffraction patterns were integrated radially, after background subtraction, to obtain intensity versus scattering angle 2θ plots.

# 3. RESULTS AND DISCUSSION

## 3.1 AQ/PET BLENDS

To determine the effect of the sulfonate group and counterion type on the degree of compatibility, several forms of the AQ polymer (ns-AQ, NaAQ, MnAQ) were blended with PET. The observed melting points of the NaAQ/PET blends were found to depress only slightly up to a blend composition of 50 wt.% NaAQ. However, a significant melting point depression was observed for both the ns-AQ/PET blends and the MnAQ/PET blends (6). This behavior may be attributed to transreactions between the AQ polymer and PET. In contrast, the melting behavior exhibited by the NaAQ/PET melt blends may be attributed to poor dispersion of NaAQ during melt-mixing due to the relative high viscosity of the $Na^+$ form of the ionomer compared to both the non-sulfonated and $Mn^{2+}$ forms.

By altering the blending procedure, the origin of the melting point depression may become evident. For 50:50 AQ/PET blends, melting point depression is only observed for melt and heat-treated samples; solution-cast samples show no melting point depression. This suggests that the AQ ionomer and PET are inherently immiscible and melting point depression occurs only after significant transesterification at elevated temperatures.

In order to compare the relative rate of ester-ester interchange for the AQ ionomer/PET blends, samples of the AQ/PET solution blends were heat-treated at 285°C for time periods ranging from 2 to 60 minutes. The data in Figures 1 and 2 show that the AQ/PET blends are initially phase separated, which is indicated by dual glass transition temperatures corresponding to the distinct $T_g$'s of the pure blend components. The NaAQ/PET solution blends, Figure 1, exhibit only a minimal change (toward convergence) in the $T_g$'s of the blend components for reaction times up to 60 minutes.

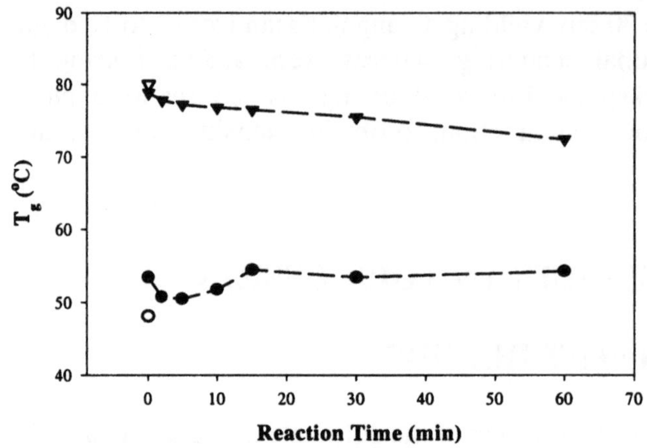


*Figure 1. Glass transition temperatures of heat-treated NaAQ/PET solution blends: (●) NaAQ phase, (▼) PET phase, (○) 100% NaAQ, and (▽) 100% PET.*

In contrast, the $T_g$'s of the MnAQ/PET solution blends, Figure 2, quickly converge to a single $T_g$ after only 5 minutes of heat-treatment. These data suggest that the counterion type can play an important role in catalyzing the desired trans-reactions.

## 3.2 AQ/N66 BLENDS

As with the AQ/PET blends, various forms of the AQ polymer (ns-AQ, NaAQ, and ZnAQ) were melt-mixed with N66, and the melting behavior was evaluated by DSC analysis. The $T_m$'s of the ns-AQ/N66 blends were found to remain effectively unchanged up to a concentration of 50 wt.% of the blend components. However, both the NaAQ/N66 blends and the ZnAQ/N66 blends exhibited a substantial melting point depression, with the largest $T_m$ depression exhibited for the ZnAQ/N66 blends (6).

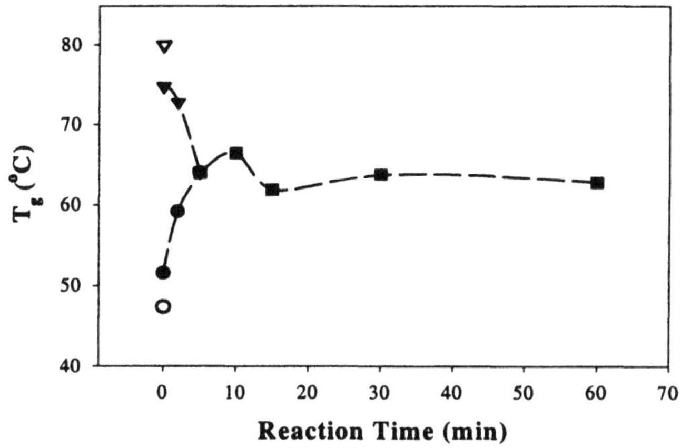


*Figure 2. Glass transition temperatures of the heat-treated MnAQ/PET solution blends: (●) MnAQ phase, (▼) PET phase, (■) MnAQ/PET mixed-phase, (○) 100% MnAQ, and (▽) 100% PET.*

For 50:50 AQ/N66 blends, negative interaction parameters are observed for melt, solution-cast and heat-treated samples containing the ionomer (see Table I). These values illustrate both the effect of the sulfonate group and counterion type on the degree of interactions between the AQ polymer and N66 and suggest that the compatibility of the ionomer with N66 is principally due to specific interactions involving the metal-counterion of the ionomer and the amide groups. Furthermore, the large negative $\chi_{12}$ values for the ZnAQ/N66 blends may be attributed to the ability of the divalent counterion to complex with two amide groups; whereas, the monovalent sodium counterion is only capable of associating with one amide group.

*Table 1. Polymer-polymer interaction parameters, $\chi_{12}$, for AQ/N66 blends.*

| Blending Method | ns-AQ/N66 | NaAQ/N66 | ZnAQ/N66 |
|---|---|---|---|
| Melt Blend | -0.1 | -0.3 | -1.2 |
| Solution Blend | 0.0 | -0.2 | -0.9 |
| Heat-treated Solution Blend | 0.0 | -0.3 | -1.2 |

## 3.3 NaSPET/PBT BINARY BLENDS

The glass transition(s) of the NaSPET/PBT blends were determined by evaluating the mechanical relaxation(s) via DMA analysis. The tan δ profiles for each of the blends and blend components were evaluated in order to determine the α-relaxation(s), as shown in Figure 3.

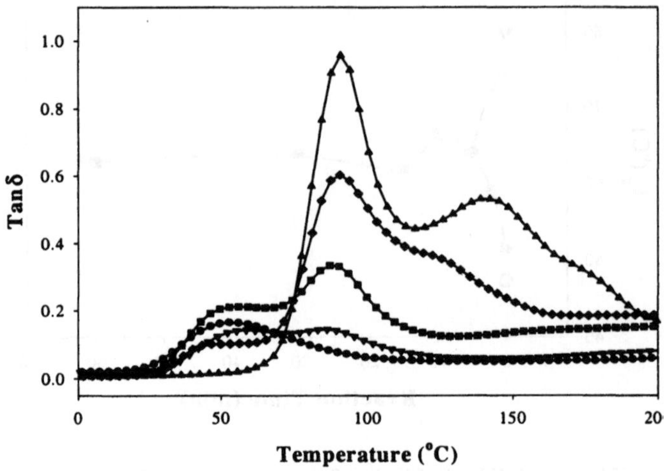


*Figure 3. a-relaxations of NaSPET/PBT Blends: (●) 100% PBT, (▼) 25NaSPET/75PBT, (■) 50NaSPET/50PBT, (◆) 75NaSPET/25PBT, and (▲) 100% NaSPET-6.*

The tan δ profiles of the NaSPET/PBT blends exhibit dual α-relaxations, indicative of an immiscible blend. The relaxation at ca. 55°C corresponds to the glass transition of the PBT phase, and the relaxation at ca. 90°C corresponds to the glass transition of the NaSPET phase. With increasing amount of NaSPET in the blends, there is little change in the temperatures of the corresponding glass transitions. Although these findings indicate inherent immiscibility between NaSPET and PBT, increasing the processing time may significantly enhance the compatibility of the NaSPET/PBT blends via ester-ester interchange.

## 3.4 NaSPET/N66 BINARY BLENDS

In order to determine the effect of blending on the dynamic mechanical relaxations, tan δ profiles of the NaSPET/N66 blends and blend components were evaluated, as shown in Figure 4. In contrast to the NaSPET/PBT binary blends, the α-relaxation of the NaSPET/N66 blends occurs at temperatures between that of pure N66 (ca. 60 °C) and pure NaSPET (ca. 90 °C). Further inspection of the α-relaxation for these blends reveals the overlap of at least two mechanical relaxations, which suggests the presence of two separate phases. The low temperature relaxation (close to that of N66) has been attributed to a N66 rich phase and the high temperature relaxation (intermediate to that of N66 and NaSPET) has been attributed to a mixed NaSPET/N66 phase.

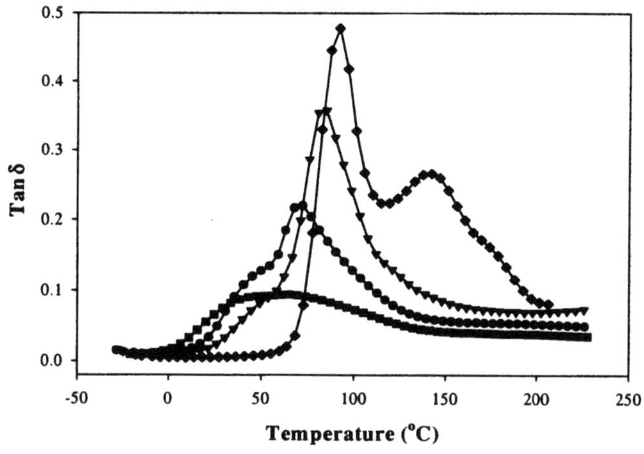


*Figure 4. Tan δ versus temperature plots for the NaSPET/N66 melt blends and blend components: (■) 100% N66, (●) 25NaSPET/75N66, (▼) 50NaSPET/50N66, and (◆) 100% NaSPET.*

Although the presence of two separate relaxations is indicative of a phase-separated morphology, the temperature of these relaxations suggests at least "partial miscibility" of the NaSPET/N66 blends. The temperature of the α-relaxation for the N66 rich phase increases and the magnitude decreases with increasing amount of NaSPET, which suggests an increase in the degree of compatibility with increasing content of NaSPET. In addition, the intermediate temperature of the α-relaxation (between that of N66 and NaSPET) for the mixed NaSPET/N66 phase suggests a high degree of compatibility (or even miscibility).

## 3.5 NaSPET/PBT/N66 COMPATIBILIZED BLENDS

The stress-strain behavior of the N66/PBT blends and blend components was evaluated in order to determine the effect of NaSPET addition on the tensile properties. Although both N66 and PBT exhibit ductile behavior (i.e., elongation without failure) as shown in Figure 5, the uncompatibilized N66/PBT blend exhibits tensile failure after only 12% elongation. However, the addition of 2 wt% NaSPET results in a significant increase in the tensile properties of the N66/PBT blends (i.e., ca. 40% increase in elongation). Increasing the amount of NaSPET to 5 wt% results in further enhancement of the tensile properties (i.e., elongation without failure) of the N66/PBT blends. Moreover, the yield stress of the compatibilized N66/PBT blends increases with an increase in the amount of NaSPET.

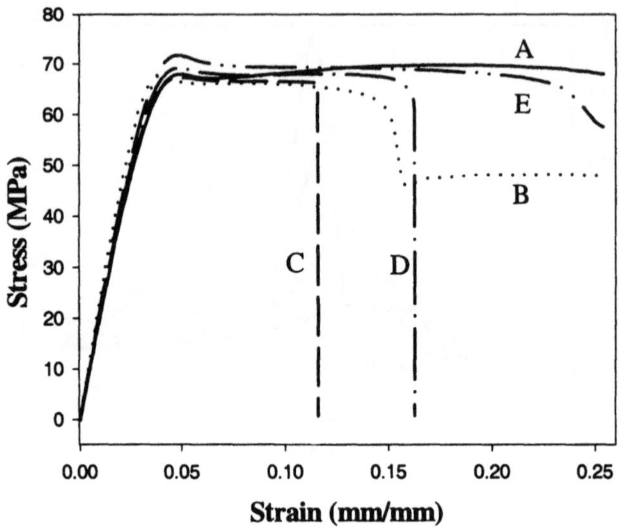


*Figure 5. Effect of NaSPET addition on the stress-strain behavior of N66/PBT blends: (A) 100% N66, (B) 100% PBT, (C) 0 wt% NaSPET, (D) 2 wt% NaSPET, and (E) 5 wt% NaSPET.*

This synergistic enhancement (while rather slight) may be attributed to strong interactions between NaSPET and N66. Similar behavior has been observed for the binary blends of NaSPET and N66, as well as for other ionomer/polyamide blends (24).

In addition to increasing the strength of interfacial adhesion, an effective compatibilizing agent typically results in a significant reduction in size of the dispersed phase (25). Therefore, the effect of NaSPET addition on the phase morphology of the N66/PBT blends was investigated by TEM, as shown in Figure 6. The uncompatibilized N66/PBT blend (i.e., 0 wt% NaSPET) exhibits a phase-separated morphology, with the size of the disperse phase on the order of 1-2 μm. However, the incorporation of 5 wt% NaSPET results in significant reduction of the disperse phase for the compatibilized N66/PBT blends (i.e., 0.3 μm – 0.6 μm). This decrease in the size of the disperse phase, upon addition of NaSPET, suggests a significant reduction in the interfacial tension between the matrix and disperse phase. This observation is consistent with the observed enhancement in mechanical properties, and may be attributed to the presence of PBT-NaSPET block copolymers formed during pre-extrusion of NaSPET/PBT blends.

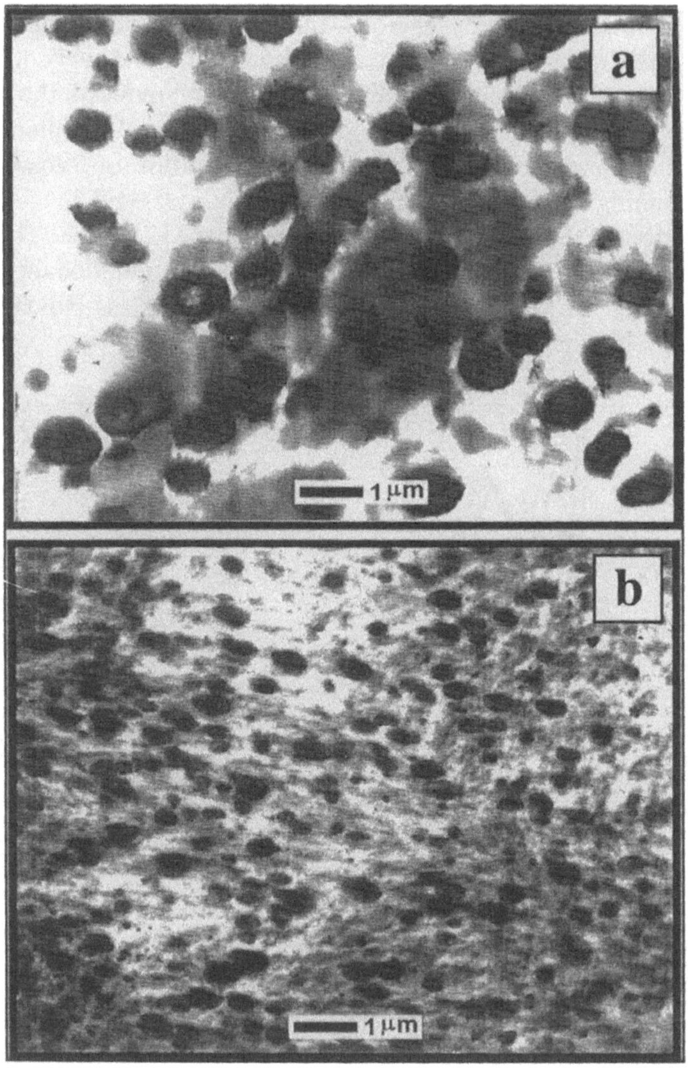


Figure 6. *Effect of NaSPET addition on the phase morphology of N66/PBT blends: (a) 20PBT/80N66 and (b) 5NaSPET/15PBT/80N66.*

## 3.6 PET NANOCOMPOSITES

The gallery spacing between clay platelets within an aggregate structure is indicative of the extent of intercalation/exfoliation and can be observed using x-ray diffraction. The incorporation of clay into PET resulted in an intercalated structure with a gallery spacing of c.a. 32.7 Å. In addition, a significant fraction of pristine clay particles (i.e., aggregates without polymer intercalation) was observed in the PET matrix, as indicated by the peak at ca. 5.6° 2θ shown in Figure 7. The addition of 9.5 wt% of Na-SPET (containing 2 mol% of ionic functionality) greatly reduced the scattering attributed to an intercalated structure and yielded only a low intensity broad shoulder in the range of 2 - 4°, 2θ. The incorporation of 9.5 wt% of Na-SPET (containing 6 mol% of ionic functionality) resulted in a very weak shoulder attributed to the intercalated structure, at c.a. 2.6°, 2θ. The addition of 9.5 wt% of Na-SPET (containing 10 mol% of ionic functionality) resulted in a featureless profile. This scattering behavior is indicative of a predominately exfoliated structure.

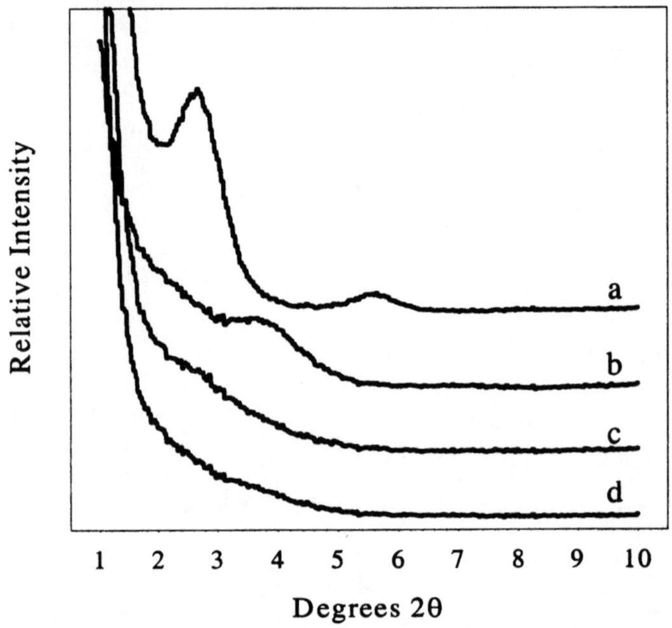


*Figure 7. X-ray diffraction of PET composite samples: a) 90% PET, 5% clay; b) 85% PET, 10% Na-SPET (2 mol %), 5% clay; c) 85% PET, 10% Na-SPET (6 mol %), 5% clay; d) 85% PET, 10% Na-SPET (10 mol %), 5% clay.*

## 3.7 PA NANOCOMPOSITES

The incorporation of clay into PA resulted in a low intensity shoulder attributed to the intercalated structure, at c.a. 2.6° 2θ shown in Figure 8. No peak was observed at larger scattering angles indicating the absence of pristine clay particles (i.e., aggregates without polymer intercalation). The addition of 9.5 and 23.75 wt% Na-SPA (2.3 mol%) respectively resulted in featureless diffraction profiles that are indicative of a predominately exfoliated structure. In addition, the scattering upturn at very small angles is significantly lower for the 23.75 wt% Na-SPA system. This behavior suggests that the long-range homogeneity of the nanocomposite is improved with increased ionomer content.

The common trends observed in these scattering data suggest that the ionic functionality of the ionomer strongly interacts with the clay platelet surface and/or organic counterions. Moreover, the clay platelets become exfoliated and homogeneously dispersed in the host matrix with increasing contributions from the ionic interactions.

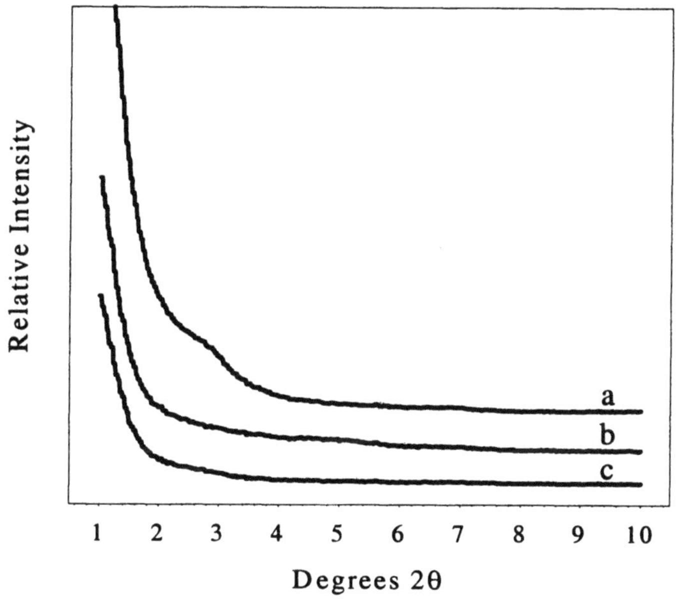


*Figure 8. X-ray diffraction of PA composite samples: a) 95% PA, 5% clay; b) 85.5% PA, 9.5% Na-SPA (2.3 mol %), 5% clay; c) 71.25% PA, 23.75% Na-SPA (2.3 mol %), 5% clay.*

# 4. CONCLUSIONS

The results obtained from this study yield the following conclusions: (i) compatibility of the AQ/PET blends may be attributed to transesterification between the polyesters which is apparently enhanced by the transition-metal form of the ionomer, and (ii) compatibility of the AQ/N66 blends may be due to the presence of specific interactions between the metal counterion of the ionomer and the amide groups. In addition, the ability of polyester ionomers, as a minor component additive, to compatibilize blends of immiscible polar polymers (e.g., PBT/N66) is demonstrated.

The incorporation of clay into PET and PA resulted in partially intercalated structures. Ionomers that are miscible with the host polymer may be used to induce strong interactions with the clay platelet surface thus compatabilizing the clay platelet within the host. With increasing ionic content and/or amount of ionomer, a greater number of specific interactions may occur yielding enhanced compatibility. Sulfonated PET containing 10 mol% ionic functionality and sulfonated polyamide containing 2.3 mol% ionic functionality are effective compatibilizers for clay in their respective homopolymers. The morphology of these systems, as observed by x-ray diffraction, is consistent with a predominantly exfoliated morphology.

The results of this study clearly demonstrate that ionomeric compatibilizers may be used to strongly interact with the inorganic clay platelets allowing for the delamination of the platelet stacks in order to produce the desired state of exfoliation in thermoplastic nanocomposites. The utilization of polyester and polyamide ionomers as compatibilizers for their respective homopolymers is under current investigation using melt extrusion as a means of dispersion.

## 5. ACKNOWLEDGEMENTS

The authors gratefully acknowledge the Eastman Chemical Company, the Office of Naval Research and the Mississippi NSF EPSCoR Program for their financial support.

## REFERENCES

1. Eisenberg, A.; Hara, M. *Polym. Eng. Sci.*, 24, 1306 (1984).
2. Serhatkulu, T.; Erman, B.; Bahar, I.; Fakirov, S.; Evstatiev, M.; Sapundjieva, D. *Polymer*, 36, 2371 (1995).
3. Kamal, M.R.; Sahto, M.A.; Utracki, L.A., *Polym. Eng. Sci.*, 22, 1127 (1982).
4. Pillon, L.Z.; Lara, J.; Pillon, D.W., *Polym. Eng. Sci.*, 27, 984 (1987).
5. Huang, C.C.; Chang, F.C., *Polymer*, 38, 2135-2141 (1997).
6. Boykin, T.L.; Moore, R.B., *Polym. Eng. Sci.*, 38, 1658 (1998).
7. Messersmith, P.B.; Giannelis, E.P. *Chem. Mater.*, 6, 1719 (1994).
8. Yano, K.; Usuki, A.; Okada, A.; Kurauchi, T.; Kamigaito, O. *J. Polym. Sci., Polym. Chem. Ed.*, 31, 2493 (1993).
9. Fischer, H.R.; Gielgens, L.H.; Koster, T.P.M. *Acta Polym.*, 50, 122 (1999).
10. Giannelis, E.P. *Adv. Mater.*, 8, 29 (1996).
11. Gilman, J.W.; Kashiwagi, T. *SAMPE J.*, 40, 40 (1997).
12. Burnside, S.D.; Giannelis, E.P. *Chem. Mater.*, 7, 1597 (1995).
13. Kyu, T.; Zhu, G.C.; Zhu, Z.L.; Tajuddin, Y.; Qutubuddin, S. *J. Polym. Sci., Poly. Phys. Ed.*, 34, 1769 (1996).
14. Tyan, H.; Liu, Y., Wei, K.H. *Polymer*, 40, 4877 (1999).

15. Ke, Y.; Long, C.; Qi, Z. *J. Appl. Sci.*, 71, 1139 (1999).
16. Kawasumi, M.; Hasegawa, N.; Kato, M.; Usuki, A.; Okada, A. *Macromolecules*, 30, 6333 (1997).
17. Liu, L.; Qi, Z.; Zhu, X. *J. Appl. Polym. Sci.*, 71, 1133 (1999).
18. Vaia, R.A.; Sauer, B.B.; Tse, O.K.; Giannelis, E.P. *J. Polym. Sci., Polym. Phys. Ed.*, 35, 59 (1997).
19. Ijdo, W.L.; Lee, T.; Pinnavaia, T.J. *Adv. Mater.*, 8, 79 (1996).
20. Usuki, A.; Kato, M.; Okada, A.; Karauchi, T. *J. Appl. Polym. Sci.*, 63, 137 (1997).
21. Hasegawa, N.; Kawasumi, M.; Kato, M.; Usuki, A.; Okada, A. *J. Appl. Polym. Sci.*, 67, 87 (1998).
22. Kato, M.; Usuki, A.; Okada, A. *J. Appl. Polym. Sci.*, 66, 1781 (1997).
23. Nishi, T.; Wang, T.T., *Macromolecules*, 8, 909 (1975).
24. Lu, X.; Weiss, R.A., *Macromolecules*, 25, 6185 (1992).
25. Favis, B.D., *Polymer*, 35, 1552 (1994).

# B. Light and Energy

Chapter 7

# SULFONATED AND CARBOXYLATED COPOLY(ARYLENESULFONE)S FOR FUEL CELL APPLICATIONS

Dirk Poppe, Torsten Zerfaß, Rolf Mülhaupt, Holger Frey
*Freiburger Materialforschungszentrum und Institut für Makromolekulare Chemie der Albert-Ludwigs Universität, Stefan-Meier-Str. 21/31, D-79104 Freiburg, Germany*

## 1. INTRODUCTION

One of the most crucial current challenges is the development of hydrogen-based technology for future energy supply. The central device for this technology is the fuel cell, which is considerably more efficient than other energy converters. In PEMFCs (proton-exchange membrane fuel cells) – and particularly in DMFCs (direct methanol fuel cells) - the polymer membrane represents the most important component (Figure 1). This membrane has to fulfill complex requirements: It has to combine electrochemical stability, workability, high ionic conductivity, low permeation of the reactants (hydrogen, methanol, oxygen) and mechanical integrity.

The membrane material has to be long-term stable under fuel cell conditions. However, for standard polymers these conditions are hard to sustain, i.e., the presence of hydrogen and oxygen at a platinum catalyst as well as the aqueous acid medium are detrimental. For instance, polymers with aliphatic groups are rapidly oxidized and consequently the membrane is destroyed under these conditions. Based on the abovementioned selection criteria, only perfluorinated polymers or materials with an aromatic backbone fulfill the strict requirements. Unfortunately, such materials often are insoluble and infusible and, hence, can not be processed.

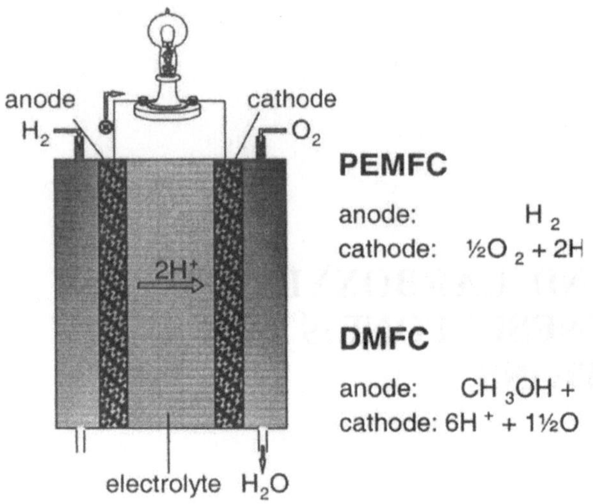


*Figure1. Schematic structure of a fuel cell*

High ionic conductivity can usually be realized by incorporation of sulfonic acid groups in the polymer membrane and subsequent swelling in water. Unfortunately, with increasing fraction of sulfonic acid groups, the polarity of the material increases, leading to undesired methanol permeation and strong swelling in water, corresponding to a deterioration of mechanical integrity.

Historically, the first materials explored in this context were cation-exchange resins based on sulfonated polystyrenes[1] or phenol-formaldehyde resins.[2] However, it soon became evident that these materials are not long-term stable against hydrolysis and oxidation of the benzylic groups.

Currently, the standard polyelectrolytes for fuel cells are copolymers derived from tetrafluoroethylene, bearing sulfonic acid functionalities. The most well-known example is Nafion® produced by Du Pont.[3, 4] In terms of mechanical and electrochemical stability, Nafion® is an excellent material for use in fuel cells. However, it possesses some disadvantages: on one hand, membrane preparation is complicated, since the protonated form is not thermoplastically processible. On the other hand, proton conduction of Nafion® is always accompanied by undesired diffusion of water through the membrane, resulting in a high electro-osmotic drag of water. Energy losses at the cathode due to floating are observed and continuous humidification of the reactants is necessary. High gas permeability and a methanol permeation of Nafion® are also disadvantageous. Only when swollen in water, Nafion® shows high ionic conductivity. Thus, application at temperatures exceeding 100 °C is not possible. Higher temperatures are desirable, since the catalysts employed at present, are less sensitive to CO poisoning at an elevated temperature of 130 °C. Even worse, in a DMFC poisoning of the Pt catalyst by CO is always observed,[5] but can be avoided at temperatures exceeding 120 °C. Therefore high ionic conductivity above 100 °C is an important target. A promising approach to this end is the incorporation of inorganic materials into the membrane, e.g., phosphotungstic acid.[6] The impregnated membranes at 110 °C show a performance comparable to Nafion® at 80 °C. Also, the synthesis of composite Nafion®/zirconium phosphate membranes with good performance up to 140-150 °C in a DMFC has been reported.[7] However, these membranes exhibit the major drawback of Nafion®: they are very expensive ($600/m$^2$).

Functional condensation polymers may represent an alternative to perfluorinated materials. In order to fulfill the abovementioned requirements for fuel cell application they have to possess an electrochemically inert structure. The incorporation of functionalities can be realized either before or after polymerization, e.g., sulfonated poly(ethersulfone)s have been functionalized by sulfonation of the polymer[8] or by sulfonation of the monomer and subsequent polymerisation.[9] These polymers and also sulfonated poly(etherketone)s[10, 11] are very polar materials showing high swelling in water and therefore limited mechanical integrity at elevated temperatures. Cross-linking of sulfonic acid groups has been employed to reduce swelling, but requires one additional synthetic step and some acid groups are no longer available as conducting groups.[8, 12] Another successful concept to reduce swelling in water is the synthesis of segmented copolymers with sulfonated poly(ethersulfone)s as hydrophilic and poly(imide)s as hydrophobic blocks,[13] but these materials still contain aliphatic structures and therefore long-term stability in fuel cells is questionable. Heterocyclic aromatic polymers functionalized[14,15,16] or doped[17,18] with acids are also a topic of recent research. A detailed overview can not be given due to the limited scope of this chapter.

Sulfonated poly(*p*-phenylene)s, solubilized by flexible side chains were proposed for fuel cell applications by Maxdem Inc.[19] Despite the high chemical stability, the side-groups necessary to process the material render the synthesis of the monomers complicated and reduce chemical stability. As it was shown by our group some years ago, sulfonated poly(*m*-phenylene-co-*p*-phenylene)s at certain compositions are soluble without additional side-groups.[20] These materials are stable under fuel cell conditions, but the synthesis is expensive. Our project is based on these results.

Here we report on the synthesis of novel functional, soluble copolyarylenes that are thermooxidatively stable and therefore promising materials for fuel cell applications. We employ the Ni-catalyzed coupling of aryl chlorides for the preparation of the copolymers. Ionic conducting groups have been incorporated in two different ways: Sulfonic acid groups ($SO_3H$)

after polymerization via sulfonation of copoly(arylenesulfone)s, carboxylic acid groups (COOH) via copolymerization of methyl 2,5-dichlorobenzoate and subsequent hydrolysis. Since to our knowledge there are no systematic investigations concerning the effect of carboxylic acid groups in fuel cell membranes, copolymers with these two functionalities as well as blends of different polyelectrolytes have been prepared. The last step has been the determination of fuel cell-relevant parameters like swelling behavior and methanol permeation.

## 2. POLYARYLENE SYNTHESIS

The polymerization method is based on the reductive coupling of aryl chlorides. The catalyst is a zero-valent nickel complex made in situ from nickel chloride ($NiCl_2$), triphenylphosphine ($PPh_3$), bipyridyl (bipy), zinc and sometimes also sodium bromide. It is sensitive to protonic groups and oxygen.

$$x\ \mathrm{Cl{-}Ar{-}Cl} \xrightarrow{\mathrm{Zn}\ [\mathrm{Ni}^0]} \mathrm{-\!\!(Ar)_x\!\!-}$$

Colon *et al.*[21] showed that biaryls can be produced in essentially quantitative yields by Ni(0)-catalyzed coupling of aryl chlorides. They were also able to use this method for the synthesis of poly(ether ether sulfone)s.[22] Ueda and co-workers optimized this procedure for the synthesis of poly(ether ketone)s[23] and applied it to the polymerization of tert-butyl 2,5-dichlorobenzoate as a precursor route to poly(*p*-phenylene) (PPP).[24] The scope of this system was extended by Sheares et al. to poly(benzophenone)s[25] and low dielectric constant materials.[26] Furthermore, McGrath and Ghassemi realized the synthesis of novel poly(arylene phosphine oxide)s.[27] Percec et al. demonstrated the usefulness of the Ni(0) catalyzed reaction for the preparation of soluble PPPs, containing $CF_3$ or $OCF_3$ substituents[28] and, more recently, PPPs with mesogenic side groups.[29] Although there is an increasing interest in using this method for the synthesis of new polyarylene architectures, to date only few publications have dealt with copolymerization.[28, 30]

The polymerization of methyl 2,5-dichlorobenzoate[31,32,33] and other dihalides or bistriflates[33] to poly(methoxycarbonyl-*p*-phenylene) is known in literature, but copolymerization has not been examined to date.

A more established method to produce polyarylene structures is the Suzuki coupling.[34, 35] A comparison of the mass balances (Figure 2) explains our motivation to develop an alternative route. As shown on the left side, the Suzuki reaction starts with aryl bromides, while the Ni(0)-coupling polymerizes aryl chlorides. Therefore the mass balance of the latter method is very advantageous. In addition, the availability of aryl chlorides is much better, compared to aryl bromides. The Pd-catalyst is more expensive than the Ni-system, and the Suzuki method requires one additional synthetic step, the preparation of the boronic acids. On the other hand, this method tolerates a broader range of reaction conditions and monomer functionalities. However, copolymerization of *m*- and *p*-dichlorobenzene by the Ni(0) catalyzed reaction failed due to the different reactivities of these monomers. A blend of two insoluble polymers was received.

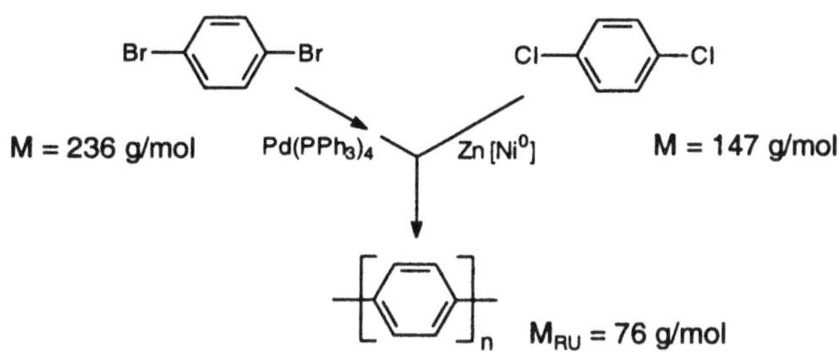


*Figure 2. Suzuki method (left) and coupling of aryl chlorides (right)*

## 2.1 Polyarylenesulfones with $SO_3H$ groups

In order to adjust the reactivities we studied the system *m*-dichlorobenzene (**M**) / 4,4'-dichlorodiphenylsulfone (**S**) (Figure 3). This copolymerization is not known in literature up to now. **S** is as electron poor as **M**, therefore its copolymerization should be more successful than of *p*-dichlorobenzene.

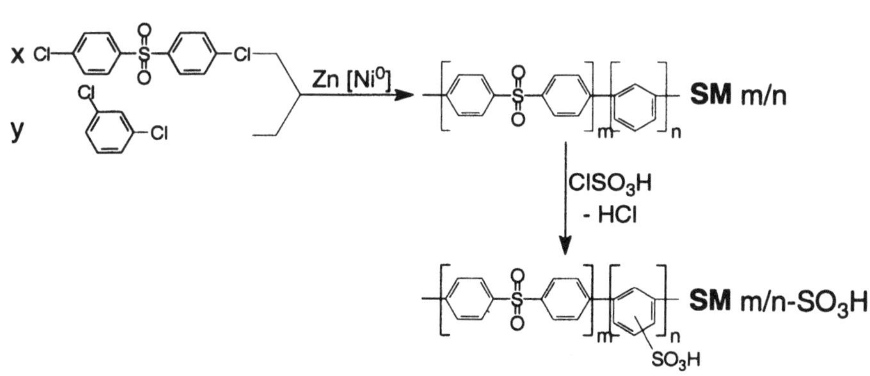


*Figure 3. Synthesis of sulfonated co(polyarylensulfone)s*

As shown in Table 1, we realized the whole range of compositions. The copolymer compositions were determined by elemental analysis and were found to correspond to the compositions of monomer mixtures. While both homopolymers (**1**, **7**) are insoluble, there is a window of copolymer composition (**3-5**) that leads to soluble materials.

The molecular weight of SM25/75 (**5**) was measured by static light scattering in DMF to $M_w$ = 8,400 g/mol ($DP_w$ = 74). Due to poor solubility we determined no other molecular weights.

*Table1. Copolymerization of 4,4'-dichlorodiphenylsulfone (S) and m-dichlorobenzene (M)*

| | molar ratio S:M | solubility | $\eta_{inh}$ / dl·g$^{-1}$ | $T_g$ / °C |
|---|---|---|---|---|
| **1** | 100:0 | - | - | - |
| **2** | 75:25 | - | - | 262 |
| **3** | 63:37 | ≈ DMF | 0.26 | |
| **4** | 50:50 | DMF, $CHCl_3$ | 0.29 | 261 |
| **5** | 25:75 | DMF, $CHCl_3$ | 0.29 | 246 |
| **6** | 11:89 | - | - | |
| **7** | 0:100 | - | - | 146 |

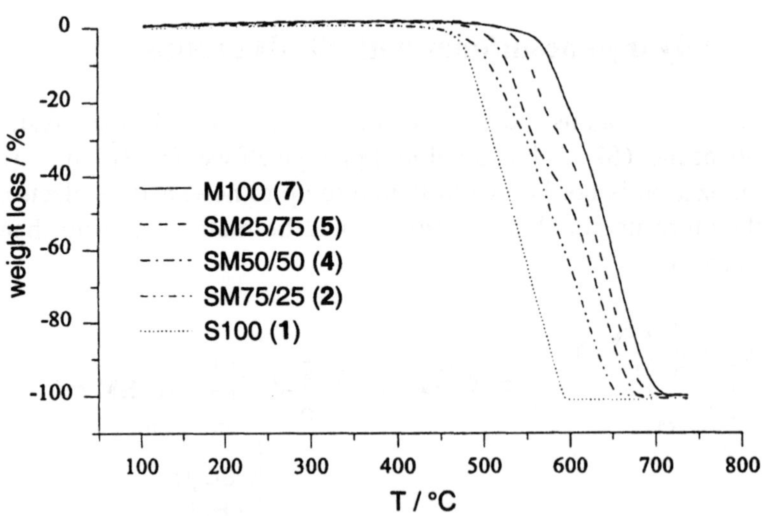


*Figure 4. TGA traces of co(polyarylenesulfone)s*

TGA showed no thermal degradation in air up to 450°C (Figure 4), evidencing the expected thermal stability of the materials. Ionic conducting groups were introduced subsequent to the synthesis of the polymers, using chlorosulfonic acid (Figure 3). The degree of sulfonation of **SM25/75 (5)** was varied between 27 and 57 %. All materials are soluble in DMF, but not in water, methanol and ethanol. Whereas membranes cast from DMF-solution with a degree of sulfonation of 38 and 43 % are brittle and break into pieces in the presence of water, the other materials (degree of sulfonation of 27, 32 and 57%) form transparent films, are flexible and mechanically stable.

## 2.2 Polyarylenes with COOH groups[36]

In order to incorporate COOH functionalities we copolymerized methyl 2,5-dichlorobenzoate (**E**) with *m*-dichlorobenzene (**M**) as well as 4,4'-dichlorodiphenylsulfone, respectively (**S**) (Fig. 5). Also these polymerizations had not been reported in literature to date. As shown in Table 2, all homo- and copolymers prepared are soluble. Molecular weights were measured by SEC calibrated to PS-standards. We were able to enhance the molecular weight of the homopolymer of methyl 2,5-dichlorobenzoate (**E100**, **8-10**) from $M_w$ = 4,600 to 7,000 g/mol by variation of reaction conditions. Unfortunately, incorporation of *m*-dichlorobenzene (**11-13**) led to a decrease of the molar masses, resulting in oligomeric products.

Since these materials are too brittle for membrane application, we had to modify the system. Again, the comonomer is dichlorodiphenylsulfone (**14-18**). The copolymer composition was determined by $^1$H-NMR spectroscopy from the comparison of the integral values of the aromatic ($\delta$ = 8.5 - 7.0 ppm) and the methoxy group ($\delta$ = 4.0 - 3.5 ppm) proton signals. Slightly increased incorporation of **M** in comparison to the monomer mixture was observed for polymers **14** to **17**, only sample **18** showed distinctively larger incorporation.

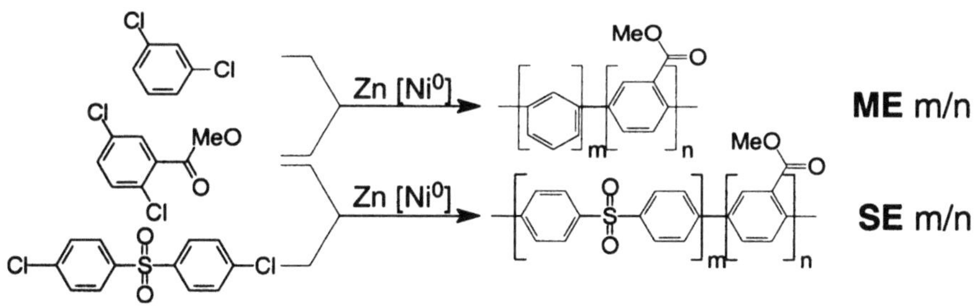


*Figure 5. Copolymerization of methyl 2,5-dichlorobenzoate*

*Table 2. Synthesis of polyarylenesulfones with ester side chains*

| | molar ratio | | | | | | solubility | SEC ($CHCl_3$) | |
|---|---|---|---|---|---|---|---|---|---|
| | monomers | | | polymers (NMR) | | | | $M_w$ / g/mol | $M_w/M_n$ |
| | S | M | E | S | M | E | | | |
| **8** | - | - | 1 | - | - | 100 | $CHCl_3$ | 4,600 | 1.88 |
| **9** | - | - | 1 | - | - | 100 | $CHCl_3$ | 5,400 | 1.90 |
| **10** | - | - | 1 | - | - | 100 | $CHCl_3$ | 7,000 | 1.92 |
| **11** | - | 1 | 1 | - | 38 | 62 | DMF, $CHCl_3$ | 2,700 | 1.71 |
| **12** | - | 1 | 2 | - | 35 | 65 | DMF, $CHCl_3$ | 1,600 | 1.53 |
| **13** | - | 1 | 10 | - | 24 | 76 | DMF, $CHCl_3$ | 1,300 | 2.33 |
| **14** | 1 | - | 3 | 17 | - | 83 | DMF, $CHCl_3$ | 7,500 | 1.83 |
| **15** | 1 | - | 2 | 25 | - | 75 | DMF, $CHCl_3$ | 36,500* | 1.70 |
| **16** | 1 | - | 2 | 25 | - | 75 | DMF, $CHCl_3$ | 35,000 | 2.56 |
| **17** | 1 | - | 1 | 47 | - | 53 | DMF, $CHCl_3$ | 3,100 | 1.81 |
| **18** | 2 | - | 1 | 10 | - | 90 | $CHCl_3$ | 18,500 | 2.55 |

* $M_w$ = 21,100 g/mol and $M_w/M_n$ = 1.49 by MALLS

In addition to SEC-characterization, for polymer **15** weight average molecular weight ($M_w$ = 21,100 g /mol; $DP_w$ = 136) and polydispersity ($M_w/M_n$ = 1.49) were determined by MALLS. As expected, molecular weights determined by SEC are overestimated due to the increased chain stiffness of polyarylenes compared to the PS-standards employed. The variance of molar masses, which is frequently observed for this synthetic method,[28] is due to the step-growth mechanism of the polymerization and the high sensitivity towards impurities. Unfortunately, these variances have a pronounced effect on the mechanical properties: transparent films were cast from $CHCl_3$ for all polymers. However, whereas **15** and **16** gave mechanically stable, flexible materials, **14** and **17** possessing considerably lower molecular weights formed brittle films. For all copolymers DSC showed no $T_g$ up to decomposition.

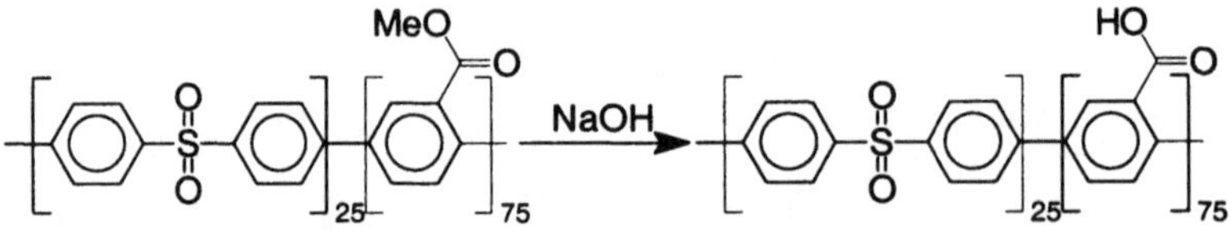


**SE25/75** **SE25/75-COOH**

*Figure 6. Hydrolysis of* ***SE25/75*** *(****15, 16****)*

Hydrolysis of **SE25/75** (Figure 6) led to a polyelectrolyte which was soluble in DMF, its deprotonated form also in water and pyridine. The protonated polymer was insoluble in methanol and water, but it was swollen in the latter without losing its mechanical integrity. TGA showed good thermal stability (Figure 7), even though no plateau was observed because of the hygroscopic nature of this polymer. No $T_g$ was observed up to decomposition by DSC. In analogy to **SE25/75**, films were cast from **SE25/75-COOH** in DMF, which still remained flexible, although the presence of hydrogen bonds should lead to a more brittle material.

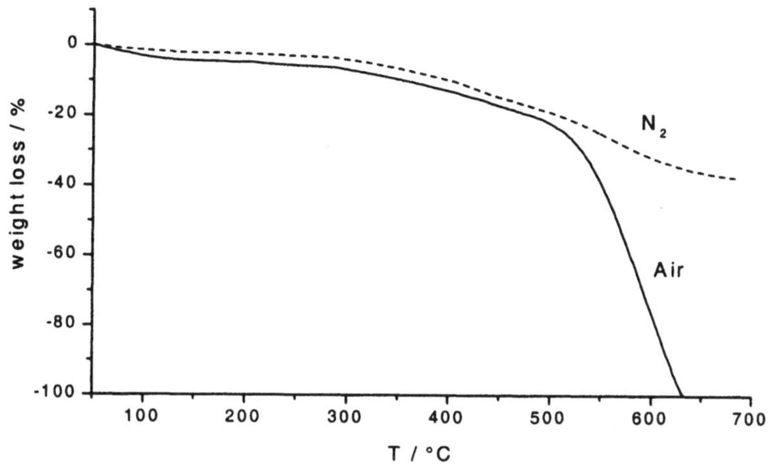


*Figure 7.* TGA thermograms of **SE25/75-COOH**

## 2.3 COOH/$SO_3$H-Blends

In addition to sulfonic (**SM25/75-$SO_3$H**) and carboxylic (e.g. **SE25/75-COOH**) acid functionalized membranes we prepared different blend membranes. To this end we mixed the mentioned materials in different ratios, solved them in DMF and cast membranes. Homogenous films with good mechanical properties were obtained. Interestingly, also a 1:1 blend of **SE25/75-COOH (15)** and the very brittle **SM25/75-$SO_3$H** (43% $SO_3$H) showed the same improved mechanical integrity as the pure carboxylic acid compound.

# 3. MEMBRANE PROPERTIES

In order to test the synthesized materials for fuel cell application, we determined the degree of swelling (*DS*) in water in dependence of the temperature (Table 3). Low swelling is favourable, since this usually corresponds to good mechanical integrity.

The degree of swelling of the sulfonic acid substituted polyarylenesulfone with 32% $SO_3$H groups is comparable to established fuel cell membrane materials (**19**). The highly sulfonated and therefore very polar second membrane (**20**) showed increased water uptake, but up to 90 °C did not lose its mechanical integrity. For the carboxylic acid based material **SE25/75-COOH** (**23**) a low *DS* is observed that remained constant with increasing temperature. Surprisingly, a 1:1 blend (**22**) exhibited approximately the same reduced swelling. Also in a 1:4 blend (**21**) the carboxylic compound caused a positive effect. Based on these results, incorporation of COOH functionalities is a concept to reduce water uptake.

*Table 3. Degree of swelling (DS) in water*

| | *DS / %* | 25 °C | 60 °C | 80 °C | 90 °C |
|---|---|---|---|---|---|
| **19** | SM25:75-$SO_3H$ (32% $SO_3H$) | 31 | | | 44 |
| **20** | SM25:75-$SO_3H$ (57% $SO_3H$) | 79 | 110 | 264 | 439 |
| **21** | COOH/$SO_3H$ blend 1:4* | 42 | 42 | 62 | 197 |
| **22** | COOH/$SO_3H$ blend 1:1** | 14 | 14 | 15 | 15 |
| **23** | SE25:75-COOH | 16 | 14 | 12 | 13 |

*57% $SO_3H$**43% $SO_3H$

For the use in a DMFC, low methanol permeation is a basic prerequisite. In order to study this property, a membrane was placed between two diffusion layers in a small fuel cell. A stream of aqueous methanol was led through one half cell, the other was rinsed with nitrogen. The methanol that permeated through the membrane was transported to a flame ionization detector by the nitrogen flow. In order to obtain comparable results, we studied three membranes. As it is shown in Table 4, in comparison to a four times thicker Nafion® 117 membrane, the sulfonic acid functionalized copolyarylene (**20**) showed a permeation reduced by one third. Unfortunately, the incorporation of carboxylic acid groups (**21**) led to increased methanol permeation. From our results we conclude that carboxylic acid functionalized polyarylenesulfone membranes are not suitable for DMFC application.

*Table 4. Methanol permeation of the polyarylenesulfone materials in comparison to Nafion®*

| | | d / μm | permeation / ppm |
|---|---|---|---|
| **20** | SM25/75-$SO_3H$ (57% $SO_3H$) | 53 | 9.2 |
| **21** | 1:4 COOH/$SO_3H$ blend | 51 | 17.8 |
| | Nafion® 117 | 200 | 13.6 |

# 4. SUMMARY

New functional copoly(arylenesulfone)s have been prepared by Ni(0)-catalyzed coupling copolycondensation. Incorporation of functionalities was achieved either by polymerization and subsequent functionalization ($SO_3H$ groups) or by polymerization of already functionalized monomers (COOH groups). For the first time it was shown that these materials are promising for fuel cell applications.

While for the polyarylenesulfone copolymers there is a range of copolymer composition that leads to soluble materials (the homopolymers are insoluble), all ester copolymers (**ME**, **SE**) are soluble. For **ME** only oligomers are observed, the system **SE** led to higher molecular weights, but whereas we were able to adjust copolymer composition, control over molecular weights was limited. Functionalization both with sulfonic and carboxylic acid groups resulted in novel polyelectrolytes with good thermal and mechanical properties.

Further investigation concerning the suitability of these materials for fuel cell applications showed that carboxylic acid functionalities in poly(arylenesulfone) blends led to reduced swelling behavior in water, but methanol permeation increases. Therefore COOH functionalized membranes are not applicable for the DMFC. Further measurements on hydrogen and oxygen permeabilities are in progress. The respective results have to be correlated with ionic conductivities and the electro-osmotic drag of water.

# 5. ACKNOWLEDGEMENT

A. Heinzel, A. Schmidt and S. Nutalapati (Fraunhofer-Institut für Solare Energiesysteme (ISE), Freiburg, Germany) are acknowledged for their help with the permeation measurements. Financial support provided by the Deutsche Forschungsgemeinschaft (DFG) in the context of the Fuel Cell Program is gratefully acknowledged. H. F. thanks the Fonds der Chemischen Industrie for valuable support.

# 6. REFERENCES

[1] G.F. D'Alelio, *US 2,366,007* (**26.12.1944**).
[2] W.T. Grubb, *US 2,913,511* (**17.11.1959**).
[3] W.G. Grot, *US 4,021,327* (E. I. Du Pont de Nemours and Comp., **03.05.1977**).
[4] W.G. Grot, *Chem. Ing. Techn.* **1972**, *44*, 167.
[5] K.D. Kreuer, *Chem. Mater.* **1996**, *8*, 610.
[6] S. Malhotra, R. Datta, *J. Electrochem. Soc.* **1997**, *144*, 2, L23.

[7] C. Yang, S. Srinivasan. A.S. Aricò, P. Cretì, V. Baglio, V, Antonucci, *Electrochem. Solid-State Lett.* **2001**, *4*, 4, A31.

[8] R. Nolte, K. Ledjeff, M. Bauer, R. Mülhaupt, *BHR Group Conf. Ser. Publ.* **1993**, *3 (Effective Membrane Processes – New Perspectives)*, 381.

[9] F. Wang, Q. Ji, W. Harrison, J. Mecham, R. Formato, R. Kovar, P. Osenar, J.E. McGrath, *Polymer Preprints* **2000**, *41*, 1, 237.

[10] C.A. Linkous, D. Slattery, *Polym. Mater. Sci. Eng.* **1993**, *68*, 122.

[11] F. Helmer-Metzmann, F. Osan, A. Schneller, H. Ritter, K. Ledjeff, R. Nolte, R. Thorwirth, *EP 0574791 A2* (Hoechst AG, **07.06.1993**).

[12] R. Nolte, K. Ledjeff, M. Bauer, R. Mülhaupt, *J. Membr. Sci.* **1993**, *83*, 211.

[13] J.B. Mecham, F. Wang, T.E. Glass, J. Xu, G.L. Wilkes, J.E. McGrath, *Polymeric Materials: Science & Engineering* **2001**, *84*, 105.

[14] X. Glipa, M. El Haddad, D.J. Jones, J. Rozière, *Solid State Ionics* **1997**, *97*, 323.

[15] M.J. Sansone, F.J. Onorato, N. Ogata, *US 5,599,639* (Hoechst Celanese Corp., **04.02.1997**).

[16] S. Kim, D.A. Cameron, Y. Lee, J.R. Reynolds, *J. Polym. Sci., Part A: Polym. Chem.* **1996**, *34*, 481.

[17] J.-T. Wang, J.S. Wainwright, R.F. Savinell, M. Litt, *J. Appl. Electrochem.* **1996**, *26*, 751.

[18] R. F. Savinell, M.H. Litt, *WO 96/13872* (Case Western Reserve University, **09.05.1996**).

[19] N. Ogato, *WO 94/24717* (Maxdem Inc., **27.10.1994**).

[20] R. Mülhaupt, H. Frey, T. Zerfaß, K. Ledjeff-Hey, R. Nolte, *DE 195 35 086 A1* (Fraunhofer-Gesellschaft, **21.09.1995**).

[21] I. Colon, D.R. Kelsey, *J. Org. Chem.* **1986**, *51*, 2627.

[22] I. Colon, G.T. Kwiatkowski, *J. Polym. Sci., Part A: Polym. Chem.* **1990**, *28*, 367.

[23] M. Ueda, F. Ichikawa, *Macromolecules* **1990**, *23*, 926.

[24] M. Ueda, M. Yoneda, *Macromol. Rapid Commun.* **1995**, *16*, 469.

[25] R.W. Phillips, V.V. Sheares, E.T. Samulski, J.M. DeSimone, *Macromolecules* **1994**, *27*, 2354.

[26] P.A. Havelka-Rivard, K. Nagai, B.D. Freeman, V.V. Sheares, *Macromolecules* **1999**, *32*, 6418.

[27] H. Ghassemi, J.E. McGrath, *Polymer* **1997**, *38*, 3139.

[28] M.C. Grob, A.E. Feiring, B.C. Auman, V. Percec, M. Zhao, D.H. Hill, *Macromolecules* **1996**, *29*, 7284.

[29] V. Percec, A.D. Asandei, D.H. Hill, D. Crawford, *Macromolecules* **1999**, *32*, 2597.

[30] G.T. Kwiatkowski, M. Matzner, I. Colon, *J. Macromol. Sci., Pure Appl. Chem.* **1997**, *A34*, 1945.

[31] V. Chaturvedi, S. Tanaka, K. Kaeriyama, *Macromolecules* **1993**, *26*, 2607.

[32] K. Kaeriyama, M.A. Mehta, H. Masuda, *Synth. Met.* **1995**, *69*, 507.

[33] V. Percec, S. Okita, R. Weiss, *Macromolecules* **1992**, *25*, 1816.

[34] M. Rehahn, A.-D. Schlüter, W.J. Feast, G. Wegner, *Polymer* **1989**, *30*, 1054.

[35] M. Rehahn, A.-D. Schlüter, W.J. Feast, G. Wegner, *Polymer* **1989**, *30*, 1060.

[36] D. Poppe, H. Frey, A. Heinzel, R. Mülhaupt, *Polymeric Materials: Science & Engineering* **2001**, *84*, 333.

# Chapter 8

# PREPARATION AND PROPERTIES OF SULFONATED OR PHOSPHONATED POLYBENZIMIDAZOLES AND POLYBENZOXAZOLES

Yoshimitsu Sakaguchi, Kota Kitamura, Junko Nakao, Shiro Hamamoto, Hiroshi Tachimori and Satoshi Takase
*Toyobo Research Center Co., Ltd.,*
*1-1 Katata 2-Chome, Ohtsu 520-0292 JAPAN*

## 1. INTRODUCTION

As a new power source for transportation, stationary application and portable power application, solid polymer electrolyte fuel cell is attracting increasing attention. Proton exchange membrane (PEM) plays a key role in this technology, and perfluorosulfonic acid polymers such as Nafion produced by DuPont have been recognized as the best materials due to their high proton conductivity and excellent chemical stability. However, Nafion has some disadvantages in limited operation temperature, high cost and high methanol permeability.

Sulfonic acid-containing aromatic condensation polymers are considered as possible candidates for the alternative materials for PEM because their backbone structures are generally heat- and solvent-resistant with good mechanical properties. Investigation for such polymers based on poly(arylene ether ketone)s (1-3), poly(arylene ether sulfone)s (4-7), polyimides (8-10), polybenzimidazoles (11 - 16), etc. have been reported so far.

In the case of polybenzazoles, Polybenzimidazoles have been mainly evaluated as a PEM. Polybenzimidazole films doped with phosphoric acid showed excellent thermal stability and high ionic conductivity at elevated temperatures (11-14). This membrane is expected to be used in high temperature fuel cells above 150°C. High ionic conductivity was also reported for polybenzimidazole system that has N-substituted alkyl or aryl sulfonic acid groups (15, 16).

Preparations of polybenzimidazoles having sulfonic acid groups on the aromatic backbone were reported in several papers (17, 18). Sulfonic acid-containing polybenzothiazoles and polybenzoxazoles were prepared in the same way (19, 20). One example of phosphonic acid-containing polybenzoxazoles was also found (21). However, evaluations as ionic conducting polymers were not reported in these literatures.

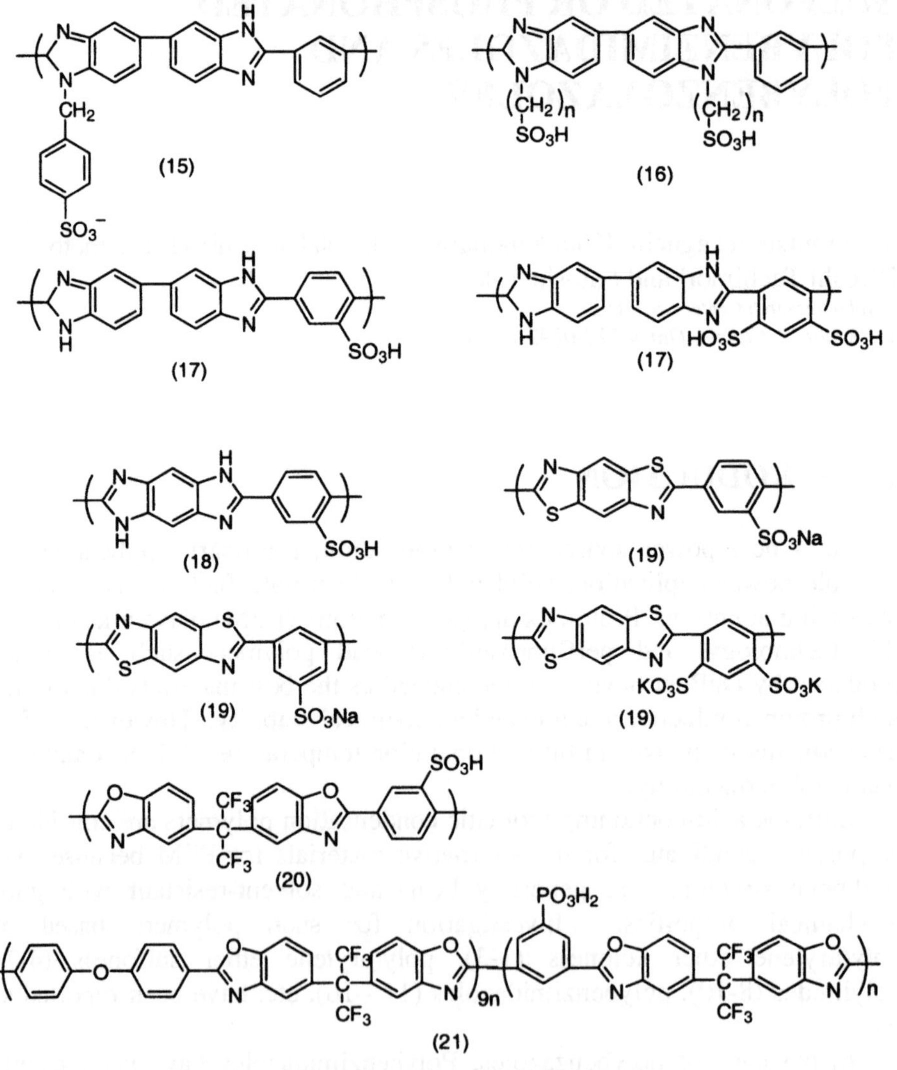


*Figure 1. Sulfonated and phosphonated polybenzazoles.*

In this study, we synthesized polybenzimidazoles and polybenzoxazoles containing sulfonic acid on the polymer backbone by using sulfoisophthalic acid and sulfoterephthalic acid as comonomers, and the basic properties of resulting polymers were characterized. In addition, phosphonic acid-containing polybenzimidazoles were synthesized as another candidate for PEM at higher temperature operation, and their properties were compared with those of sulfonic acid-containing corresponding polymers.

## 2. EXPERIMENTAL

*Materials.* 3,3',4,4'-Tetraaminodiphenylsulfone (TAS, Konishi Chemical Ind.), 3,3'-dihidroxybenzidine (HAB, Wakayama Seika Kogyo), terephthalic acid (TPA, Mitsui Petrochemical Ind.), 5-sulfoisophthalic acid monosodium salt (SIA, Tokyo Kasei), 3,5-dicarboxyphenylphosphonic acid (DCP, Nissan Chemical Ind.), phosphorus pentoxide (nakalai Tesque) and 75.5% polyphosphoric acid (Nakalai Tesque) were used without further purification. 2-Sulfoterephthalic acid monosodium salt (STA) was used as highly purified monomer.

*Polymerization.* As an example, copolymerization of polybenzimidazole with mixed dicarboxylic acids of STA and TPA is described as follows.

In a polymerization vessel equipped with a nitrogen inlet and a stirring rod, 1.500g of TAS ($5.390 \times 10^{-3}$ mol), 0.3044g of TPA ($1.836 \times 10^{-3}$ mol), 0.9540g of STA ($3.557 \times 10^{-3}$ mol), 20.48g of polyphosphoric acid and 16.41g of polyphosphoric acid were mixed under nitrogen flow. The reaction temperature was raised stepwise, 100°C, 150°C and 200°C, then kept at that temperature until the reaction mixture became highly viscous enough. After being cooled to room temperature, the reaction mixture was mixed with water to precipitate the polymer. The polymer obtained was washed with water by using a blender until the washings were completely neutralized.

Other copolymers were prepared by the same procedure except changing the comonomers and feed ratio.

*Characterization.* Inherent viscosities were measured at a concentration of 0.5 g/dl in concentrated sulfuric acid at 30°C. Infrared spectra were measured on a Biorad FTS-40 spectrometer with a Biorad UMA-300A microscope. Thermogravimetric analysis (TGA) was conducted at a heating rate of 10°C/min under flowing air or argon, using a Shimadzu TGA-50 thermogravimetric analyzer. Ionic conductivities of the films were measured at 80°C and 95%RH with a Solartron 1250 Frequency Response Analyzer over the frequency range from 1Hz to 65KHz, and calculated based on the complex impedance plot.

## 3. RESULTS AND DISCUSSION

The structures and abbreviations of monomers are listed in Figure 2. Polymerizations were carried out with changing the molar ratios of dicarboxylic acids, TPA/SIA, TPA/STA and TPA/DCP in polyphosphoric acid as shown in Figure 3. The results of Polymerizations for sulfonic acid- and phosphonic acid-containing polybenzimidazoles and polybenzoxazoles with 66mol% of ionic monomer content are summarized in Table 1.

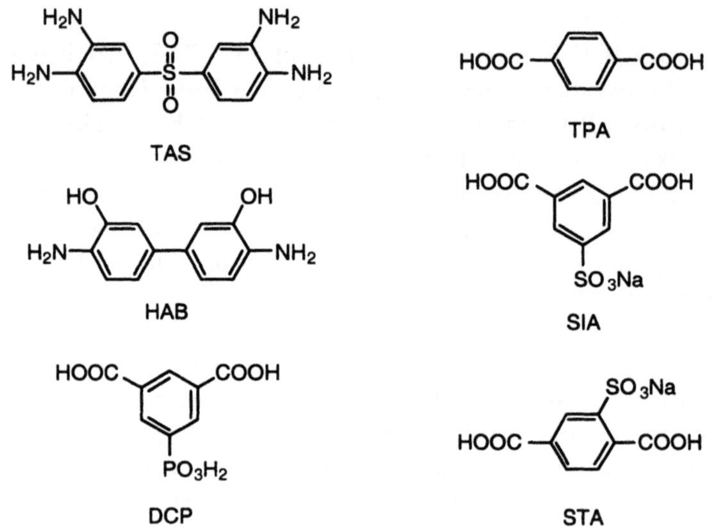


*Figure 2. Chemical structures of monomers*

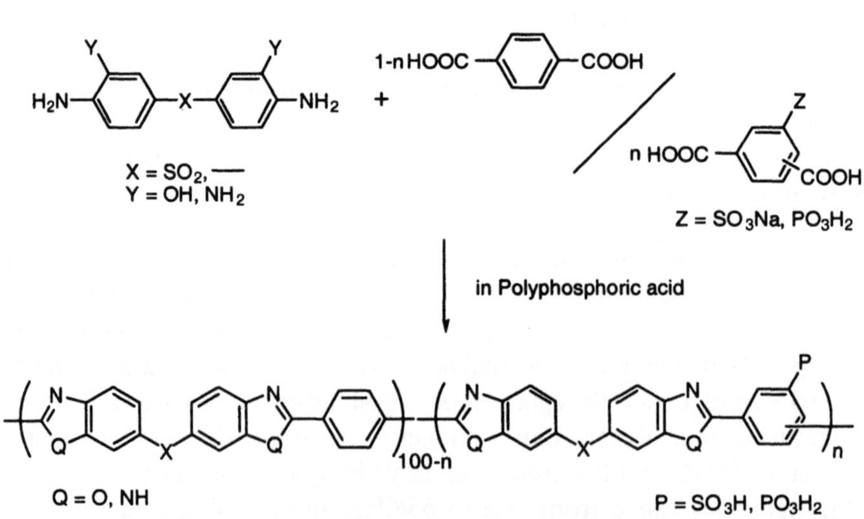


*Figure 3. Synthesis of sulfonic acid- and phosphonic acid-containing polybenzimidazoles and polybenzoxazoles.*

*Table 1. Preparation of Sulfonic Acid- and Phosphonic Acid-Containing Polybenzimidazoles and Polybenzoxazoles*

| Polymer | Monomer | | | B/C | $\eta_{inh}$ 1) | NMP |
|---|---|---|---|---|---|---|
| | A | B | C | | (dl/g) | solubility 2) |
| HTI66 | HAB | TPA | SIA | 34/66 | 0.82 | No |
| HTT66 | HAB | TPA | STA | 34/66 | 0.42 | No |
| TTI66 | TAS | TPA | SIA | 34/66 | 0.50 | Yes |
| TTT66 | TAS | TPA | STA | 34/66 | 1.46 | Yes |
| TTP66 | TAS | TPA | DCP | 34/66 | 1.28 | Yes |

(1)Measured in concentrated sulfuric acid at a concentration of 0.5g/dl at 30°C.

(2)Determined by heating with NMP at 170°C for 3h.

Sulfur content in the resulting polybenzimidazoles were 90-95% of theoretical values that consisted of one sulfone linkage in TAS monomer and the sulfonic acid group in the ionic monomer. It is considered that the desulfonation during the polymerization can be ignored in this system. Although the sulfonic acid-containing monomers were used as sulfonic acid sodium salt in the dicarboxylic acids, elemental analysis showed that almost all the sulfonic acid group in resulting polymers were free from the salt. IR and NMR spectrum also confirmed incorporation of sulfonic acid groups into the polybenzazoles. IR spectrum showed characteristic absorptions of sulfonic acid group at around 1200 and 1000cm$^{-1}$. IR spectrum of the polybenzimidazole (TAS//TPA/STA=100//34/66, abbreviated as TTT66) is shown in Figure 4 as an example.

The sulfonic acid-containing polybenzimidazole copolymers showed relatively good solubility in NMP, and films were obtained by casting the NMP solutions and evaporation of the solvent. The sulfonic acid-containing polybenzoxazole copolymers, HTI and HTT, were not soluble in NMP. Therefore, films from these polymers were prepared by wet process with the solution in methanesulfonic acid, in which the cast film was coagulated and washed with water. Reinforced films containing small amount of polyparaphenylenebenzbisoxazole (PBO) (intrinsic viscosity = 24) were prepared from the mixed solution of polybenzoxazole copolymers and PBO. This method was effective to obtain a good film even if the film made from a single component was brittle. In the case of Polybenzimidazole copolymers, although films were prepared successfully from the both solutions of NMP and methanesulfonic acid, the NMP solution gave more uniform and transparent films. Better solubilities for the polybenzimidazoles were probably due to the existence of sulfone linkages in the polymer backbone.

In general, sulfonic acid groups on the aromatic rings are not stable at high temperature due to thermal desulfonation. For example, the thermal

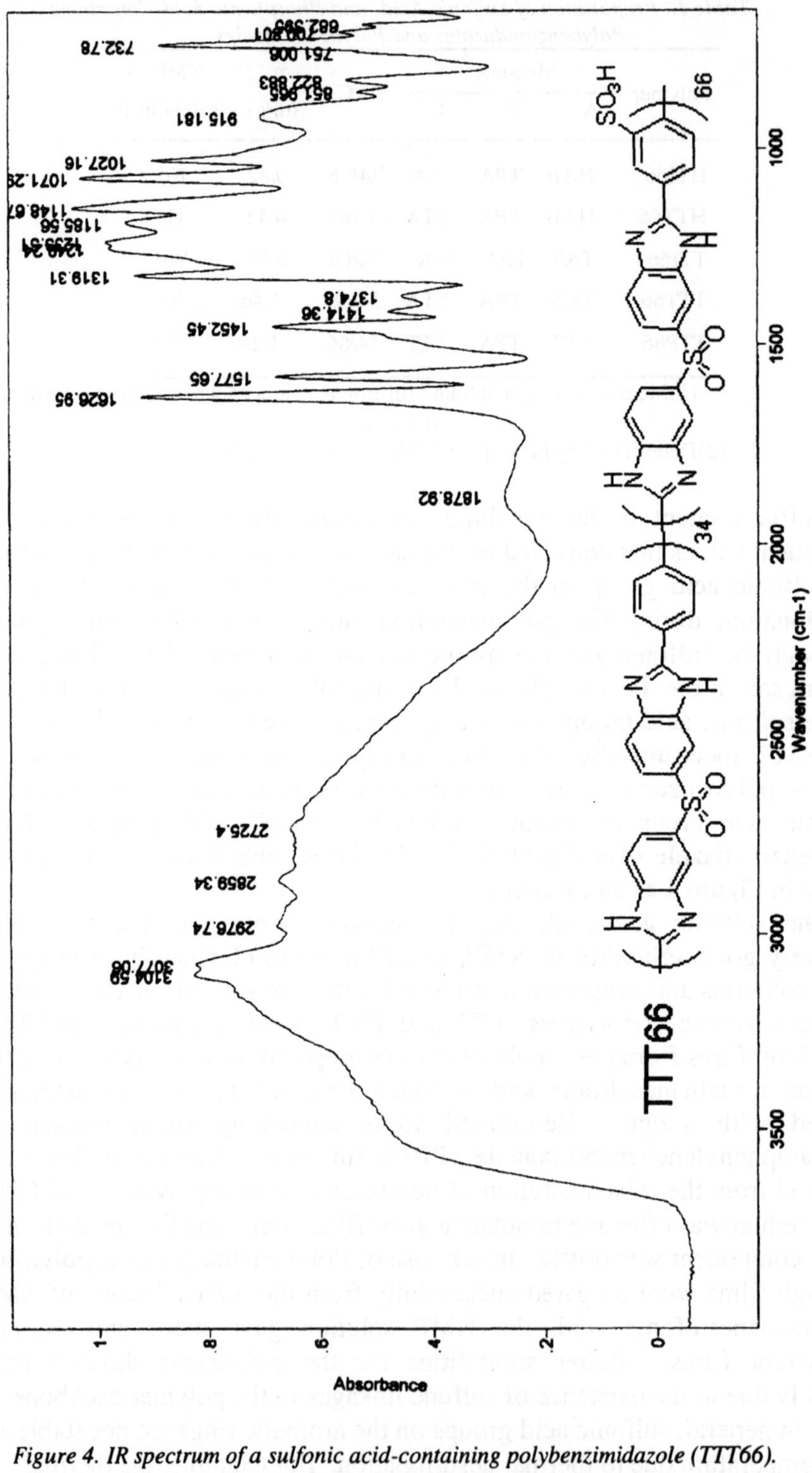


*Figure 4. IR spectrum of a sulfonic acid-containing polybenzimidazole (TTT66).*

decomposition of sulfonated polyether ether ketone was reported around 315°C (22). However, the sulfonic acid-containing polybenzimidazoles and polybenzoxazoles in this work showed good thermal stability. This effect is attributed to the position of the sulfonic acid groups which is situated in the deactivated part in the polybenzazoles due to electron withdrawing character of azole rings. **Figure 5** shows the TGA curve of TTT66. After the weight loss by water desorption with heating, the weight did not change until the temperature exceeding 400°C. As a whole, the polybenzimidazole copolymers, TTI and TTT, were stabler than the polybenzoxazole copolymers, HTI and HTT, in TGA analysis.

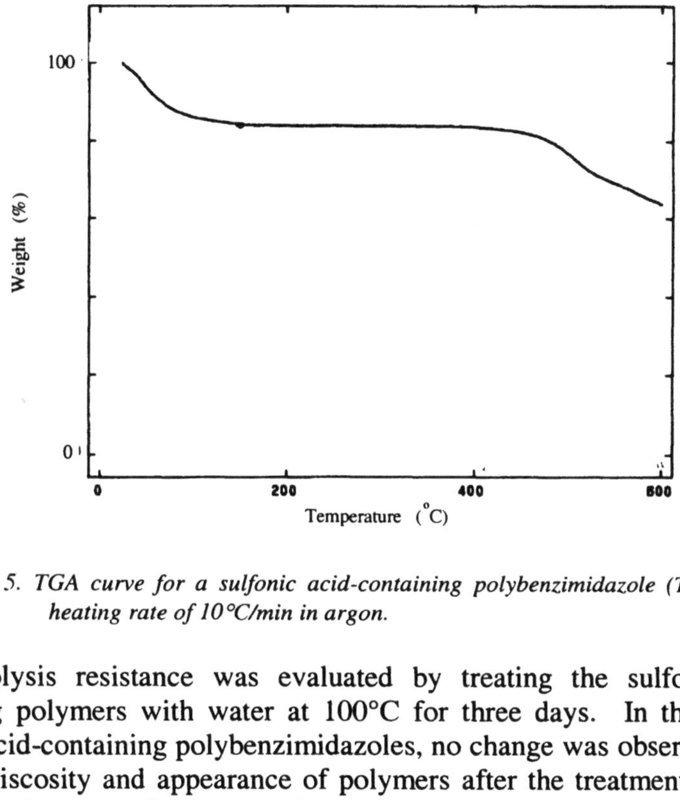


*Figure 5. TGA curve for a sulfonic acid-containing polybenzimidazole (TTT66) at a heating rate of 10°C/min in argon.*

Hydrolysis resistance was evaluated by treating the sulfonic acid-containing polymers with water at 100°C for three days. In the case of sulfonic acid-containing polybenzimidazoles, no change was observed in the inherent viscosity and appearance of polymers after the treatment. On the other hand, the inherent viscosity of sulfonic acid-containing polybenzoxazoles decreased and the polymers swelled to some extent.

Ionic conductivity was measured on the films at 80°C at 95%RH. The sulfonic acid-containing polybenzoxazoles showed higher conductivity than the polybenzimidazoles as shown in Figure 6. In both systems, the copolymers consisted of sulfonic acid introduced by meta-catenated dicarboxylic acid showed higher conductivity than those by para-catenated.

Phosphonic acid-containing polybenzimidazoles were also prepared, and an example of the results (TTP66) is shown in Table 1. TTP66 was soluble

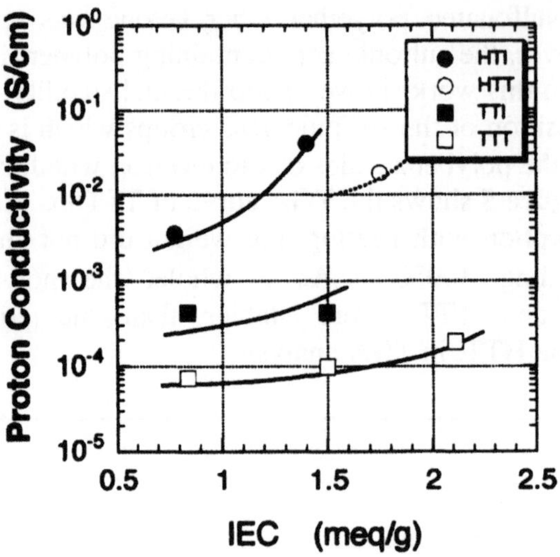


*Figure 6. Proton conductivity of sulfonic acid-containing polybenzazoles.*

in NMP and a film was prepared from the NMP solution. As expected, TTP copolymers showed similar or higher thermal stability than TTT copolymers. Proton conductivity of TTP copolymers was one order lower than that of TTT copolymers. This is attributed to a lower acidity of phosphonic acid compared to sulfonic acid.

Comparing the properties of polybenzimidazoles and polybenzoxazoles, it seems that difference in the basicity between the azole rings is a key point. Figure 7 shows the protonated structures of imidazole and oxazole rings. As basicity of imidazole ring is stronger than oxazole ring and protonated salt forms a symmetrical structure in imidazole ring, the salt of imidazole ring with sulfonic acid is much stabler than that of oxazole ring. This characteristics leads to the higher thermal stability and hydrolysis resistance in polybenzimidazole copolymers. On the other hand, as protons are easily trapped on the imidazole rings, ionic conductivity in the polybenzimidazoles stayed at lower level even if the content of sulfonic acid groups was increased.

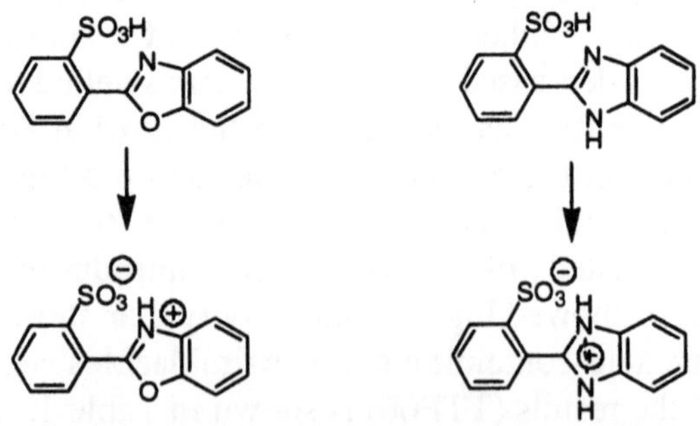

*Figure 7. Protonated structure of sulfonic acid-containing oxazole and imidazole.*

# 4. CONCLUSIONS

Five series of copolybenzazoles, polybenzimidazoles and polybenzoxazoles, containing sulfonic acid or phosphonic acid were prepared and their basic properties were compared, although the backbone structures were not exactly corresponding each other. The sulfonic acid-containing polybenzimidazoles showed higher thermal stability and hydrolysis resistance than the sulfonic acid-containing polybenzoxazoles. However, Proton conductivity of the sulfonic acid-containing polybenzoxazoles was superior to that of the sulfonic acid-containing polybenzimidazoles. The phosphonic acid-containing polybenzimidazoles showed similar or higher thermal stability and lower proton conductivity than the sulfonic acid-containing correspondents. The characteristics of the polybenzimidazole systems seem to be explained by the salt formation between ionic functional groups and imidazole rings.

# 5. REFERENCES

1. Kobayashi, T.; Rikukawa, M.; Sanui, K.; Ogata, N., Solid State Ionics., 106, 219 (1998).
2. Roziere, B.; Bauer, B.; Jones, D.J.; Tchicaya, L.; Alberti, G.; Casciola, M.; Massinelli, L.; Peraio, A.; Besse, S.; Ramunni, E., J. New Mater. Electrochem. Syst., 3, 93 (2000)
3. Kaliaguine, S.; Zaidi, S.M.; Mikhailenko, S.D.; Robertson, G.P.; Guiver, M.D., J. Memb. Sci., 173, 17 (2000).
4. Ulrich, H.; Rafler, G., Angew. Makromol. Chem., 263, 71 (1998).
5. Kerres,J.; Zhang, W.; Cui, W., J. Polym. Sci., Polym. Chem., 36, 1441 (1998).
6. Wang, F.; Ji, Q.; Harrison, W.; Mecham, J.; Formato, R.; Kovar, R.; Osenar, P.; McGrath, J.E., ACS Polym. Prepr., 41(1), 237 (2000).
7. Lufrano, F.; Squadrito, G.; Patti, A.; Passalacqua, E., J. Appl. Polym. Sci., 77, 1250 (2000).
8. Zhang, Y.; Litt, M.; Savinell, R.F.; Wainright, J.S.; Vendramini, J., ACS Polym. Prepr., 41(2), 1561 (2000).
9. Gunduz, N.; McGrath, J.E., ACS Polym. Prepr., 41(2), 1565 (2000).
10. Genies, C; Mercier, R.; Silion, B.; Cornet, N.; Gebel, G.; Pineri, M., Polymer, 42, 359 (2001).
11. Wainright, J.S.; Wang, J.; Weng, D.; Savinell, R.F.; Litt, M., J. Electrochem. Soc., 142, L121 (1995)
12. Wang, J.; Savinell, R.F., Wainright, J.; Litt, M.; Yu, H., Electrochim. Acta, 41, 193 (1996)
13. Samms, S.R.; Wasmus, S.; Savinell, R.F., J. Electrochem. Soc., 143, 1225 (1996)
14. Kawahara, M.; Morita, J.; Rikukawa, M.; Sanui, K.; Ogata, N., Electrochim. Acta, 45, 1395 (2000)
15. Glipa, X.; Haddad.M.E.; Jones, D.J.; Roziere, J., Solid State Ionics, 97, 323 (1997).
16. Rikukawa, M.; Sanui, K., Prog. Polym. Sci., 25, 1463 (2000)
17. Uno, K.; Niume, K.; Iwata, Y.; Toda, F.; Iwakura, Y., J. Polym. Sci., Polym. Chem., 15, 1309 (1977)
18. Dang, T.D.; Bai, S.J.; Heberer, D.P.; Arnold, F.E.; Spry, R., J. Polym. Sci., Polym. Phys., 31, 1941 (1993)

19. Kim, S.; Cameron, A.; Lee, Y.; Reynolds, J.R.; Savage, C.R., J. Polym. Sci., Polym. Chem., 34, 481 (1996)
20. Dang, T.D.; Chen, J.P.; Arnold, F.E., **ACS Symposium Series 585 Hybrid Organic-Inorganic Composites,** ACS, Washington, DC, P.280 (1995).
21. Ahmad, Z.; Wang, S.; Mark, J.E.; Chen, J.P.; Arnold, F.E., Polym. Mat. Eng. Sci., 70, 303 (1993).
22. Venkatasubramanian, N.; Dean, D.J.; Arnold, F.E., ACS Polym. Prepr., 37(2), 354 (1996)

Chapter 9

# DESIGN OF CONJUGATED POLYMERS FOR SINGLE LAYER LIGHT EMITTING DIODES

Zhonghua Peng
*University of Missouri-Kansas City, Departments of Chemistry, Kansas City, Missouri 64110*

## 1. INTRODUCTION

The discovery of polymer light-emitting diodes (LEDs) in 1990 (1) has stimulated an enormous amount of research in the field of polymer LEDs (2-6). Compared to conventional semiconductor LEDs, polymer LEDs offer a variety of advantages, such as low fabrication cost, excellent mechanical properties, and tunable emission colors. It is widely expected that polymer LEDs will have a significant impact on display technologies.

The light emission from a LED comes from the radiative decay of excitons resulting from the recombination of electrons and holes injected from the opposite electrodes. An efficient LED device thus requires both high photoluminescence (PL) quantum efficiency of the emitter and balanced injection and transport of electrons and holes. Unfortunately, most of the conjugated polymers developed so far do not have high PL quantum efficiency (usually less than 30%) in the solid state, and do not transport electrons and holes with comparable efficiency (1,7,8). Although a multi-layer device structure, comprised of the emissive polymer layer plus additional layers to facilitate electron and hole transport, has been proven to be able to tackle the second problem, fabrication of such a multi-layer device is often difficult, time-consuming, and more expensive than single-layer devices. Efficient single-layer devices may be fabricated with polymer blends (9), and yet, such devices may not be stable due to the tendency of phase separation. Thus, it is desirable to develop a polymer system which can be fabricated as efficient single-layer LED devices. The challenge for a material chemist is to design a highly emissive polymer, transporting electrons and holes with comparable efficiencies (10).

We have been pursuing polymers suitable for fabricating single-layer LEDs. In this chapter, we first report the approaches we have taken to improve the solid-state PL quantum efficiency of conjugated polymers. We then focus on our efforts in balancing the charge injection and transport. Finally, approaches towards novel polymers with both high PL efficiency and bipolar charge-transporting properties are presented.

## 2. CONJUGATED POLYMERS EXHIBITING HIGH SOLID STATE PL EFFICIENCIES

One chief reason for the low PL efficiency of conjugated polymers in the solid state is that conjugated backbones tend to stack cofacially with each other due to the favorable inter-chain $\pi$–$\pi$ interactions: this leads to a self-quenching process of excitons (11,12). Introducing appropriate substituents to the PPV backbone to prevent such $\pi$–$\pi$ stacking should therefore increase its PL efficiency (13). We have found that 2,5-di-(2-biphenyl)-1,4-phenylene (DBPP) was a good pack-blocking unit (14). The DBPP unit adopts a three-dimensional geometry (the two pendant phenyl rings twisted away from the center phenyl ring) due to the steric interactions. Thus, if this unit is incorporated into a PPV backbone, it should prevent the PPV backbone from stacking.

A DBPP-type monomer with dialdehyde functional groups, 2,5-di-(2-biphenyl)-1,4-benzenedicarboxaldehyde, was synthesized by the Suzuki Coupling reaction (15) of 2,5-dibromo-1,4-benzenedicarboxaldehyde and 2-biphenylboronic acid, both synthesized in one step from commercially-available starting materials (16). Three polymers (Polymers **I-III**) have been synthesized by the Horner-Wittig-Emmons (HWE) reaction (17). Detailed synthetic procedures were reported elsewhere (14).

Polymers **II** and **III** exhibit excellent solubility in common organic solvents such as THF and chloroform, while Polymer **I** is only partially soluble in these solvents. As shown in Table **1**, Polymers **II** and **III** exhibit high molecular weights and significantly improved solid-state PL quantum efficiencies (14). For comparison, the PL efficiencies of a PPV polymer (poly(2,5-dioctyloxy-1,4-phenylene vinylene)(DO-PPV) were measured under the same conditions. Its PL efficiencies in dilute solution and as solid film are both significantly lower than that of Polymers **I-III**. Compared to solutions, PL spectra of films are slightly red-shifted by less than 20 nm for Polymers **I-III**, in contrast to a nearly 40 nm red-shift for DO-PPV. This difference can be clearly seen in Figure 1, where the PL spectra of Polymer II and DO-PPV in solution and as solid films are shown. These results indicate that the DBPP unit can indeed significantly block the interchain interactions, thus limiting the exciplex formations. Improvement in PL efficiencies of other conjugated polymers, such as poly(*p*-phenylene) (PPP), poly(phenylene thiophene) (PPS), etc. is expected by introducing the DBPP unit into the respective polymer backbone.

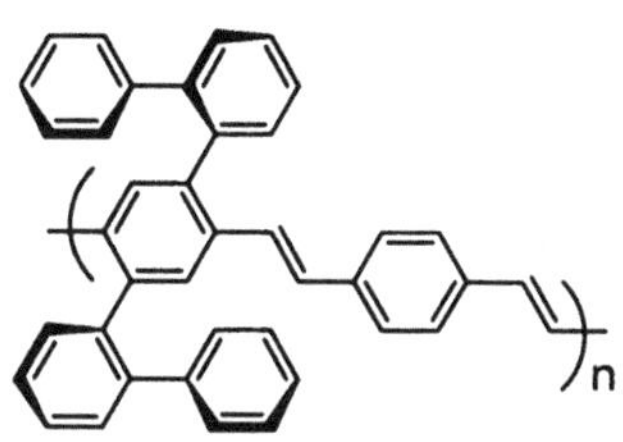


I

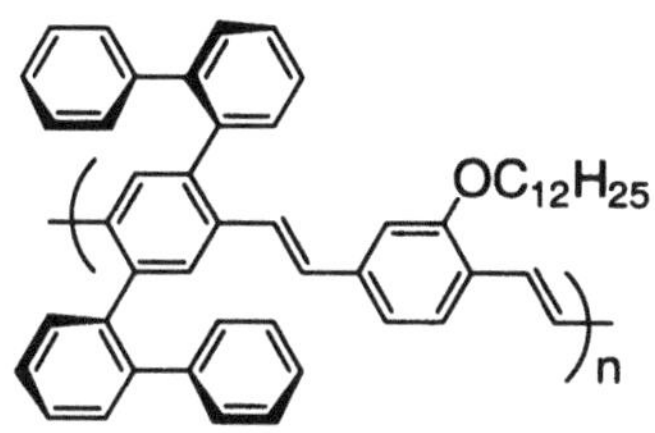


II

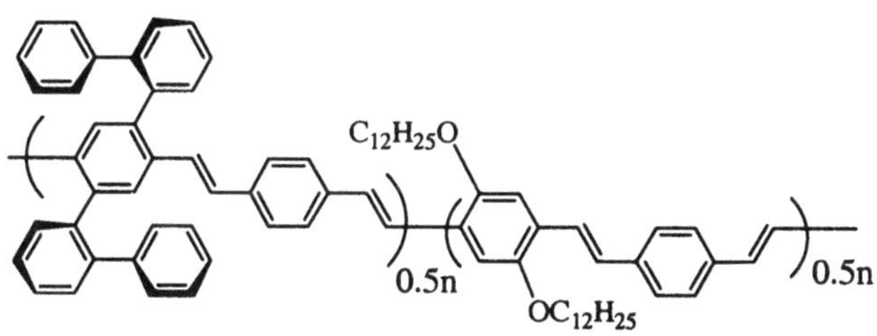


III

*Table 1. Optical Properties of Polymers **I-III***

| Polymer | Mw (kDa) | PDI | fluorescence: THF solution $\lambda_{max}$ (nm) | THF solution $\Phi_{pl}$[a] | film $\lambda_{max}$ (nm) | film $\Phi_{pl}$[b] |
|---|---|---|---|---|---|---|
| I | 4.5 | 1.5 | 488 | 0.77 | 504 | 0.82 |
| II | 151 | 2.5 | 503 | 0.73 | 520 | 0.61 |
| III | 235 | 2.8 | 510 | 0.53 | 522 | 0.77 |
| DO-PPV | 25 | 2.3 | 507 | 0.32 | 546 | 0.10 |

[a]*Using a dilute quinine sulfate solution in 1N $H_2SO_4$ as the standard (18).*
[b]*Using diphenylanthracene (dispersed in PMMA film with concentration less than $10^{-3}$M, assuming PL efficiency of 0.83) as the standard (19).*

The solid-state PL efficiency of a conjugated polymer may also be improved by designing polymers exhibiting certain three-dimensional structures such as helices. We envision that the π–π interaction between two helices will be much weaker than that of two linear conjugated chains. Based on this idea, we have synthesized a new polymer as shown as Polymer **IV** (20).

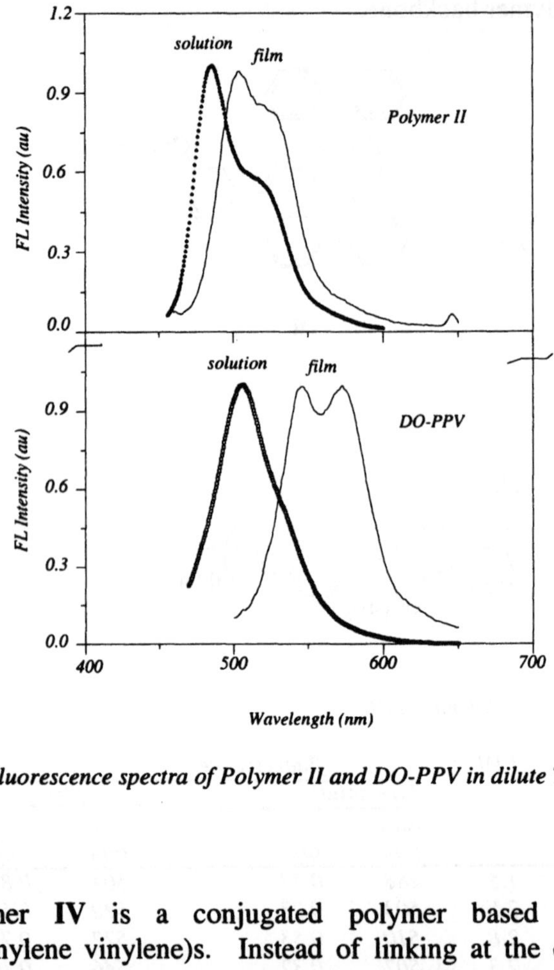


*Figure 1. Fluorescence spectra of Polymer II and DO-PPV in dilute THF solution and as solid films.*

Polymer **IV** is a conjugated polymer based on biphenyl-linked oligo(phenylene vinylene)s. Instead of linking at the commonly used 4,4' position of the biphenyl, the 2,2' positions are utilized. Due to steric reasons, the biphenyl adopts a twisted structure (two phenyl rings not on the same plane). It is hoped that the continuous twisting of the adjacent phenylene vinylene units may result in a helical type of structure, which not only limits the effective conjugation length of the polymer backbone, but also decreases the inter-chain interactions.

Polymer **IV**, a yellow powder, is soluble in common organic solvents such as THF, chloroform, etc. GPC measurements show that the polymer contains two parts, one with a molecular weight of 11k Dalton and a polydispersity of 1.33, and the other with a much lower molecular weight of 2.6k Dalton. $^{1}$H NMR and $^{13}$C NMR spectra of the polymer show no signals corresponding to the end groups. For example, no chemical shift corresponding to aldehyde protons is noticeable in the $^{1}$H NMR. Similarly, no chemical shifts corresponding to carbonyl carbons are observed in the $^{13}$C NMR. These results indicate that the low molecular weight portion is likely cyclic oligomers. While significant oligocyclics exist in the polymer, good quality films can be spin-cast from polymer solutions.

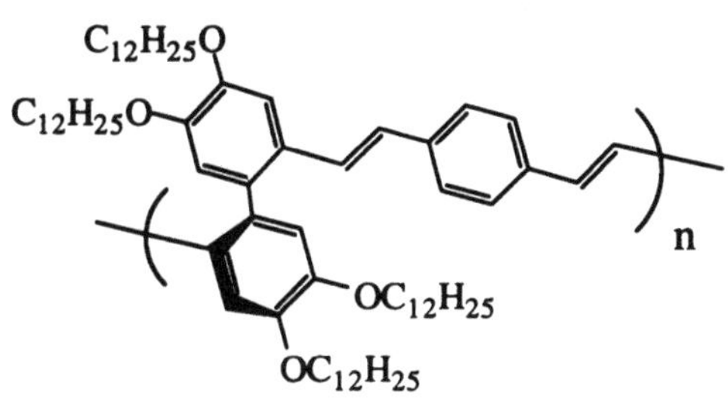


**IV**

PL spectra of Polymer **IV** in solution and as solid film are measured by a Shidmazu 5301PC Photoluminescence Spectrometer. In solution, the polymer shows intense blue emissions with two emission peaks, at 440 nm and 465 nm, and a quantum efficiency of 87%. Compared to that of the solution, the PL spectrum of a spin-cast polymer film shows two slightly red-shifted peaks, at 452 nm and 477 nm. While the emission edge is only slightly red-shifted (less than 15 nm), the PL spectrum of the polymer film is significantly broadened into the longer wavelength range. As a result, the polymer film shows bluish-green emissions. Nevertheless, the quantum efficiency of the polymer film was estimated to be 76% (19, 20). These results indicate that the biphenyl linkage can indeed disrupt both conjugation and inter-chain interactions, resulting in highly efficient PL emissions.

## 3. EXPLORING APPROACHES TOWARD BALANCED CHARGE INJECTION AND TRANSPORT

Most conjugated polymers transport electrons much less efficiently than they transport holes. One approach to improve the electron-transporting properties of a conjugated polymer is to introduce electron-withdrawing substituents into the polymer backbone. CN-substituted PPV is one example (21,22). The electron-withdrawing substituents lower the LUMO of the backbone, thus decreasing the electron injection barrier. We have recently utilized oxadiazole rings as main-chain substituents (23). The structures of these polymers are shown as Polymers **V-VII**. There are two unique features to this system. First, it is hoped that the orthogonal arrangement of the two rigid units may limit the $\pi$-stacking or excimer formation of the PPVs, thus increasing the radiative decay quantum yield and photo-stability of the polymer (24). Second, the electron-deficient oxadiazole unit will facilitate electron transport by lowering the LUMO of the backbone, much like the CN substitutent does in CN-PPVs (21).

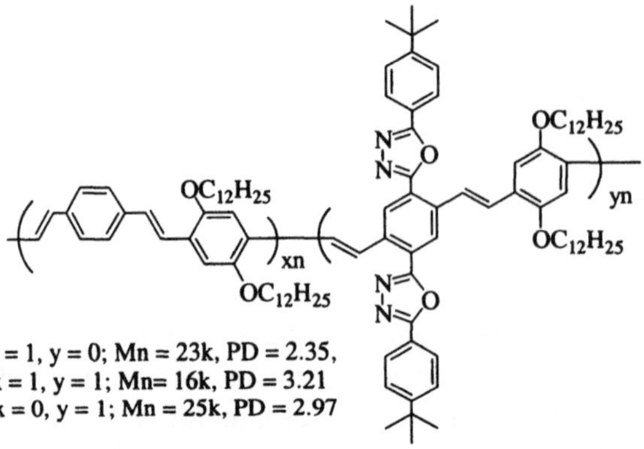


**V.** x = 1, y = 0; Mn = 23k, PD = 2.35,
**VI.** x = 1, y = 1; Mn= 16k, PD = 3.21
**VII.** x = 0, y = 1; Mn = 25k, PD = 2.97

The UV/Vis spectra of Polymers **VI** and **VII** show two strong absorption peaks: one at ca. 465 nm is the absorption peak of the PPV backbone, while the other absorption peak at 320 nm can be assigned to the orthogonal conjugated oxadiazole unit. When polymer films are excited at 460 nm, both Polymer **VI** and Polymer **VII** show one broad emission peak at 640 nm. In dilute THF solutions, both polymers show an emission maximum of 550 nm. No emissions are observed from the oxadiazole unit at 395, even when polymers are excited at 320 nm where oxadiazole units absorb strongly, indicating the existence of efficient energy transfer from the oxadiazole moiety to the PPV backbone.

Single-layer LED devices of all three polymers were fabricated and the external efficiencies and turn-on voltages of these polymer devices are listed in Table **2**. The effect of the oxadiazole-substitutents on the LED efficiency can be clearly seen from Table **2**. As the number of oxadiazole substituents increases, the EL efficiency increases and the operating voltage decreases, especially for the

LED device with Al as the cathode. The difference between the turn-on voltage of light and that of current also decreases from Polymer **V** to **VI** and to **VII**, indicating an improvement in balanced charge-injection.

*Table 2. External Quantum Efficiencies of Single-Layer LEDs*

| | ITO/Polymer/Al | | ITO/Polymer/Ca | |
|---|---|---|---|---|
| *Polymer* | $\eta$% | $V_{on}$/V ($V_{on}$'/V)[a] | $\eta$% | $V_{on}$/V |
| V | 0.002 | 15 (10) | 0.01 | 10 |
| VI | 0.018 | 10 (8.5) | 0.068 | 7.5 |
| VII | 0.041 | 8 (8) | 0.066 | 7.5 |

[a] *$V_{on}$ is the turn-on voltage of light with $V_{on}$' the turn-on voltage of current.*

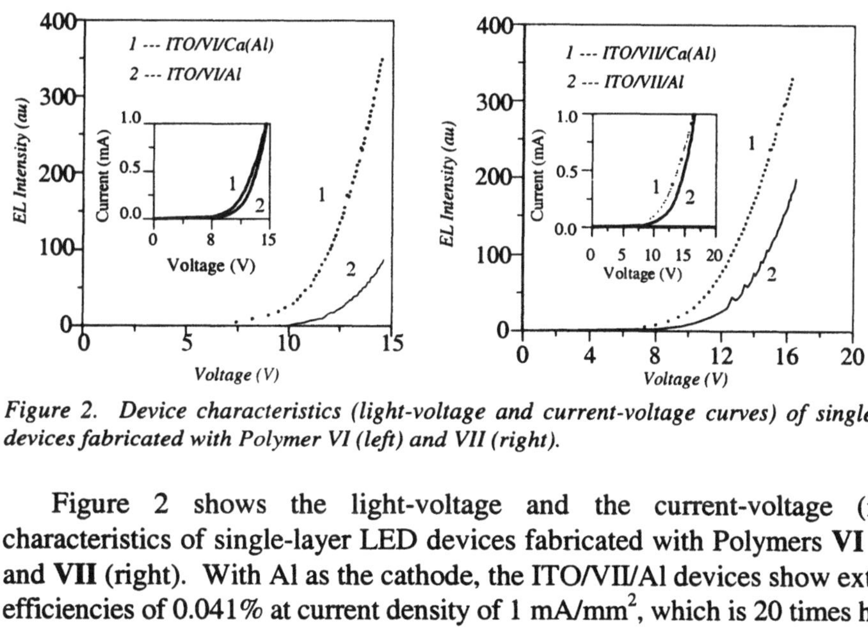


*Figure 2. Device characteristics (light-voltage and current-voltage curves) of single-layer devices fabricated with Polymer VI (left) and VII (right).*

Figure 2 shows the light-voltage and the current-voltage (inset) characteristics of single-layer LED devices fabricated with Polymers **VI** (left) and **VII** (right). With Al as the cathode, the ITO/VII/Al devices show external efficiencies of 0.041% at current density of 1 mA/mm$^2$, which is 20 times higher than that of PPV Polymer **V**. The efficiency decreases slightly when the current density increases. The devices also show the same turn-on voltage (8 V) for both the light and the current, indicating a reasonably balanced charge injection. With Ca as the cathode, a slightly higher efficiency (0.066%) and a lower turn-on voltage were obtained. Polymer **VI**, which has half the oxadiazole content as that of Polymer **VII**, gives less efficient devices (0.018%) when Al is used as the cathode. However, the efficiency is still one order of magnitude higher than that of PPV Polymer **V**. With Ca as the cathode, Polymer **VI** shows the same EL efficiency as Polymer **VII**. From Al cathode to Ca cathode, the improvement in efficiency is 61% for Polymer **VII** and 280% for Polymer **VI**. These results indicate two points: 1) the electron injection properties are improved from Polymer **V** to **VI** and to **VII**. 2) The balanced injection of both charge carriers has yet to be achieved even in Polymer **VII**. This may be due to the limited electronic interactions of the cross-conjugated oxadiazole units with the conjugated backbone. The LED efficiencies of Polymers **VI** and **VII** are higher than that of side-chain oxadiazole PPVs(25), but lower than that of main-chain oxadiazole-PPV, Polymer **VIII** (26,27). This seems consistent with the strength of electronic interactions between oxadiazole units and the conjugated backbones in these different types of oxadiazole-containing PPVs.

To find out whether it is possible to further improve the device efficiency, a new PPV (Polymer **IX**) containing oxadiazole units in both the main chain and as main-chain substitutents has recently been synthesized.

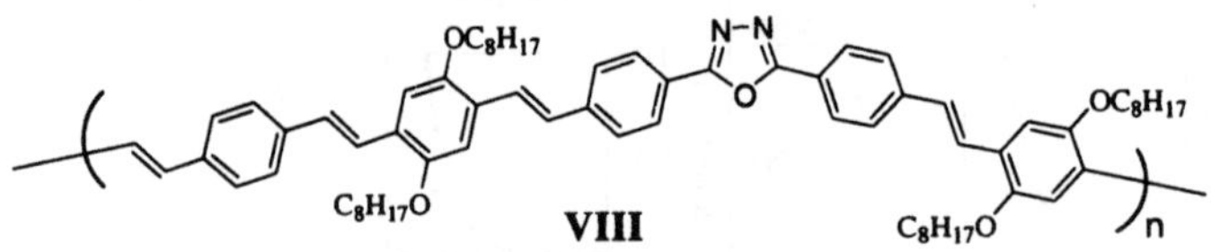


**VIII**

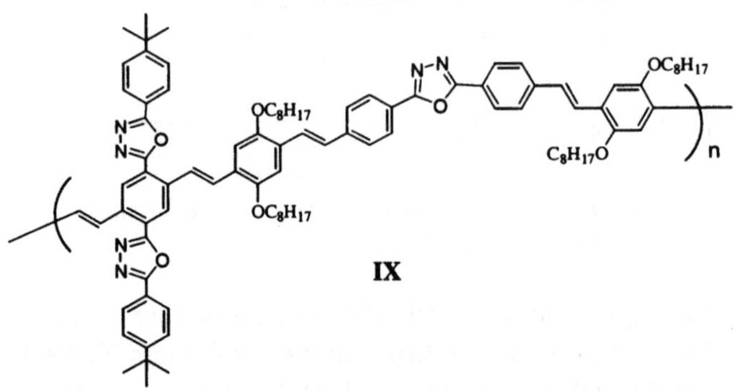


**IX**

The synthesis of Polymer **IX** is reported elsewhere (28). Polymer **IX** is soluble in common organic solvents and shows an absorption plateau from 320 nm to 470 nm in the UV-vis spectrum. In dilute THF solutions, Polymer **IX** shows strong orange-red emissions with an emission maximum at 525 nm. When polymer films are excited at 460 nm, Polymer **IX** shows a broad emission with a maximum at 620 nm.

Devices fabricated with Polymer **IX** show uniform red-orange emissions, easily observable under normal room light, under forward bias. The EL spectrum shows identical features to that of the PL spectrum of the polymer film, although its maximum (600 nm) is blue–shifted by 20 nm compared to the PL spectrum. Figure **3** shows the current-voltage (solid line) and light-voltage (dotted line) characteristics of single-layer LED devices fabricated with Polymer **IX**. With Al as the cathode, the devices showed external efficiencies of 0.07%. This value is 30 times greater than Polymer **V** and more than 2 times greater than Polymer **VII**, but again lower than Polymer **VIII**. The device can sustain current densities as high as 15 mA/mm$^2$, where the brightness of the device is 3000 cd/m$^2$, based upon the EL spectra, the relative sensitivity of a normal eye to light of varying wavelengths, and the measured output light intensity. With Ca as the cathode, a higher external efficiency (ca. 0.15%) and a lower turn-on voltage (7 V) are observed. The significant decrease in turn-on voltage may be due to the reversal of dominant charge carriers: holes are the dominant charge carriers in Al-as-cathode devices, while electrons are the dominant charge carriers in Ca-as-cathode devices.

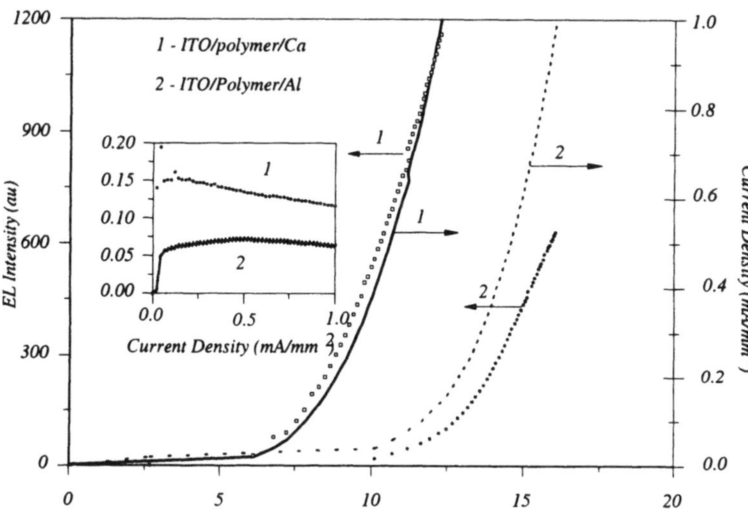


*Figure 3. Current-voltage (solid line) and light-voltage (dotted line) characteristics of single-layer LED devices fabricated with Polymer IX. The inset is the curves of external efficiencies over current densities.*

The single-layer devices fabricated with Polymer **IX** show reasonable stability under ambient conditions. The stability of the device (ITO/Polymer/Al) was monitored by measuring the light intensity under a constant driving voltage. The results are shown in Figure 4. At a driving voltage of 14.5V, the device drops to within 50% of its initial brightness in 3 minutes. However, at 12V, where the initial brightness of the device is 25 Cd/m$^2$, the device maintains over 50% of its brightness after two hours of continuous operation. Devices fabricated with Polymer **V** on the other hand usually short-circuit in less than one minute under operating voltages even close to its threshold voltage. Similar improvements in device stability are also observed for Polymer **VIII** (26).

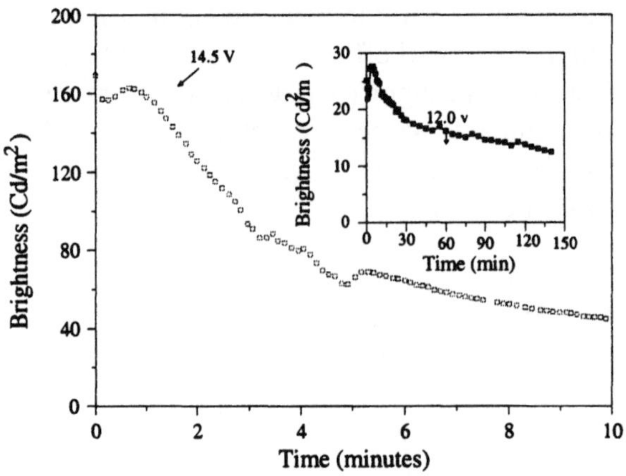


*Figure 4. Stability measurements of a single-layer device (ITO/Polymer X / Al) operating at 14.5 V or 12.0 V (inset) under ambient conditions.*

It is rather disappointing, however, that the EL efficiency of Polymer **IX** is not higher than that of Polymer **VIII** when aluminum is used as the cathode. One of the major reasons we have found during photoluminescence studies of polymers in solutions and as solid films is that this polymer, like other oxadiazole-substituted PPVs (**VI** and **VII**), exhibits very strong co-facial packing. The PL emission wavelengths of polymer films are red-shifted by nearly 100 nm compared to that of polymer solutions. The large red shift indicates strong interchain interactions. As a result, the PL quantum efficiencies for polymers **VI**, **VII**, and **IX** are one order of magnitude lower than that of Polymer **VIII**. The strong inter chain interaction may result from the alternating electron-withdrawing and electron-donating substitutions on the PPV phenyl rings. Such a substitution pattern will induce, in addition to the aromatic $\pi$–$\pi$ interaction, an additional electrostatic interaction among conjugated backbones.

In addition to combining oxadiazole units in both the main chain and side chains, we also explored the possibility of increasing electron injection by introducing electron-withdrawing groups to the cross-conjugated oxadiazole segments in **VII**. Polymers **X**, **XI**, **XII** and **XIII** are four examples (29,30).

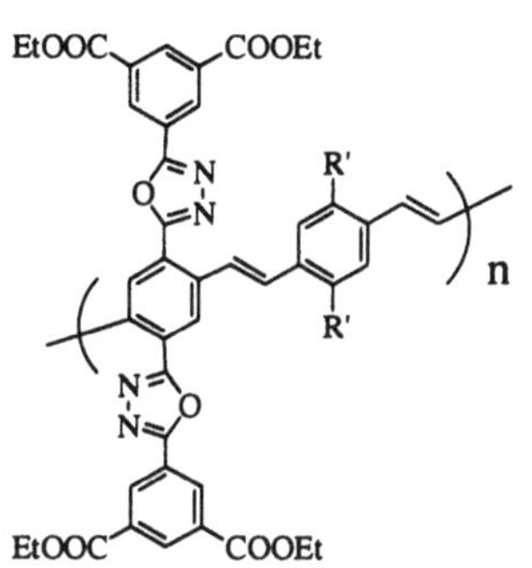


**X:** R' = $OC_6H_{12}$ **XI:** R' = $OC_{12}H_{25}$
**XII:** R' = $O(CH_2CH_2O)_3CH_3$

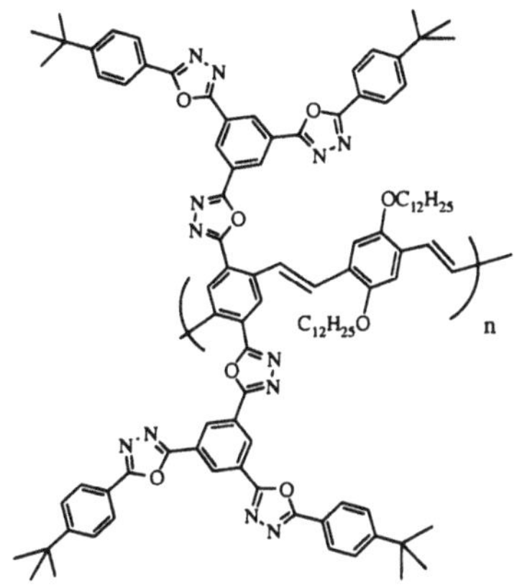


**XIII**

*Table 3. Characterization of the polymers*

| | *GPC* | | *Absorption* | *PL Emission* | |
|---|---|---|---|---|---|
| | *Mn (kDa)* | *Pd* | *$\lambda_{max}$ (nm)* | *$\lambda_{max}$ (nm)* | *$\Phi$* |
| *VII* | *25* | *2.97* | *324, 464* | *540* | *0.291* |
| *X* | *135* | *4.18* | *312, 508* | *412, 545, 628* | *0.021* |
| *XI* | *49* | *2.83* | *312, 506* | *389, 550, 628* | *0.034* |
| *XII* | *91* | *3.18* | *312, 504* | *389, 540, 623* | *0.057* |
| *XIII* | *19* | *2.1* | *310, 500* | *560, 595* | *0.06* |

All four polymers, synthesized by the Heck coupling reaction, possess high molecular weights and exhibit excellent solubility in common organic solvents. PL measurements of these polymers, however, show surprising results. As shown in Table 3, the PL quantum efficiencies of these polymers in dilute solutions are significantly lower than that of **VII**. Detailed photoluminescence and cyclovoltametry studies showed that a charge transfer transition from the polymer backbone to the cross-conjugated oxadiazole unit exists in these four polymers (29). The introduction of multiple electron-withdrawing groups to the oxadiazole segment appears to lower the $\pi^*$ state of the oxadiazole segments to such an extent that it has a lower energy than the $\pi^*$ state of the PPV backbone. Consequently, a charge-transfer excited state from the HOMO of the PPV backbone to the LUMO of the oxadiazole segment becomes the lowest-lying excited state in these polymers. The existence of such charge transfer transitions significantly lowers the PL efficiency. These results remind us that when energy levels are manipulated to optimize charge-injection barriers, care must be taken to avoid the formation of charge-transfer transitions if efficient single-layer light-emitting diodes are to be fabricated.

## 4. POLYMERS WITH BOTH HIGH PL EFFICIENCY AND BALANCED CHARGE-INJECTION PROPERTIES.

To fabricate an efficient single-layer device, a polymer must combine high photoluminescence quantum efficiency with balanced charge injection and transporting properties. Often, however, conjugated polymers with high solid-state PL quantum efficiencies do not possess bipolar charge-transporting properties. On the other hand, as shown earlier and by other researchers (31,32), a bipolar charge-transporting polymer often does not have high solid-state PL efficiency. It is thus quite a challenge to design a polymer combining both properties. We have shown that incorporating DBPP units into the main chain of a PPV can dramatically improve its PL quantum efficiency at solid state, while incorporating oxadiazole units into the main chain or as main-chain substituents can significantly enhance its electron injection and transporting properties. It seems logical that if we incorporate both units into a single PPV polymer, the resulting polymer might be both highly photoluminescent and highly electron-transporting. Polymer **XIV** is one example where the DBPP unit is incorporated into the main chain while oxadizole units are introduced as main-chain substituents.

The synthesis of Compound **5** was previously reported[23]. To synthesize Compound **4**, Compound **1** was first synthesized by the Suzuki coupling reaction of 2,5-dibromo-*p*-xylene with 2-biphenylboronic acid in 78% yield. Compound **1** was then converted to **2** by benzyl-bromination with NBS (85% yield), followed by the reaction with triethylphosphite (68% yield). HWE reaction of **2** with **3** gave the desired Monomer **4** in 55% yield. Polymer **XIV** was synthesized by the Heck coupling reaction.

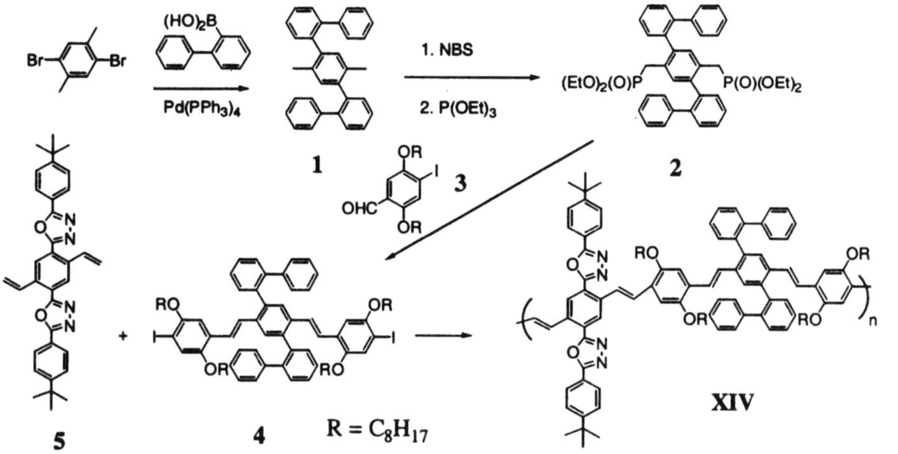


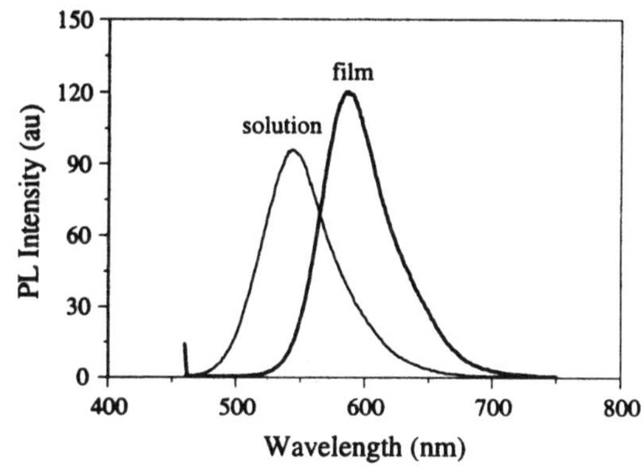


*Figure 5. Fluorescence spectra of Polymer XIV in THF solution and as a solid film.*

Polymer **XIV**, with a number average molecular weight over 56k Dalton, is soluble in common organic solvents such as THF and chloroform. Fluorescence measurements show that Polymer **XIV** exhibits a PL quantum efficiency of 38% in the solid state. This value is significantly higher than other oxadiazole-substituted PPVs, such as Polymers **VI**, **VII** and **IX**. As shown in Figure 5, the fluorescence spectrum of Polymer **XIV** shows a maximum at 544 nm in dilute THF solution and 585 nm in the solid state. A red shift of around 40 nm is observed, in sharp contrast to the nearly 100 nm red-shift observed for other oxadiazole-substituted PPVs (**VI**, **VII**, and **IX**). This result indicates again that the DBPP unit can indeed block the interchain interaction, thus improving its solid-state PL efficiency. With the oxadiazole substituents, Polymer **XIV** is expected to possess improved charge-injection balance as well, endowing

Polymer **XIV** with both high PL efficiency and balanced charge-injection properties. We are currently evaluating its LED performance.

## 5 CONCLUSIONS

We have shown that incorporating DBPP units into the backbone of a PPV results in significant enhancement of its solid-state PL efficiency. We have also demonstrated that introducing oxadiazole units as main-chain substitutents imparts bipolar charge-transporting properties. Combining both DBPP and oxadiazole units into a single conjugated polymer may result in a polymer exhibiting both high PL quantum efficiency and bipolar charge-transporting properties, making such a polymer a promising candidate for single-layer LED applications.

## 6. REFERENCES


1. Burroughes, J. H.; Bradley, D. D. C.; Brown, A. R.; Marks, R. N.; Mackay, K.; Friend, R. H.; Burn, P. L.; Holmes, A. B. *Nature* **1990**, 347, 539.
2. Kraft, A.; Grimsdale, A. C.; Holmes, A. B. *Angew. Chem., Int. Ed. Eng.* **1998**, 37, 402.
3. Greiner, A. *Polymers for Advanced Technologies* **1998**, 9(7), 371.
4. Bradley, D. *Curr. Opin. Solid State Mater. Sci.* **1996**, 1 (6), 789.
5. Feast, W. J.; Tsibouklis, J.; Pouwer, K. L.; Groenendaal, L.; Meijer, E. W. *Polymer* **1996**, 37, 5017.
6. Yang, Y. *MRS Bull.* **1997**, 22(6), 31.
7. Hide, F.; Diazgarcia, M. A.; Schwartz, B. J.; Heeger, A. J. *Acc. Chem. Res.* **1997**, 30 (20), 430.
8. Gustafsson, G.; Cao, Y.; Treacy, G. M.; Klavetter, F.; Colaneri, N.; Heeger, A. J. *Nature* **1992**, 357, 477.
9. Wu, C. C.; Sturm, J. C.; Register, R. A.; Tian, J.; Dana, E. P.; Thompson, M. E. *IEEE Trans. Electron Dev.* **1997**, 44, 1269.
10. Liu, Yunqi; Ma, Hong; Jen, Alex K-Y. *Chem. Commun.* **1998**, 24, 2747-2748.
11. McBranch, D. W.; Sinclai, M. B.; in *The Nature of The Photoexcitations in Conjugated Polymers*; Sariciftci, N. S., Ed.; World Scientific Publishing: Singapore, **1997**; Chapter 20, p 608.
12. Rothberg, L. J.; Yan, M.; Papadimitrakopoulos, F.; Galvin, M. E.; Kwock, E. W.; Miller, T. M. *Synth. Met.* **1996**, 80, 41.
13. Hsieh, B. R.; Yu, Y.; Forsythe, E. W.; Schaaf, G. M.; Feld, W. A. *J. Am. Chem. Soc.* **1998**, *120*, 231.
14. Peng, Z.; Zhang, J.; Xu, B. *Macromolecules* **1999**, 32, 562.
15. Suzuki, A. *Pure Appl. Chem.* **1991**, 63, 419.
16. Peng, Z.; Galvin, M. E. *Acta Polymer.* **1998**, 49 (5), 244.
17. Wadsworth, W. S. Jr. *Org. React.* **1977**, 25, 73.

18. Demas, J. N.; Crosby, G. A. *J. Phys. Chem.* **1971**, 75, 991.
19. Guilbault, G. G., Ed. *Practical Fluorescence*, 2nd ed.; Marcel Dekker Inc.: New York, **1990**.
20. Xu, B.; Zhang, J.; Peng, Z. *Synth. Met.* **2000**, 113, 35.
21. Greenham, N. C.; Moratti, S. C.; Bradley, D. D. C.; Friend, R. H.; Holmes, A. B. *Nature* **1993**, 365, 628-631.
22. Moratti, S. C.; Cervimi, R.; Holmes, A. B.; Baigent, D. R.; Friend, R. H.; Greenham, N. C.; Gruner, J.; Hamer, P. J. *Synth. Met.* **1995**, 71, 2117.
23. Peng, Z.; Zhang, J. *Chem. Mater.* **1999**, 11(4), 1138.
24. Yang, J. S.; Swager, T. M. *J. Am. Chem. Soc.* **1998**, 120, 5321.
25. Bao, Z.; Peng, Z.; Galvin, M. E.; Chandross, E. A. *Chem. Mater.* **1998**, *10*, 1201.
26. Peng, Z.; Bao, Z.; Galvin, M. E. *Adv. Mater.* **1998**, 10, 680.
27. Peng, Z; Bao, Z.; Galvin, M. E. *Chem. Mater.* **1998**, 10, 2086.
28. Peng, Z.; Zhang, J. *Synth. Met.* **1999**, 105, 73.
29. Xu, B.; Zhang, J; Peng, Z. *Synth. Met.* **2000**, 114, 337.
30. Peng, Z.; Zhang, J.; Xu, B. *Polym. Prepr.* **2000**, 41, 881.
31. Sarnecki, G. J.; Friend, R. H.; Holmes, A. B.; Moratti, *Synth. Met.* **1995**, 69, 545.
32. Lux, A.; Holmes, A. B.; Cervini, R.; Davies, J. E.; Moratti, S. C.; Gruner, J.; Cacialli, F.; Friend, R. H. *Synth. Met.* **1997**, 84, 293.

Chapter 10

# SYNTHESIS AND CHARACTERIZATION OF NOVEL BLUE LIGHT-EMITTING POLYMERS CONTAINING DINAPHTHYLANTHRACENE

Shiying Zheng and Jianmin Shi
*Eastman Kodak Company, Research & Development, Rochester, NY 14650*

## 1. INTRODUCTION

Electroluminescent devices are opto-electronic devices where light emission is produced in response to an electrical current through the device. The physical model for EL is the radiative recombination of electrons and holes (1). The term light emitting diode (LED) is commonly used to describe an EL device where the current-voltage behavior is non-linear, meaning that the current through the EL device is dependent on the polarity of the voltage applied to the EL device. Both organic and inorganic materials have been used for the fabrication of LEDs. Inorganic EL materials such as ZnS/Sn, GaAsP, and GaAs have been known and used for many years. The first commercial GaAs LEDs were introduced in 1962. However, the drawbacks of inorganic materials include difficulties to process and to obtain large surface areas and efficient blue light.

EL from an organic material was first discovered on anthracene crystals by Pope et al. in the 1960s (2). These organic EL devices required high operating voltages ranging from 400-2000 V and had low quantum efficiency. In the mid-1980s Tang and Van Slyke at Eastman Kodak Company made a breakthrough by using multi-layer structures to achieve organic LED (OLED) with high brightness, enhanced quantum efficiency, and low operating voltages (3, 4). This discovery spurred tremendous research interest both in academia and industry. A new barrier was crossed in 1990 when scientists at Cambridge University discovered the EL in conjugated poly(vinylene vinylene) (PPV) (5). These discoveries have provided a new impetus to the development of OLED for display and other purposes. Organic EL displays represent an alternative to the well-established display technologies based on cathode-ray tubes and liquid crystal displays (LCDs), particularly with respect to large-area displays for which existing methods are not well suited. The principal interest in the use of polymers for LEDs lies in the scope for low-cost manufacturing, using solution-processing of film-forming polymers such as spin-coating, inkjet delivery, and reel to reel coating.

Moreover, the rapid growing market of portable electronic devices such as laptops, hand-held computers, media players, pagers, and cellular phones has driven the search for a new generation of display technology. Portable display technology is constrained by the need for high efficiency, full color, low cost, battery-compatible driving voltage, reasonable lifetime, and resistance to the temperature extremes for outdoor or automobile use. Most approaches, such as plasma displays, vacuum florescence, and inorganic thin-film EL, face some combination of these obstacles. As a result, the market has been largely left to LCDs, which require backlighting and have small viewing angle.

OLEDs have been demonstrated to be brighter, thinner, lighter, and faster than LCDs. They also require less power to operate, offer higher contrast and wide viewing angle (>165 degree), and have great potential to be cheaper to manufacture, especially polymer-based LEDs. OLEDs are threatening to challenge LCDs as dominant flat panel display in a broad range of portable electronics. According to Stanford Resources, a San Jose, California-based consulting group, the global market for OLED displays is currently valued at $84 millions. The group expects the market to be $700 million by 2005, and $1.6 billion by 2007.

Tremendous amount of research activities have been on going in the discovery of new polymers and in the performance of the related LED devices (6-8). The colors polymers emit can be easily tailored by chemical modification of the polymer structures. Up to date, the colors emitted by the current polymers span almost the whole range of the visible spectrum with the exception of blue. Stable, efficient, and high-brightness blue light-emitting materials are desirable for full-color display applications. These materials can also serve as energy-transfer donors in the presence of lower energy fluorophores (9). Recent material research has been focused on the synthesis of polymers with large energy band gaps to emit blue light. The first report of blue-emission from a conjugated polymer LED was for polydialkylfluorene (PF) (10) followed by poly(p-phenylene) (PPP) (11). Other conjugated blue-light-emitting polymers include PPP ladder and stepladder polymers (12), poly(pyridine) (13), poly(thiophenyloxadiazolylphenylene) (14).

An effective approach to achieve large energy band gaps for blue light is to control the conjugation length. For example, well-defined conjugated chromophores can be incorporated in the side chains (15) or the main chains of the polymers. In the latter case, the conjugation length of the chromophores can be controlled by introducing flexible non-conjugated spacer groups to tune the optical properties of the polymers. The spacer groups usually prevent the extended conjugation and contribute to the solubility and film-forming properties of the polymers. This approach has been demonstrated in several examples such as PPV, polythiothene, PPP, and poly(oxadiazoles) (16-19).

In this chapter, we present the design, synthesis, and characterization of blue light-emitting polymers, **P1-P3,** containing novel blue chromophores, 9,10-di(2-naphthyl)anthracene (Figure 1). **P1-P3** are polymers alternating with the blue chromophores and the flexible spacer groups. The chromophores have been demonstrated to be particularly useful for the fabrication of efficient and stable blue EL devices (20). Polymers incorporating such chromophores are expected to be highly flurorescent and emit blue light. For comparison, a model compound **M1** was synthesized.

## 2. EXPERIMENTAL

Common reagents were purchased from Aldrich, Acros, or Eastman Kodak Company and used as received unless otherwise specified. The EL grade PEDOT aqueous solution was purchased from Bayer Incorporation under the trade name Baytron P and was filtered through a 2 μm of filter before spin coating. The ITO glass substrate was obtained from Donnelly Applied Films and the ITO thickness is about 300 Å with a sheet resistance of about 68 ohms/square. Prior to the spin-coating of polymer solution, the substrate glass was thoroughly cleaned by scrubbing, ultrasonication in a detergent, vapor degreasing, and irradiation in an UV-ozone chamber. The polymers were dissolved in toluene as 1 wt% solution and filtered through 2 μm filters before spin coating. The cathode metals for the devices, Mg:Ag (10:1 in volume ratio), were deposited by co-evaporation with a deposition rate of 10 Å/s, and the thickness of the cathode is 2000 Å.

$^{1}$H NMR and $^{13}$C NMR spectra were recorded on a Varian XL-300 spectrometer using TMS as an internal standard. Molecular weights for the polymers were determined via size exclusion chromatography (SEC) in THF with polystyrene standards. TGA analysis was performed on TA instruments,

TGA 2950, and DSC was carried out on DSC 2920. Both TGA and DSC data were obtained at a heating rate of 10 °C/min. Melting points were measured with an Electrothermal 9200 melting point apparatus and are uncorrected. UV-vis and photoluminescence spectra were recorded on a FluoroMax-2. The thickness of the polymer coating was about 700 Å, which was measured by a TENCOR P-10 surface profiler. The active area is 0.1 $cm^2$. The luminance was measured using a Photo Research PR650 spectrophotometer. A DC voltage or current source was used to trace the current-voltage characteristics.

**Compound 1:** 1st step: Synthesis of 1,3-di(2-(6-bromonaphthoxy))-2-propanol. 6-Bromo-2-hydroxy-naphthalene (66.7 g, 0.299 mol) and NaOH (7.4 g, 0.185 mol) were suspended in 500 mL of dioxane and heated for half hour until the reaction became homogeneous. Epichlorohydrin (11.5 g, 0.124 mol) was added dropwise and the reaction was heated at reflux for 20 h. The brown reaction mixture was poured into 1 L of water and the gray precipitate was filtered, washed with water and methanol, and dried under vacuum. The crude product was recrystallized from acetone to give 41.3 g of pure product (67% yield). $^1$H NMR ($CDCl_3$) δ (ppm): 4.28-4.35 (m, 5H), 4.50-4.55 (m, 1H), 7.15 (d, J = 2.2 Hz, 2H), 7.18 (dd, $J_1$ = 8.9 Hz, $J_2$ = 2.4 Hz, 2H), 7.48 (dd, $J_1$ = 8.7 Hz, $J_2$ = 1.8 Hz, 2H), 7.56 (d, J = 8.8 Hz, 2H), 7.64 (d, J = 8.9 Hz, 2H), 7.91 (d, J = 1.1 Hz, 2H); $^{13}$C NMR ($CDCl_3$): 66.07, 67.87, 105.32, 115.04, 118.36, 126.84, 127.04, 127.70, 127.76, 128.24, 131.32, 155.40; M.p. 148-150 °C; FD-MS: m/z 500 ($M^+$).

2nd step: To a 500 mL round-bottomed flask were added the above compound 1,3-di(2-(6-bromonaphthoxy))-2-propanol (30.0 g, 0.060 mol), NaOH (7.2 g, 0.18 mol), $NaHSO_4$ (20.4 g, 0.060 mol), 1-bromododecane (44.9 g, 0.18 mol), water 30 mL, and THF 160 mL. The reaction was heated under reflux for two days. After cooling to room temperature, 250 mL of water was added, and the reaction was extracted with ether (6x200 mL). The organic phase was combined, dried over $MgSO_4$, and concentrated. The crude product was purified by column chromatography on silica gel using $CH_2Cl_2$/hexane (15/85) as an eluent to give 25.2 g pure product as a white powder (63% yield). $^1$H NMR ($CDCl_3$) δ (ppm): 0.87 (t, J=6.9 Hz, 3H), 1.25-1.39 (m, 18H), 1.60-1.70 (m, 2H), 3.73 (t, J=6.6 Hz, 2H), 4.13 -4.18 (m, 1H), 4.26 -4.37 (m, 4H), 7.16 (d, J=2.2 Hz, 2H), 7.19 (dd, $J_1$ = 8.9 Hz, $J_2$ = 1.8 Hz, 2H), 7.48 (dd, $J_1$=8.8 Hz, $J_2$ = 1.8 Hz, 2H), 7.57 (d, J=8.8 Hz, 2H), 7.63 (d, J = 8.9 Hz, 2H), 7.91 (d, J=1.4 Hz, 2H); $^{13}$C NMR ($CDCl_3$): 14.10, 22.67, 26.05, 29.33, 29.44, 29.61, 29.65,

30.03, 31.90, 67.76, 71.16, 76.40, 106.95, 117.22, 119.85, 128.35, 128.51, 129.64, 130.13, 132.94, 156.92; M.p. 99-101 °C; FD-MS: m/z 668 ($M^+$).

**Compound 2:** 1st step: Compound **1** (12.0 g, 0.018 mol) was dissolved in 150 mL of anhydrous THF and cooled to -78 °C. To this solution was added slowly nBuLi solution (33.8 mL, 1.6 M in hexane, 0.054 mol) to keep temperature lower than -60 °C. The mixture was stirred at -78 °C for 1 h and trimethyl borate (8.1 mL, 0.072 mol) was added slowly. The reaction mixture was slowly warmed up to room temperature and stirred at room temperature overnight. The reaction was quenched with dilute HCl solution and stirred under nitrogen for 1 h. The reaction mixture was then extracted with ether 5 times and dried over $MgSO_4$. Solvent was evaporated and the crude product was used for preparation of boronic diester without purification.

2nd step: Synthesis of bis-2,2-dimethyltrimethylene diboranate. The crude diboronic acid and 2,2-dimethylporpane-1,3-diol (neopentyl glycol) (3.8 g, 0.037 mol) were dissolved in toluene and heated under vigorous reflux under a Dean-Stark trap. After the completion of the reaction, toluene was evaporated and the crude product was recrystallized from hexane to give 7.7 g of off-white needle-like crystals (57% yield). $^1$H NMR ($CDCl_3$) δ (ppm): 0.85 (t, J = 6.9 Hz, 3H), 1.02 (s, 12H), 1.24-1.34 (m, 18H), 1.58-1.65 (m, 2H), 3.71 (t, J = 6.6 Hz, 2H), 3.79 (s, 8H), 4.11-4.16 (m, 1H) , 4.24-4.36 (m, 4H), 7.17 (s, 2H), 7.14 (d, J = 8.5 Hz, 2H), 7.67 (dd, J = 8.2 Hz, 2H), 7.75 (d, J = 8.8 Hz, 2H), 7.8 (d, J = 8.3 Hz, 2H), 7.91 (s, 2H); $^{13}$C NMR ($CDCl_3$): 14.11, 21.93, 22.68, 26.07, 29.35, 29.47, 29.63, 29.66, 30.06, 31.92, 67.83, 71.15, 72.37, 76.47, 106.77, 118.61, 125.77, 128.62, 130.26, 130.66, 134.80, 136.13, 157.38; M.p. 59-61 °C; FD-MS: m/z 736 ($M^+$).

**Compound 3:** 2,6-Dihydroxyanthraquinone (100.0g, 0.42 mol) and 2-ethylhexyl bromide (165.0 g, 0.86 mol) were dissolved in 1 L of DMF. To this solution was added anhydrous $K_2CO_3$ (120.0g, 0.87 mol). The reaction was heated at 90 °C overnight. Most of DMF was removed and 500 mL of water was added. The reaction was extracted with ether (3x400 mL), washed with brine (1x200 mL), and dried over $MgSO_4$. Solvent was removed and the crude product was recrystallized from methanol to give yellow powdery product 125.2 g (65% yield). $^1$H NMR ($CDCl_3$) δ (ppm): 0.92-0.98 (m, 12H, $CH_3$), 1.34-1.54 (m, 16H), 1.75-1.81 (m, 2H, C$\underline{H}$($CH_3$)), 4.02 (d, J = 5.5 Hz, 4H, $OCH_2$), 7.19 (d, J = 8.4 Hz, 2H), 7.70 (s, 2H), 8.19 (d, J = 8.5 Hz, 2H); $^{13}$C NMR ($CDCl_3$): 11.12, 14.06, 23.04, 23.88, 29.08, 30.51, 39.34, 71.34, 110.64, 120.84, 127.00, 129.62, 135.88, 164.29, 182.27. M.p. 49-51 °C; FD-MS: m/z 464 ($M^+$).

**Compound 4:** To a 1 L round bottom flask was added tin (80.0 g, 0.67 mol), compound **3** (75.0 g, 0.16 mol), and 375 mL of acetic acid. The reaction was refluxed for 2 h during which the reaction became a slurry. The reaction was cooled to room temperature and the top layer was decanted. The solid was washed with $CH_2Cl_2$. The combined organic phase was washed with water, saturated $NaHCO_3$ solution, and brine and dried over $MgSO_4$. Solvent was removed to yield 72.0 g of yellow solid. The yellow solid was dissolved in 200 mL of isopropanol and added dropwise to a solution of $NaBH_4$ (6.5 g, 0.17 mol) in 300 mL of isopropanol. The reaction was heated at reflux overnight. After cooled to room temperature, the reaction was quenched with dilute HCl solution and then poured into water. The yellow precipitate was collected by filtration, washed with water and ethanol and dried to give 55.2 g of pure product as an yellow powder (78% yield in two steps). $^1$H NMR ($CDCl_3$) δ (ppm): 0.92-1.62 (m, 14H, alkyl), 1.79-1.87 (m, 1 H, alkyl), 3.99 (d, J = 5.7 Hz, 2H, $OCH_2$), 7.14 (d, J = 9.4 Hz, 2H), 7.17 (s, 2H, 1 and 5 of anthracene), 8.17 (s, 2H, 9 and 10 of anthracene); $^{13}$C NMR ($CDCl_3$): 11.19, 14.10, 23.10, 24.07, 29.18, 30.72, 39.44, 70.48, 104.58, 120.85, 124.09, 128.71, 129.06, 131.30, 156.22. M.p. 60-62 °C; FD-MS: m/z 436 ($M^+$).

**Compound 5:** Compound **4** (13.50 g, 0.031 mol) was added to 150 mL of DMF and cooled down to 0 °C. To this suspension was added NBS (11.60 g, 0.065 mol) in 60 mL of DMF. Upon the addition of NBS, the reaction became clear and turned to dark green color. The reaction was stirred at room temperature under nitrogen overnight. The reaction was poured into 200 mL of water, and extracted with methylene chloride (3x300 mL). The combined organic phase was washed thoroughly with water (3x100 mL) and brine (1x100 mL), and dried over $MgSO_4$. After removal of the solvent, the dark brown residue was washed with hexane to collect greenish yellow crystals. The crude crystals were recrystallized from acetone to give flaky greenish yellow fluorescent product. The filtrates were combined and purified by chromatography on silica gel with hexane as eluent. Total yield: 5.5 g (30% yield). $^1$H NMR ($CDCl_3$) δ (ppm): 0.93-1.70 (m, 14H, alkyl), 1.81-1.89 (m, 1 H, alkyl), 3.12 (d, J = 5.4 Hz, 2H, $OCH_2$), 7.34 (d, J = 9.2Hz, 2H), 8.00 (d, J = 9.2 Hz, 2H), 8.71 (s, 2H, 1 and 5 of anthracene); $^{13}$C NMR ($CDCl_3$): 11.12, 14.10, 23.08, 23.93, 29.15, 30.52, 39.88, 72.76, 107.74, 117.02, 125.27, 129.51, 129.75, 130.12, 152.87. M.p. 103-105 °C; FD-MS: m/z 590 ($M^+$).

**Model compound M1:** Magnesium turnings (5.55 g, 231 mmol), 30 mL

THF, and two crystals of iodine were placed into a round-bottomed flask under nitrogen. The flask was heated to 50 °C and dibromoethane was used to help initiate the Grignard reaction. 2-Bromo-6-methoxy naphthalene (50.0 g, 211 mmol) in 400 mL of THF was added dropwise over 90 minutes during which time heat was increased to allow for gentle reflux. After addition, the reaction mixture was allowed to reflux for an additional 2 hours.

9,10-Dibromoanthracene (23.63 g, 70.3 mmol), 300 mL of THF and a catalytic amount of dichlorobis(triphenylphosphine)palladium (II) were all placed under nitrogen in a round-bottomed flask. The Grignard reagent prepared above was then added via a double-end needle transfer while still warm. After the addition, the reaction mixture was allowed to reflux overnight. After cooling to room temperature, the precipitated solid was collected by vacuum filtration and washed with ether, THF, and water. The collected solid was placed into a beaker and HCl (6M) was added followed by ethanol. This mixture is stirred for 60 minutes and the solid is again collected by vacuum filtration. The product was washed until the washings are neutral. After drying, the solid is refluxed gently in $CH_2Cl_2$ for 60 minutes. The pure product was filtered and dried to give 85.2 g solid (82% yield). $^1$H NMR ($CDCl_3$) δ (ppm): 3.97 (s, 6H), 7.29 (dd, $J_1$=8.9 Hz, $J_2$=2.4 Hz, 2H), 7.40 (d, J=6.9 Hz 2H), 7.41 (d, J=6.9 Hz, 2H), 7.53 (d, J 2.2 Hz, 2H), 7.57 (d, J=8.0 Hz, 2H), 7.64 (d, J=6.8 Hz, 2H), 7.65 (d, J=6.8 Hz, 2H), 7.94 (d, J=9.0 Hz, 2H), 7.97 (s, 2H), 8.10 (d, J=8.4 Hz, 2H); FD-MS: m/z 490 ($M^+$).

**Polymerization:** Equal mole of dibromide and boronic diester were dissolved in toluene. To this solution was added 2 M $Na_2CO_3$ aqueous solution (3 eq to mononer) and phase transfer catalyst Aliquat® 336 (0.13 eq to monomer). The reaction mixture was bubbled with dry nitrogen for 15 min and catalyst tetrakis(triphenylphosphine)palladium (0.03 eq to monomer) was added. The reaction was heated under vigorous reflux for 24 h, and small amount of phenylboronic acid was added for end-capping of bromo group. The reaction was heated at 95 °C for 13 h and bromobenzene was added to end-cap boronic ester group. The reaction was heated for another 6 h and then poured into 200 mL of methanol. The precipitated polymer was washed with diluted HCl solution, and dried. The polymer was extracted with acetone with a Sohxlet setup overnight to removed oligomer and residual catalyst. Polymer was then extracted with THF and re-precipitated into methanol three times.

## 3. RESULTS AND DISCUSSION

The synthesis of the monomers and polymers is illustrated in Figure 1. The key compound to the polymers is boronic diester monomer **2**, incorporating the flexible spacer group. The boronic diester was synthesized instead of boronic acid is due to the ease of the purification to obtain pure monomer. It is in general more difficult to purify boronic acid than ester which can be recrystallized from an organic solvent such as hexane, heptane, or toluene. Compound **1** was synthesized in 2 steps according to a modified literature procedure (21). Monomer **5** is a highly fluorescent crystalline material and was synthesized from dihydroxyanthraquinone in 3 steps. Only the last step gave very poor yield because of the competitive bromination of the ortho position of the alkoxy group. The synthesis of the analog monomer di-*t*-butylanthracene monomer was reported elsewhere (22). The model compound **M1** was synthesized via a cross-coupling reaction catalyzed by a palladium catalyst. The polymers were synthesized via the Suzuki coupling reaction in the presence of a phase transfer reagent (23). The characterization of the polymers is summarized in Table 1. The model compound and polymers are highly fluorescent materials. **M1**, **P1** and **P2** are blue fluorescent while **P3** is blue-green fluorescent. Polymers are readily soluble in common organic solvents such as THF, chloroform, toluene, and 1,2-dichloroethane. A pinhole-free thin film was obtained easily by spin-coating. All the polymers are thermally stable and show onset decomposition temperature, $T_d$, above 390 °C.

*Table 1. Characterization of polymers.*

| *Polymer* | $M_w$ | *PDI* | $T_d$ (°C) | $T_g$ (°C) | *Absorption (nm)* | *PL (solution) (nm)* | *PL (film) (nm)* |
|---|---|---|---|---|---|---|---|
| **P1** | 15,700 | 2.08 | 418 | 90 | 379, 398 | 430 | 436 |
| **P2** | 10,000 | 1.62 | 412 | 112 | 379, 395 | 440 | 449 |
| **P3** | 17,600 | 1.60 | 390 | 46 | 416 | 457, 468 | 464, 488 |

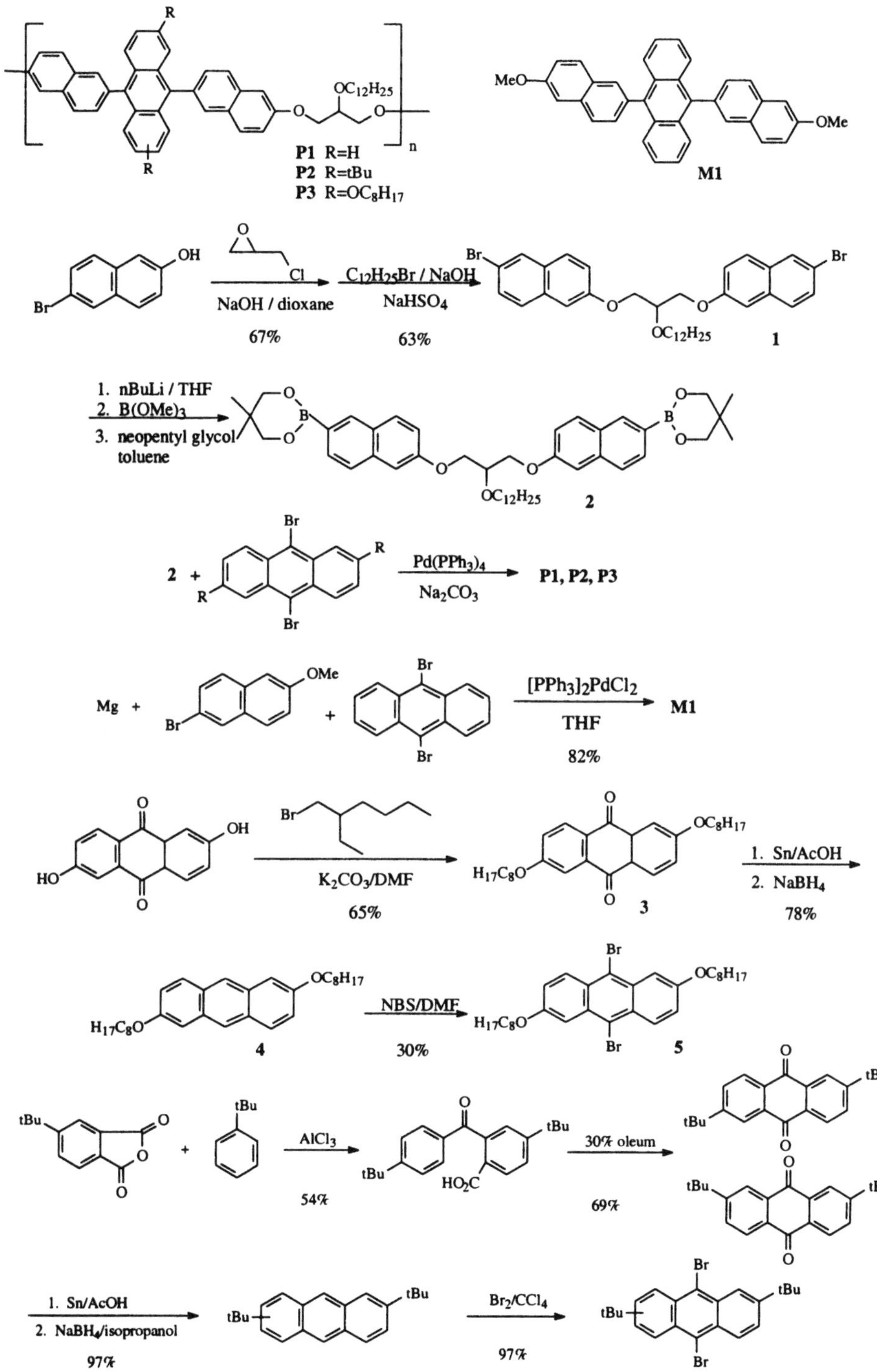


*Figure 1. Synthesis of monomers, polymers and model compound.*

The absorption and emission spectra of the model compound and polymers in toluene are shown in Figure 2. **P1** and **P2** show similar absorption spectra in dilute solutions, and the spectra are similar to that of **M1**. **M1** shows strong absorption peaks at 379 and 398 nm, which are attributed to the anthracene moiety. However, in **P3**, the absorption peaks are red-shifted due to the strong electron-donating characteristic of the alkoxy side chain. **M1** emits blue light at 432 nm when excited at 370 nm, and **P1** shows almost identical emission peaks at 430 nm. The almost identical absorption and emission spectra of **M1** and **P1** suggests that the spacer groups in **P1** have no effect on the optical properties of the polymer and the active chromophore is dinaphthylanthracene. The emission peaks, 440 nm for **P2** and 457 nm for **P3,** are red-shifted due to the electron-donating nature of alkyl and alkoxy side chains on the anthracene moiety. In solution, polymers show high quantum efficiency, 72% and 75% for **P1** and **P2** respectively. In solid thin films, the emission peaks of the polymers shift to longer wavelength. **P1** shows photoluminescence (PL) at 436 nm, **P2** at 449 nm, and **P3** at 464 nm. This might be due to the intrachain and/or interchain mobility of the excitons and excimers generated in the solid state. The quantum efficiencies of the thin films are rather high. **P1** shows an efficiency of 22%, and **P2** has a 37% efficiency.

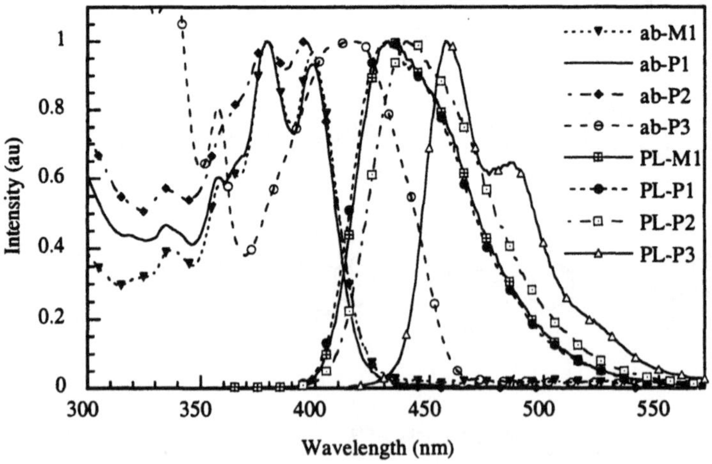


*Figure 2. Absorption and emission spectra of model compound and polymers in solution.*

The EL properties of the polymers were investigated by evaluation of the single-layer LED devices, ITO/polymer/Mg:Ag. The devices were fabricated by spin-coating the polymer solution onto an ITO coated glass substrate. The current and light output versus the voltage were measured in forward bias (a positive voltage was applied at the ITO electrode and a negative bias for the Mg/Ag electrode). All the devices exhibited typical diode behavior. The forward current was observed to increase superlinearly with the increase of applied voltage after exceeding the turn-on voltage. The absorption, emission, and EL spectra of **P2** are shown in Figure 3. **P1** shows blue EL at 444 nm, with CIE coordinate x=0.18 and y=0.16, very close to the PL in solid thin film. Similarly, **P2** emits blue light at 468 nm (x=0.17, y=0.20). The I-V characteristics of the device is shown in Figure 4. The threshold voltage of the device from **P2** for light emission is as low as 4.9 V. The results suggest that our polymer show relatively low turn-on voltages compared with other polymers with flexible spacer groups (24). A common issue related to the blue light emitting polymer is the difficulty of charge injection into the polymer leading to lower device efficiency. One strategy is to modify the contact between the polymer and ITO. For example, a thin layer of conducting polymer such as polyaniline (25) or poly(3,4-ethylene dioxythiophene) (PEDOT) (26) can be coated on ITO. The device efficiency can be improved at least an order of magnitude in the presence of PEDOT inter-layer. The luminous efficiency of device from **P2** with PEDOT is 0.4 Cd/A at 20 mA/cm$^2$. Considering the device was fabricated in the air and the device structure was not optimized, these results are very encouraging and promising.

## 4. CONCLUSIONS

We have synthesized and characterized novel blue-light emitting polymers containing blue chromophore 9,10-di(2-naphthyl)anthracene. These polymers show excellent solubility and thermal stability. The absorption and emission spectra indicate that the active chromophores in the polymers are 9,10-di(2-naphthyl)anthracenes. Polymers show blue photoluminescence in solution and solid state. The single layer LED devices emit blue light with low turn-on voltages. The device efficiency can be significantly improved with the proper engineering of polymer-electrode interface. Future work will focus on improving device efficiency by optimizing device structures in respect to hole- and electron injection materials.

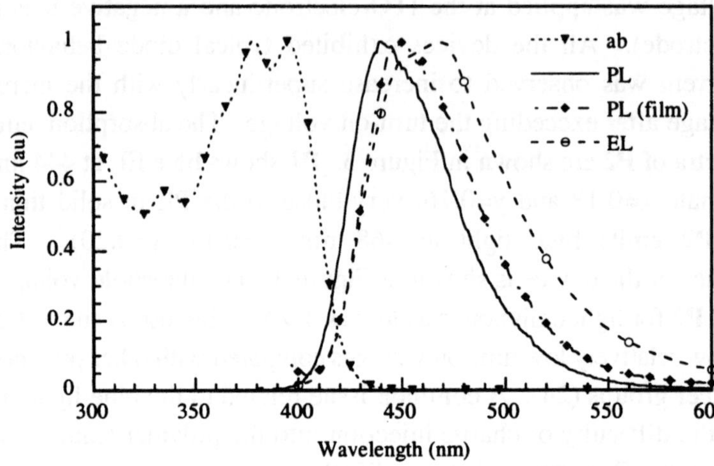


*Figure 3. Absorption, emission, and EL spectra of **P2**.*

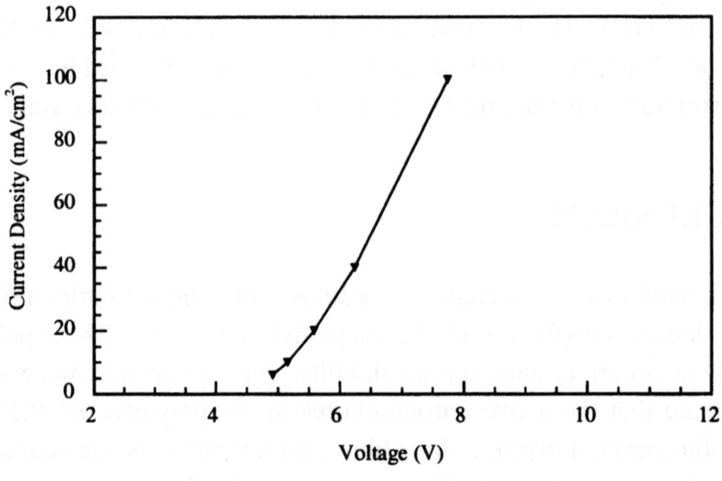


*Figure 4. Current-voltage characteristics of single layer device from **P2**.*

## 5. REFERENCES

1. Sze, S. M. Physics of Semiconductor Devices, John Wiley & Sons, New York, 1981.
2. Pope, M.; Kallmann, H. P.; Magnate, P. J. Chem. Phys., 38, 2042 (1963).
3. Tang, C. W.; Van Slyke S. A. Appl. Phys. Lett. 51, 913 (1987).
4. Tang, C. W.; Van Slyke S. A.; Chen, C. H. Appl. Phys. Lett. 63, 3610 (1987).
5. Burroughes, J. H.; Bradley, D. D. C.; Brown, A. R.; Marks, R. N.; Mackay, K.; Friend, R. H.; Burns, P. L.; Holmes, A. B. Nature, 347, 539 (1990).
6. Sheats, J. R.; Antoniadis, H.; Hueschen, M.; Leonard, W.; Miller, J.; Moon, R.; Roitman, D.; Stocking, A. Science, 273, 884 (1996).
7. Spreitzer, H.; Becker, H.; Kluge, E.; Kreuder, W.; Schenk, H.; Demandt, R.; Schoo, H. Adv. Mater., 10, 1340 (1998).
8. Bernius, M. T.; Inbasekaran, M.; O'Brien J.; Wu, W. Adv. Mater. 12, 1737, (2000).
9. Kido, J.; Hongawa, K.; Nagai, K. Macromol. Symp., 84, 81 (1994).
10. Ohmori, Y.; Uchida, M.; Muro, K.; Yoshino, K. Jpn. J. Appl. Phys., Part 2, 20, L1941 (1991).
11. Grem, G.; Leditzky, G.; Ullrich, B.; Leising, G. Adv. Mater., 4, 36 (1992).
12. Grem, G.; Paar, C.; Stampfl, J.; Leising, G.; Huber, J.; Scherf, U. Chem. Mater., 7, 2 (1995).
13. Gebler, D. D.; Wang, Y. Z.; Blatchford, J. W.; Jessen, S. W.; Lin, L. B.; Gustafson, T. L.; Wang, H. L.; Swager, T. M.; MacDiarmid, A. G.; Epstein, A. J. J. Appl. Phys., 78, 4264 (1995).
14. Huang, W.; Meng, H.; Yu, W.-L.; Gao, J.; Heeger, A. J. Adv. Mater., 10, 593 (1998).
15. Lee, J.-K.; Schrock, R. R.; Baigent, D. R.; Friend, R. H. Macromolecules, 28, 1966 (1995).
16. Zheng, S.; Shi, J.; Mateu, R. Chem. Mater., 12, 1814 (2000).
17. Andersson, M. R.; Berggren, M.; Inganaes, O.; Gustafsson, G.; Gustafsson-Carlberg, J. C.; Selse, D.; Hjertberg, T.; Wennerstroem, O. Macromolecules, 28, 7525 (1995).
18. Pei, Q.; Yang, Y. Adv. Mater., 7, 559 (1995).
19. Hilberer, A.; Brouwer, H.-J.; van der Scheer, B.-J.; Wildeman, J.; Hadziioannou, G. Macromolecules, 28, 4525 (1995).
20. Shi, J.; Tang, C. W.; Chen, C. H. U.S. Patent 5,935,721, (1998).
21. Scherf, U.; Mullen, K. ACS Symp. Ser., 672, 358 (1997).
22. Zheng, S.; Shi, J.; Klubek, K. P. Eur. Pat. Appl. 1088875 A2 (2001).
23. Inbasekaran, M.; Wu, W.; Woo, E. P. U.S. Patent 5,777,070, (1998).
24. Kim, H. K.; Ryu, M.-K.; Kim, K.-D.; Lee, S.-M.; Cho, S.-W.; Park, J.-W. Macromolecules, 31, 1114 (1998).
25. Yang, Y.; Heeger, A. Appl. Phys. Lett. 64, 1245 (1994).
26. Groenendaal, L.; Jonas, F.; Freitag, D.; Pielartzik, H.; Reynolds, J. Adv. Mater. 12, 481 (2000).

# Chapter 11

# Novel Two-Photon Absorbing Polymers

Kevin D. Belfield*[1], Alma R. Morales[1,2], Stephen Andrasik[1], Katherine J. Schafer[1], Ozlem Yavuz[1], Victor M. Chapela[2] and Judith Percino[2]
*[1]Department of Chemistry, University of Central Florida, [2]Centro de Quimica, Benemerita Universidad Autonoma de Puebla, Mexico*

## 1. INTRODUCTION

Multiphoton absorption has a number of inherent characteristics that make it attractive for use in a number of fast emerging technologies, including multiphoton fluorescence imaging, microfabrication, and optical power limiting to name but a few. To meet the needs of these, and other, applications, the design and synthesis of polymeric materials that exhibit high multi- or two-photon absorption is critical.

Two-photon absorption (TPA) is the nonlinear process in which two photons can be simultaneously absorbed in the material at high input irradiance. Although multiphoton absorption processes have been known since 1931, through the theory of simultaneous absorption of two-photons developed by Goeppert-Mayer (*1*), this field remained undeveloped until the advent of the pulsed laser providing very high-intensity light. It was not until the early 1960s that the two-photon absorption process was experimentally verified by Kaiser and Garrett (*2*), using pulsed lasers that provided very high intensity. In the presence of intense laser pulses, molecules can simultaneously absorb two or more photons mediated by a so-called 'virtual state,' a state of the molecule that has no classical analogue (*3*). The transition probability for absorption of two identical photons is proportional to $I^2$, where I is the intensity of the laser pulse. The combined energy of the two photons accesses a stable excited state of the molecule. If the two photons are of the same energy (wavelength), the process is referred to as degenerate TPA. On the other hand, if the two photons are of different energy (wavelength), the process is non-degenerate TPA.

As light passes through a molecule, the virtual state may form, persisting for a very short duration (of the order of a few femtoseconds). TPA can result if a second photon arrives before decay of this virtual state, and for this reason, the probability of TPA is proportional to the square of the light intensity. Thus, TPA involves the concerted interaction of both photons that combine their energies to produce an electronic excitation analogous to that conventionality caused by a single photon of a correspondingly shorter wavelength. Unlike single-photon absorption, whose probability is linearly proportional to the incident intensity, the TPA process depends on both a spatial and temporal overlap of the incident photons and takes on a quadratic (nonlinear) dependence on the incident intensity. The linear absorption of material may be very weak in the wavelength range where the TPA occurs, facilitating the potential to excite materials at greater depth than might be possible via one-photon excitation

Two-photon transitions can be described by two different mechanistic types. In non-polar molecules with a low-lying strongly absorbing state $\langle g|$ near the virtual level, only excited states that are optically forbidden by single photon selection dipole rules can be populated via two photon absorption (type 1) (*3*). The probability that this low-lying state can contribute to the virtual state is predicted by Heisenberg's uncertainty principle, with a virtual state lifetime approximated as $h/(4\pi\Delta E)$, where h is Planck's constant and $\Delta E$ is the energy difference between the virtual and actual states. Using this equation, it is predicted that an allowed state can contribute to formation of the virtual sate for time $t_{virtual}$ which is equal to about $h/(4\pi\Delta E)$ with the transition probability proportional to $\Delta\mu^2$. In polar molecules, strong TPA can occur by a different mechanism (type 2) in which a large change in dipole moment ($\Delta\mu > 10D$) occurs upon excitation of the ground to an excited state (*3*), in this case the lifetime of virtual state is proportional to $\Delta\mu^2$. An allowed two-photon transition can be viewed a a sequence of two one-photon allowed transitions $\langle g| \leftarrow \langle u| \leftarrow \langle g|$ (*4*). Two-photon absorptivity, $\delta$, is expressed in Goeppert-Mayer units (GM), with $1GM = 1 \times 10^{-50}$ $cm^4$ s $molecule^{-1} photon^{-1}$.

Particular organic molecules can undergo upconverted fluorescence trough non-resonant two-photon absorption using near-IR radiation, resulting in an energy emission greater than that of the individual photons involved (up-conversion). The use of a longer wavelength excitation source for fluorescence emission affords advantages not feasible using conventional UV or visible fluorescence techniques, e.g. deeper penetration of the excitation beam and reduction of photobleaching. The wavelength used for two-photon excitation is roughly twice that for one-photon excitation.

Thus, there is a need to developed new materials wherein improved TPA as well as other molecular properties or process for example, large fluorescence quantum yield. To meet the needs of these, and other, applications, the design and synthesis of polymeric materials that exhibit high multi- or two-photon absorption is critical. We have previously reported the synthesis and characterization of compounds that exhibit high two-photon absorptivity (*5 - 7*). Based on these findings, we have designed polymers bearing chromophores that exhibit high two-photon absorptivity. Herein, we describe the synthesis, structural characterization and photophysical characterization of a new class of fluorenylbisbenzothiazole-based polymer and maleic anhydride modified polymers produced via polycondensation. Photophysical properties including, linear absorption, single photon fluorescence, and two-photon upconverted fluorescence emission spectra are re reported.

## 2. RESULTS AND DISCUSSION

Preparation of 2, 7-diiodofluorene **1** and its subsequent dialkylation to yield 9,9-didecyl-2,7-diiodofluorene **2** was performed using conditions reported previously (*5*). The dicyano monomer (2,7-dicyano-9,9-didecylfluorene) **3** was synthesized through nucleophillic substitution of **2** with CuCN in pyridine in high yield (*8*). The synthesis of poly[benzo[1,2-d:4,5-d']bisthiazole-9,9-didecylfluorene] **5** was accomplished via polycondensation of dicyano monomer **3** with 2,5-diamino-1, 4-benzenedithiol (*9*). The preparation was conducted in polyphosphoric acid (PPA) (*10*). The polymerization process entailed the following stages: (a) dehydrochlorination of 2,5-diamino-1,4-benzenethiol dihydrochloride (DABTD·2HCl) in the presence of the dinitrile monomer in 77% PPA at 50-65 °C for 24 h; (b) addition of $P_2O_5$ to raise the $P_2O_5$ content of the medium; (c) increasing the temperature to 100-145 °C to promote chain propagation and cyclodehydration processes Figure 1.

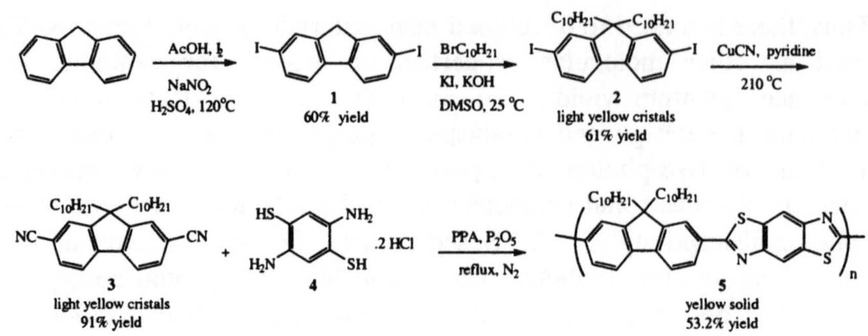

*Figure 1. Synthesis of poly[benzo[1,2-d:4,5-d']bisthiazole-9,9-didecylfluorene] 5.*

The linear UV-visible absorption spectrum and fluorescence emission spectrum for the polymer **5** are presented in Figure 2. Three major absorption bands are noted, with high absorption in the shorter wavelength with $\lambda_{maxima}$ at ca. 275 and 340 nm, from the bisbenzothiazole moiety, while the longer absorption, with its $\lambda_{max}$ at 390 nm, likely due to the fluorenyl ring system. The fluorescence spectrum obtained upon excitation at 389 nm revealed three major emission bands with $\lambda_{maxima}$ at ca. 420, 450, and 480 nm, with a Stoke's shift of ca. 30 nm from the region of the absorption maxima, providing some semblance to the longer absorption band profile. The fluorescence quantum yield (0.75) for polymer **5** was remarkably high in $CHCl_3$.

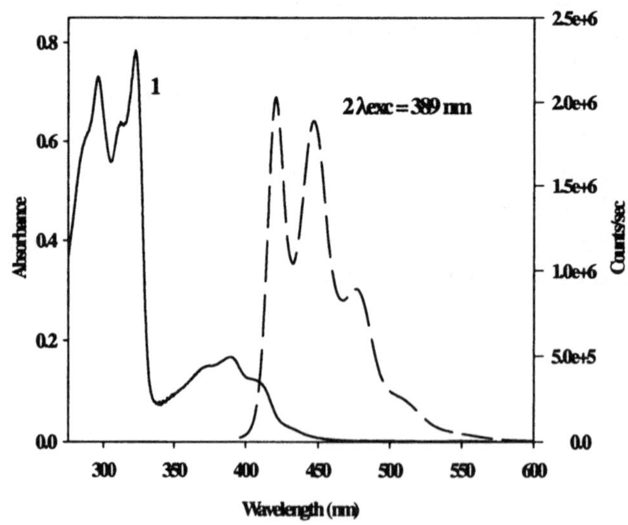


*Figure 2. Linear absorption (1) and fluorescence emission (2) ($\lambda_{exc}$ = 389 nm) spectra of fluorenylbisbenzothiazole polymer 5 in $CHCl_3$.*

Two-photon upconverted fluorescence spectra were recorded for **5** over the wavelength range of 650 to 870 nm. Shown in Figure 3 are the two-photon upconverted fluorescence spectra at the excitation wavelengths of 650, 680, 700, and 870 nm. As expected, the two-photon upconverted emission spectrum was similar in line-shape to the single-photon (conventional) emission spectrum. A slight difference was noted in that a reduction of intensity of the short wavelength emission band ($\lambda_{max}$ = 420 nm) was observed in the single-photon emission spectrum due to overlap with the long wavelength absorption band ($\lambda_{max}$ = 390 nm). The shorter wavelength emission band was more pronounced in the two-photon upconverted emission spectrum, with the one and two-photon spectra differences attributable to concentration effects.

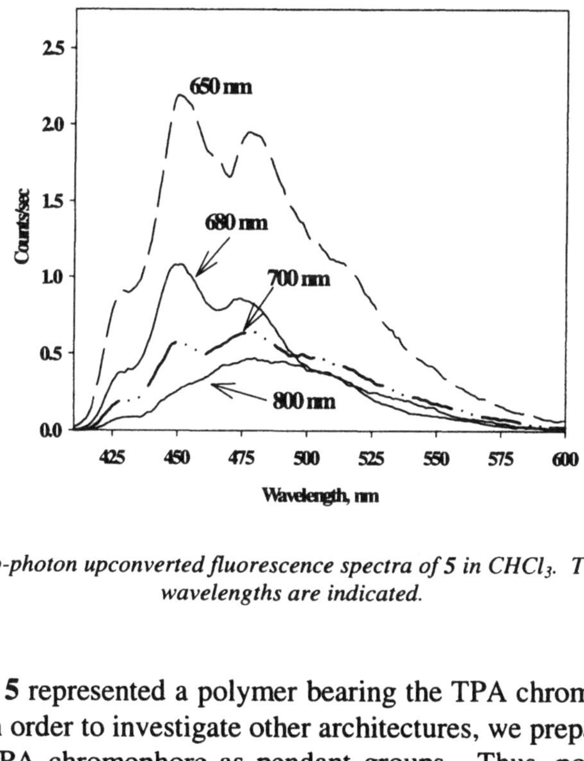


***Figure 3.*** *Two-photon upconverted fluorescence spectra of 5 in $CHCl_3$. The fs excitation wavelengths are indicated.*

Polymer **5** represented a polymer bearing the TPA chromophore in the main chain. In order to investigate other architectures, we prepared polymers bearing the TPA chromophore as pendant groups. Thus, poly(styrene-*co*-maleic anhydride) was modified by condensation with fluorenylamine **6** (*5*), as illustrated in Figure 4. HMDS and $ZnCl_2$ were used as condensation agents (*11*), affording the modified copolymer **7** in 36% yield after purification.

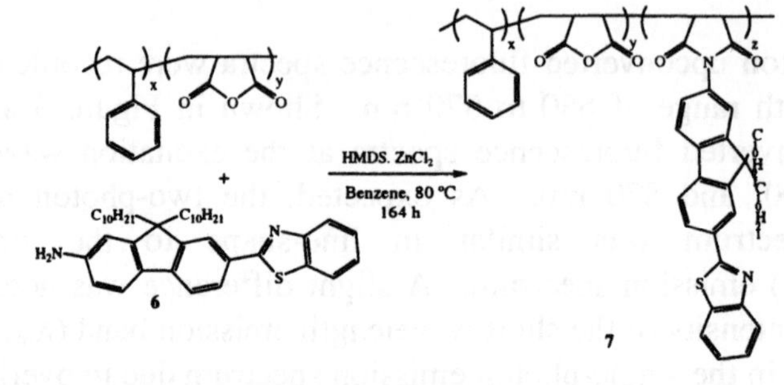


***Figure 4.*** *Synthesis of poly(styrene-co-maleic anhydride) 7.*

The linear absorption spectrum and fluorescence emission spectrum for 7 are presented in Figure 5. A Stoke's shift of >100 nm was observed with a quantum yield of 0.61 (DMF). Upon excitation at two different wavelengths 370 and 400 nm, two components are clearly evident in the emission spectrum. One species has an emission slightly red-shifted relative to the other Figure 6. The longer wavelength emission may result from a charge transfer complex between the pendant fluorophore and maleic anhydride moieties in the backbone of polymer.

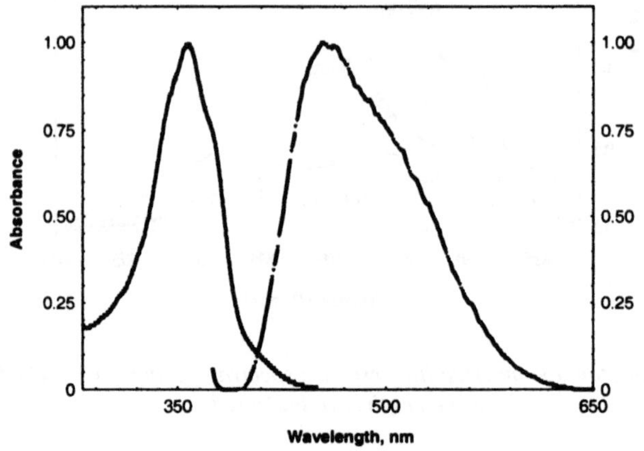


***Figure 5.*** *Linear absorption (0.27 mg/mL in THF) and fluorescence emission spectra ($\lambda_{exc}$ = 361 nm) of modified poly(styrene-co-maleic anhydride) 7 in DMF.*

Two-photon upconverted fluorescence spectra for 7 were recorded over the wavelength range of 650 to 870 nm. Shown in Figure 7 are the two-photon upconverted fluorescence spectra at excitation wavelengths of 660, 790, and 870 nm. Again two components were observed. One component with the shorter wavelength emission centered at 450 nm, is attributable to the fluorenylimide fluorophore. An additional component produced emission at longer wavelengths upon longer wavelength excitation, likely due to a charge transfer complex, a subject of further investigation. Emission lifetime measurements were consistent with the steady state spectra, revealing a short-lived species (< 100 ps), and a longer lived species with a lifetime of ca. 3 ns.

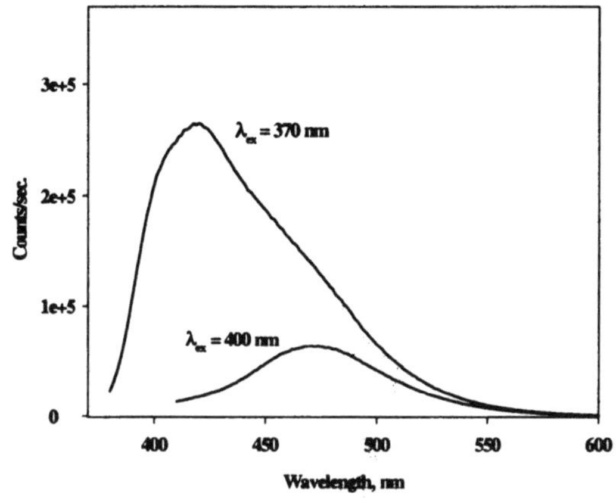


*Figure 6. Fluorescence emission spectra at ($\lambda_{exc}$ = 370 and 400 nm) of modified poly(styrene-co-maleic anhydride) 7 in DMF.*

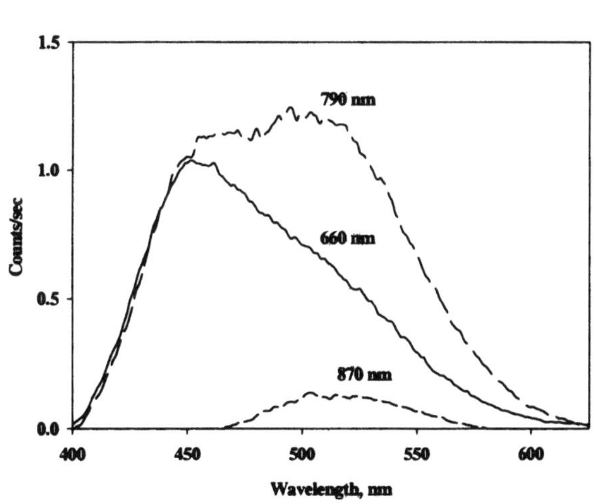


*Figure 7. Two-photon upconverted fluorescence spectra of modified poly(styrene-co-maleic anhydride) 7 in DMF. The fs excitation wavelengths are indicated.*

Poly(ethylene-*g*-maleic anhydride) was modified by condensation with fluorenylamine **6**, as illustrated in Figure 8. HMDS and $ZnCl_2$ were used as condensation agents, affording the modified copolymer **8** in 52% yield after purification.

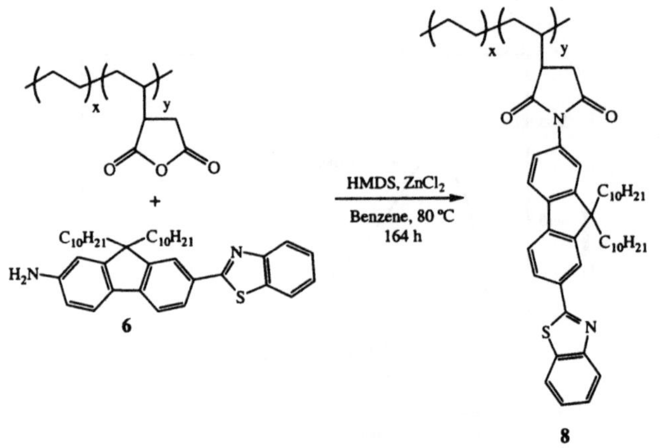


***Figure 8.*** *Synthesis of poly(ethylene-g-maleic anhydride)* ***8.***

The linear absorption $\lambda_{max}$ was ca. 350 nm for **8** with a Stoke's shift of its emission of ca. 100 nm (benzene). As can be seen in Figure 9, when excited with fs near-IR radiation (830 nm), the polymer exhibited upconverted fluorescence with an emission maximum at 450 nm.

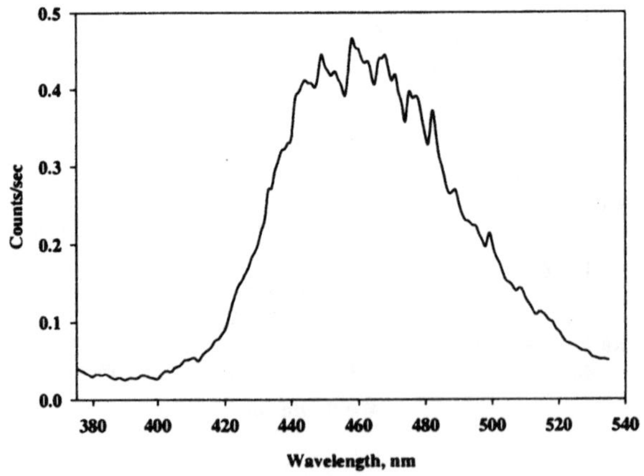


***Figure 9.*** *Two-photon upconverted fluorescence spectrum of modified poly(ethylene-g-maleic anhydride)* ***8*** *in benzene ($\lambda_{exc}$ = 830 nm).*

# 3. EXPERIMENTAL

## 3.1 Measurements

$^{1}H$ and $^{13}C$ NMR spectra were recorded on Varian 300 NMR spectrometer at 300 MHz. FT-IR spectra were recorded on Perkin Elmer spectrophotometer. UV-Visible spectrophotometric measurements were recorded on Varian Cary 3 spectrophotometer. Elemental analyses were performed at Atlantic Microlab. Steady-state fluorescence measurements were recorded with PTI Quantamaster spectrofluorometer using a Xe lamp and excitation monochromator. Two-photon upconverted fluorescence studies were performed using a Clark-MXR model CPA-2001 Ti:sapphire laser that provides 775 nm, 150 fs pulses with 900 μJ/pulse at 1 KHz repetition rate. Half of the laser power was split to pump a Quantronix OPG/OPA system, tunable from 500 to 1600 nm. In our experiments, we used second harmonic signal and second harmonic idler to accommodate the range from 600 to 950 nm for two-photon upconverted florescence experiments. Fluorescence lifetime measurements were obtained with a PTI TimeMaster system. The excitation source was the output of a nitrogen laser (337 nm, 500 ps pulsewidth, 10 Hz repetition rate). Fluorescence quantum yields, Q, were measured by a standard method relative to Rhodamine 6G in ethanol (Q = 0.94) (*12*).

## 3.2 Synthesis

### 3.2.1 General

Reactions were conducted under $N_2$ atmosphere. Pyridine was distilled over barium oxide before use. All other reagents and solvents were used as received from commercial suppliers. 7-Benzothiazole-2-yl-9,9-didecylfluorene-2-ylamine has been synthesized as previously described (*5*).

### 3.2.2 2,7-Dicyano-9,9-didecylfluorene (3)

Fluorene was diiodinated using iodination conditions employed previously to yield 2,7-diiodofluorene **1** (*5*). Recrystallization (60:40 cyclohexane/ethyl acetate) afforded a crystalline solid (60% yield, mp = 205-206 °C). A mixture of **1** (3 g, 7.17 mmol), DMSO (20 mL), 1-bromodecane (3.163 g, 14.35 mmol), and KI (0.127 g, 0.769 mmol) were stirred at room temperature, followed by slow addition of powdered KOH under $N_2$. The reaction was complete in 3.5 h, after which the mixture was poured into water and extracted with hexane. The organic extract was washed with water, dried over $MgSO_4$, and concentrated, resulting in an orange oil. 1-Bromodecane and DMSO were removed in vacuo and 9,9-didecyl-2,7-diiodofluorene **2** was isolated via column chromatography using silica gel (hexanes), providing 3.06 g of yellow crystals (61 % yield, mp = 45-46 °C). Compound **2**, CuCN, and pyridine were placed into a three-neck, 50 mL flask fitted with reflux condenser and protected from moisture with a calcium chloride tube. The mixture was heated in a silicone oil bath at 210 °C for 17 h, after which time it was poured while still hot, into a flask containing $NH_4OH$ (6 mL). A mixture of benzene and ether (1:1.6, 13 mL) was added and filtered. The organic layer was separated and washed successively with the $NH_4OH$ solution, 5 mL of 6N HCl, 5 mL of water, and 5 mL of 6N NaCl. The organic extract was concentrated, affording a brown oil. Purification was accomplished by column chromatography (80:20 hexanes/methylene chloride), providing 2,7-dicyano-9,9-didecylfluorene **3** as yellow crystals (1.57 g , 91% yield, mp = 46-47 °C). Anal. Calcd for $C_{35}H_{48}N_2$: C, 84.62%, H, 9.74%, N, 5.64%. Found: C, 84.27%, H, 10.08%, N, 5.54%. $^1H$ NMR (300 MHz, $CDCl_3$) δ: 7.85 (d, 2H), 7.69 (m, 4H), 2.02 (m, 4H), 1.19 (m, 28H), 0.93 (m, 6H), 0.53 (m, 4H). $^{13}C$ NMR (300 MHz, CDCl3) δ: 152.2, 143.6, 131.8, 126.9, 121.6, 119.4, 112.1, 56.2, 40.0, 32.0, 29.9, 29.6, 29.4, 29.4, 23.9, 22.8, 14.3. FT-IR (KBr, $cm^{-1}$): 3056 (νArCH), 2926, 2854 (νalCH), 2226 (νC≡N), 1609 (νArC=C).

### 3.2.3 Poly(benzo[1,2-d:4,5-d']bisthiazole-9,9-didecylfluorene) (5)

Into a three-neck, 100 mL reaction flask were placed 2,7-dicyano-9,9-didecylfluorene **3** (0.727 g, 0.0014 mol), 2,5-diamino-1,4-benzenethiol dihydrochloride **4** (0.362 g, 0.0014 mol), and polyphosphoric acid (3.75 g). The reaction vessel was fitted with a mechanical stirrer, flushed with nitrogen and then heated to 45 °C under vacuum and stirred for 16 h. The temperature was then gradually raised to 60 °C for 4 h, and 100 °C for 2 h, resulting in the reaction mixture turning orange. The mixture was cooled to room temperature and 1.83 g of phosphorus pentoxide ($P_2O_5$) was added. The solution was then slowly heated to 100 °C and stirred for 16 h (reddish-orange solution), followed by heating to 130 °C for another 16 h, then at 145 °C for 6 h. The reaction mixture was immediately poured into water, resulting in a yellow-brown precipitate. The polymer was neutralized with $NH_4OH$ (20%) and washed with water in a soxhlet extractor for 32 h. The polymer was dried and again washed with hexane in a soxhlet extractor, yielding a yellow solid (0.49 g, 53% yield). mp = 169-170.5 °C). $^1$H NMR (300 MHz, $CDCl_3$) δ: 7.79 (d, 5H), 6.15 (bs, 3H), 2.03 (m, 4H), 1.01 (bm, 34H), 0.53 (bs, 4H). FT-IR (KBr, $cm^{-1}$): 2925, 2853(νalCH), 1646, 1609 (νC=N imino), 1606 (νArC=C).

### 3.2.4 7-Benzothiazol-2-yl-9,9-didecylfluoren-2-ylamine-modified poly(styrene-co-maleic anhydride) (7)

Poly(styrene-co-maleic anhydride) obtained from Aldrich (styrene to maleic anhydride mole ratio 1.3:1) (2.8013 g) was dissolved in 2 mL THF. Distilled benzene (6 mL) was than added and the polymer solution was placed in a 25 mL two- necked round bottom flask fitted with a reflux condenser under $N_2$. While stirring 7-benzothiazole-2-yl-9,9-didecylfluorene-2-ylamine (0.470 g, 0.79 mmol, in 3.2 mL dry benzene) was added dropwise at r.t. to the reaction flask, resulting in a yellow-white precipitate upon addition. The reaction mixture was stirred for 1 h, after which recrystallized $ZnCl_2$ (0.1208 g, 0.79 mmol) added in one portion, followed by heating to 80 °C. Hexamethyldisilazane (HMDS) (0.25 mL, 1.18 mmol) in dry benzene (2.5 mL) was added slowly over a period of 30 min, and the reaction mixture was then refluxed for 80 h. During the reflux period two additional portions of $ZnCl_2$ and HMDS were added at 24 and 48 h, and the reaction followed by TLC. Upon completion, the reaction mixture was cooled to room temperature and poured into 0.5 N HCl (30 mL). The aqueous phase was extracted with ethyl acetate (3 x 25 mL), and the combined organic extracts were washed successively with 30 mL of saturated $NaHCO_3$ and brine solution and dried over $MgSO_4$. The solution was concentrated under reduced pressure and the crude polymer dissolved in a minimum amount of N,N-dimethylacetamide (DMAc) and precipitated in cold methanol twice then dried at reduced pressure. The final product was a bright yellow powder (36 wt% yield).

### 3.2.5 7-Benzothiazol-2-yl-9,9-didecylfluoren-2-ylamine-modified poly(ethylene-g-maleic anhydride) (8)

Poly(ethylene-g-maleic anhydride) obtained from Aldrich (0.5 wt. % maleic anhydride) (2.0052 g) was dissolved in a minimal amount of THF (12 mL) at 110 °C. Distilled benzene (6 mL) was then added and the polymer solution placed into a 25 mL two-necked round bottom flask fitted with a reflux condenser under $N_2$. While stirring, 7-benzothiazole-2-yl-9,9-didecylfluorene-2-ylamine (0.05 g, 0.084 mmol, in 2.0 mL dry benzene) was added dropwise into the reaction flask. The reactants were stirred for 1 h, then recrystallized $ZnCl_2$ (0.0114 g, 0.084 mmol) was added in one portion. The resulting reaction mixture was kept at 110 °C. HMDS (0.03 mL, 0.1261 mmol) in dry benzene (0.3 mL) was added slowly over a period of 30 min. The reaction mixture was then refluxed for 164 h, during which time it became a dark-brown color. During the reflux period, three additional portions of $ZnCl_2$ and HMDS were added at 24, 72, and 96 h. TLC indicated the reaction was complete. The reaction mixture was cooled to room temperature and poured into 0.5 N HCl (110 mL). The aqueous phase was extracted with ethyl acetate (3 x 75 mL). The combined organic extracts were washed successively with 100 mL of saturated $NaHCO_3$ and brine, then dried over anhydrous $MgSO_4$. The solution was concentrated under reduced pressure. The polymer was dissolved in chloroform and precipitated in cold methanol twice, then dried at reduced pressure. The final product was a light green solid (52 wt% yield).

## 4. CONCLUSIONS

The ease of synthesis, high two-photon absorptivity, luminescence fluorescence properties, and solubility makes these polymer a good candidates for two-photon based applications, such as optical power limiting and two-photon fluorescence imaging, aspects currently under investigation. Also, luminescence properties of fluorenyl bisbenzothiazole-based polymer suggest possible application as blue light emitting diode material.

## 5. ACKNOWLEDGMENTS

We wish to acknowledge the donors of the Petroleum Research Fund of the ACS, Research Corporation (Cottrell College Science Award), National

Science Foundation (DMR-9975773, ECS-9970078, and ECS-9976630), National Research Council (COBASE program), and the University of Central Florida for support for this work. The authors also wish to thank Profs. Eric W. Van Stryland and David J. Hagan along with Mr. Joel M. Hales for assistance with two-photon fluorescence experiments.

## 6. REFERENCES

1. Goeppert–Mayer, M., Ann. Phys., 9, 273 (1931).
2. Kershaw, S., **Characterization Techniques and Tabulations for Organic Nonlinear Optical Materials** (M. G. Kuzyk, C. W. Dirk, Eds.), Marcel Dekker, NY, 1998, Chapter 7.
3. Birge, R. R.; Parsons, B.; Song, Q. W.; Tallent, J. R., **Molecular Electronics** (J. Jortner, M. Ratner, Eds.), Blackwell Science, London, 1997, Chapter 15.
4. Birge, R. R., Acc. Chem. Res., 19, 138 (1986).
5. Belfield, K. D.; Schafer, K. J.; Mourad, W.; Reinhardt, B. A., J. Org. Chem., 65, 4475 (2000).
6. Belfield, K. D.; Schafer, K. J.; Hagan, D. J.; Van Stryland, E. W., Negres, R. A., Org. Lett., 1, 1575 (1999).
7. Belfield, K. D.; Schafer, K. J.; Liu, Y.; Liu, J.; Ren, X.; Van Stryland, E. W., J. Phys. Org. Chem., 13, 837 (2000).
8. Newman, M. S., J. Am. Chem. Soc., 59, 2473 (1937), **Organic Synthesis** (E. C. Horning, Ed.), Wiley, NY, 1955, Vol. 3, p. 631.
9. Tan, L.-S.; Srinivassan, K. R.; Bai, S. J., J. Polym. Sci., A, Polym. Chem,, 35, 1909 (1997).
10. Wolfe, J. F.; Sybert, P. D.; Sybert, J. R., U. S. Patent 4,533,693.
11. Reddy, P. W.; Kondo, S.; Toru, T.; Ueno, Y., J. Org. Chem., 62, 2652 (1996).
12. Lakowicz, J. R. **Principles of Fluorescence Spectroscopy**, Kluwer Academic/Plenum, NY, 1999.

# C. Bioactivity and Biomaterials

# Chapter 12

# NATURAL FUNCTIONAL CONDENSATION POLYMER FEEDSTOCKS

Charles E. Carraher, Jr.
*Florida Atlantic University, Department of Chemistry and Biochemistry, Boca Raton, FL 33431, and Florida Center for Environmental Studies, Palm Beach Gardens, FL 33410*

## 1. INTRODUCTION

Condensation polymer organic feedstocks or starting materials can be divided into natural and synthetic materials. While the bulk of synthetic polymers are vinyl in nature, condensation polymers overwhelm non-condensation materials in nature. He we will briefly review the vast storehouse of natural condensation polymers.

Biological polymers represent successful strategies that are being studied by scientists as avenues to different and better polymers and polymer structure control. Sample "design rules" and approaches that are emerging include the following.

*Identification of mer sequences that give materials with particular properties.

*Identification of mer sequences that key certain structural changes.

*Formation of a broad range of materials with a wide variety of general/specific properties and function (such as proteins/enzymes) through a controlled sequence assembly from a fixed number of feedstock molecules (proteins-about 20 different amino acids; five bases for nucleic acids and two sugar units).

*Integrated, in situ (in cells) polymer production with precise nanoscale control.

*Repetitive use of proven strategies with seemingly minor structural differences but resulting in quite divergent results (protein for skin, hair, and muscle).

*Control of polymerizing conditions that allow steady state production far from equilibrium.

There often occurs a difference in "mind-set" between the nucleic acid and protein biopolymers and other biopolymers such as polysaccharides. Nucleic acids and proteins are site specific with one conformation. Nucleic acids and proteins are not a statistical average, but rather a specific material with a specific chain length and conformation. By comparison, synthetic and many other biopolymers are statistical averages of chain lengths and conformations. The distributions are often kinetic/thermodynamic driven.

This difference between the two divisions of biologically important polymers is also reflected in the likely hood that there are two molecules with the exact same structure. For molecules such as polysaccharides the precise structures of individual molecules vary, but for proteins and nucleic acids the structures are identical from molecule to molecule. This can be considered a consequence of the general function of the macromolecule. For polysaccharides the major, though not the sole functions, are energy and structural. For proteins and nucleic acids, main functions include memory and replication, in additional to proteins sometimes also serving a structural function.

Another difference between proteins and nucleic acids and other biopolymers and synthetic polymers involves the influence of stress/strain activities on the materials properties. Thus, application of stress on many synthetic polymers and some biopolymers encourages realignment of polymer chains and regions often resulting in a material with greater order and strength. By comparison, application of stress to certain biopolymers, such as proteins and nucleic acids, causes a decrease in performance (through denaturation, etc.) and strength. For these biopolymers, this is a result in the biopolymer already existing in a compact and "energy favored" form and already existing in the "appropriate" form for the desired performance. The performance requirements for the two classifications of polymers is also different. For one set, including most synthetic and some biopolymers, performance behavior involves response to stress/strain application with respect to certain responses such as chemical resistance, absorption enhancement, and other physical properties. By comparison, the most cited performances for nucleic acids and proteins involves selected biological responses requiring specific interactions occurring within a highly structured environment that demands a highly structured environment with specific shape and electronic requirements.

## 2. POLYSACCHARIDES

Polysaccharides constitute the largest organic polymer feedstock . The vast majority of these polysaccharides are renewed in some cycle, often the cycle being one year or a growing cycle. On a bulk basis they constitute a vast storehouse of largely untapped materials. On a structural basis they represent materials where only a general structure is typically present with each chain structure varying in length, number of branches, length of branching, connective linkage, etc. These variations are seasonal, age-dependent, nutrient dependent source dependent, etc. The variations are satisfactorily encompassed into general behavioral categories for bulk use and modification, but for areas where the exact structure is important, such as in biomedical areas, there is a need for precise structural knowledge and control.

While there is a wide variety of polysaccharides we will look superficially at only several of these, utilizing the examples as illustrations.

By bulk, the largest general groupings of polysaccharides are the celluloses and starches. Cellulose was originally "discovered" by Payen in 1838. For thousands of years impure cellulose formed the basis of much of our fuel and construction systems in the form of wood, lumber (cut wood) and dried plant material; served as the vehicle for the retention and conveying of knowledge and information in the form of paper; and clothing in the form of cotton, ramie, and flax. Much of the earliest research was aimed at developing stronger materials with greater resistence to the natural elements (including cleaning) and to improve dyeability so that the color of choice by common people for their clothing material could be other than a drab off-white. In fact, the dyeing of textile materials, mainly cotton, was a major driving force in the expansion of the chemical industry in the latter part of the $19^{th}$ century.

The general repeat unit of cellulose is given below where the D-glucose units are joined by a beta-acetal linkage

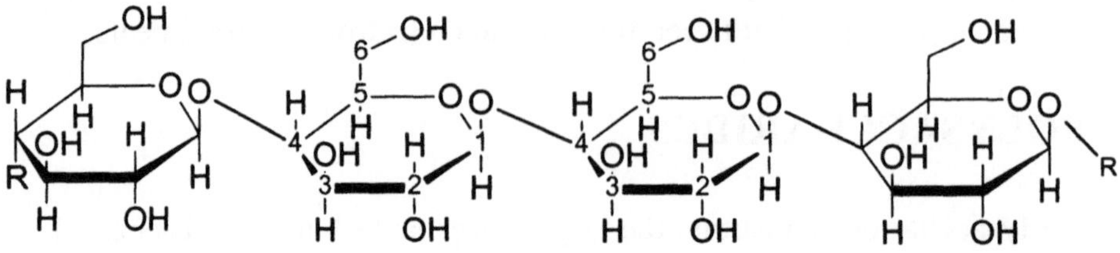

Cellulose

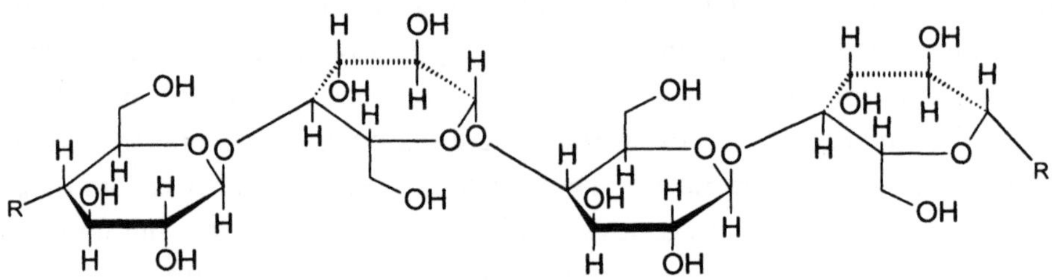

Cellulose 3d structure

The top structure is most commonly employed as a description of the repeat unit of cellulose but the lower structure more nearly represents the actual three-dimensional structure with each D-glucosyl unit rotated 180 degrees. We will employ a combination of these two structural representations. Numbering is also shown above and the type of linkage is written as 1->4 since the units are connected through oxygens contained on carbon 1 and 4 as below.

By agreement with the anometic nature of the particular carbons involved in linking together the glucosyl units, this unit is described as being a β-linkage. Thus, for cellulose this linkage is a β 1->4 linkage. The other similar 1->4 linkage found in starch is called an α-linkage. The geometric consequence of this difference is great. The linear arrangement of cellulose with the β linkage gives an arrangement where the OH groups reside somewhat uniformly on the outside of the chain allowing close contact and ready hydrogen bond formation between chains. This arrangement results in a tough, insoluble, rigid, and fibrous material that is well suited as cell wall material for plants. By comparison, the α-linkage of starch (namely amylose) results in a helical structure where the hydrogen bonding is both interior and exterior to the chain allowing better wettability. This difference in bonding also results in one material being a "meal" for humans (the α linkage) whereas the other is a meal for termites.

Structural and physical information is important in designing appropriate modifying systems and in treatment of the feedstock to ready it for the particular treatment.

The various crystalline structures of cellulose have different physical properties and chemical reactivities. These variations are a consequence of the properties varying according to plant source, location in the plant, plant age, season, seasonal conditions, treatment, etc.

In the cellulose regenerating process, sodium hydroxide is initially added such that approximately one hydrogen, predominately the hydroxyl group on carbons 2 and 3, is replaced by the sodium ion. This is followed by treatment with carbon disulfide forming cellulose xanthate which is eventually re-changed back again, regenerated, to cellulose. This sequence is depicted below.

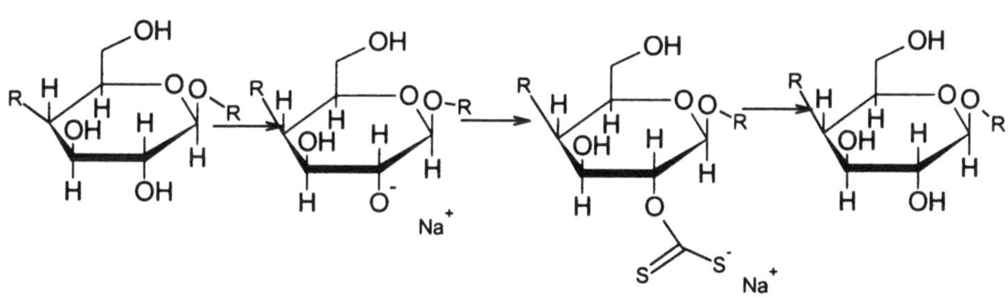

Cellulose ------->Sodium Salt ------>Cellulose Xanthate ------>Regenerated Cellulose- rayon or cellophane

There are a number of products that have been formed from cellulose. Here the cellulose is the feedstock or functionalized polycondensation reactant. These include inorganic and organic esters and organic ethers. Following is a brief description of some of these.

## 2.1 Inorganic Esters

The most widely used so-called inorganic ester of cellulose is cellulose nitrate (CN), also called nitrocellulose and gun cotton. Celluloid is produced from mixtures of cellulose nitrate and camphor. Cellulose nitrate was first made about 1833 when cellulose-containing linen, paper, or sawdust was reacted with concentrated nitric acid. It was the first "synthetic" cellulose recognized product. Initially, CN was used as a military explosive and improvements allowed the manufacture of smokeless powder. A representation of CN is given below.

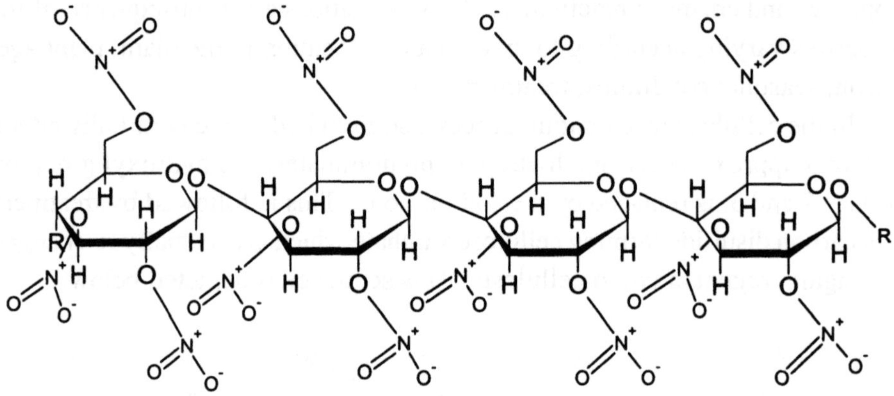

Cellulose Nitrate

The development of solvents and plastizing agents for cellulose nitrate led to the production of many new and useful non-explosive uses. Celluloid was produced in 1870 from a mixture of CN and camphor. Films were cast from solution and served as the basis for the original still and motion pictures. After World War I the development of stable CN solutions allowed the production of fast-drying lacquer coatings.

## 2.2 Organic Esters

The most important cellulose ester is cellulose acetate because of its use in fibers and plastics. They were first made in 1865 by heating cotton with acetic anhydride. During World War I a cellulose acetate replaced the highly flammable CN coating on airplane wings and fuselage fabrics. Below are illustrative structures of these materials showing the predominately so-called mono-substituted material(top), and predominately tri-substituted material (bottom).

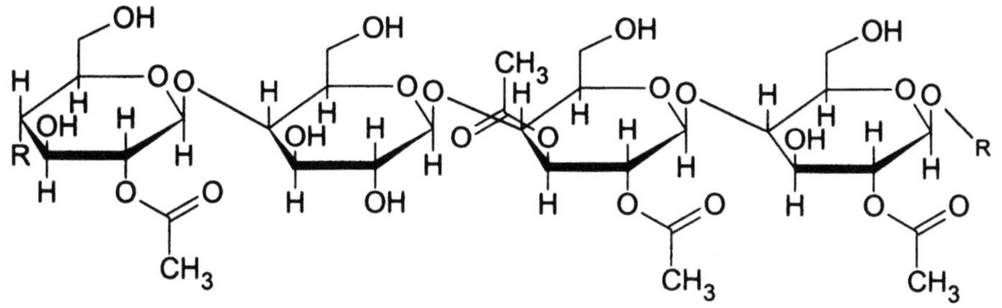

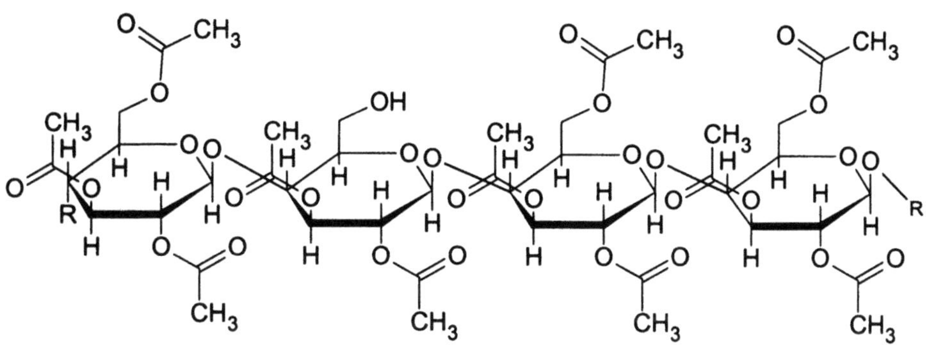

Reaction occurs differently since there are two "types" of hydroxyl groups, the two ring hydroxyls and the methylene hydroxyl. In the typical formation of esters such as cellulose acetate the ring hydroxyl groups are acetylated initially (top structure) prior to the C-6 exocyclic hydroxyl. Under the appropriate reaction conditions reaction continues to almost completion with almost all three of the hydroxyl groups esterified (bottom structure). In triacetate products only small amounts (on the order of 1 %) of the hydroxyls remain free and of these generally about 80 % are the C-6 hydroxyl..

Cellulose esters are used as plastics for the formation by extrusion of films and sheets and by injection molding of parts. They are thermoplastics and can be fabricated employing most of the usual techniques of (largely compression and injection) molding, extrusion and casting . Cellulose esters plastics are noted for their toughness, smoothness, clarity, and surface gloss.

Acetate fiber is the generic name of a fiber that is partially acetylated cellulose. They are also know as cellulose acetate and triacetate fibers. They are nontoxic and generally non allergic so are ideal from this aspect as clothing material.

While acetate and triacetate differ only moderately in the degree of acetylation, this small difference accounts for differences in the physical and chemical behavior for these two fiber materials. Triacetate fiber is hydrophobic and application of heat can bring about a high degree of crystallinity that is employed to "lock-in" desired shapes (such as permanent press). Cellulose acetate fibers have a low degree of crystallinity and orientation even after heat treatment. Both readily develop static charge and thus anti-static surfaces treatments are typically employed to clothing made from them.

While cellulose acetates are the most important cellulose esters they suffer by their relatively poor moisture sensitivity, limited compatibility with other synthetic resins, and a relatively high processing temperature.

As noted before, starch can be divided into two general structures, branched amylopectin and largely linear amylose.

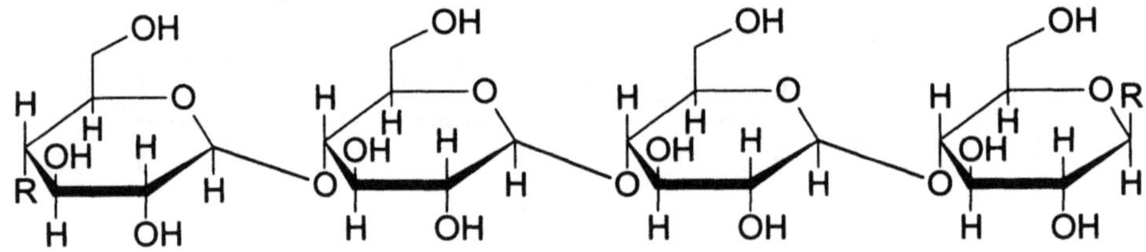

Linear Amylose

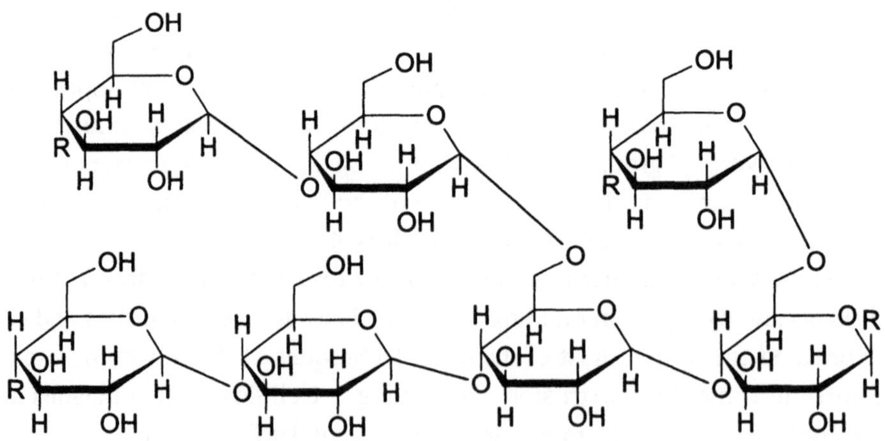

Branched Amylopectin

Most starches contain about 10 to 20 % amylose and 80 to 90 % amylopectin thought the ratio can vary greatly.

Amylose typically consists of over 1000 D-glucopyranoside units. Amylopectin is a larger molecule containing about 6,000 to 1,000,000 hexose rings essentially connected with branching occurring at intervals of 20 to 30 glucose units. Branches also occur on these branches giving amylopectin a fan or tree-like structure similar to that of glycogen. Thus, amylopectin is a highly structurally complex material. Unlike nucleic acids and proteins where specificity and being identical are trademarks, most complex polysaccharides can boast of having the "mold broken" once a particular chain was made so that the chances of finding two exact molecules is very small.

Starch granules are insoluble in cold water but swell in hot water, first reversible until gelationization occurs at which point the swelling is irreversible. At this point the starch loses its birefringence, the granules burst, and some starch material is leached into solution. As the water temperature continues to increase to near $100^0$ C a starch dispersion is obtained. Oxygen must be avoided during heating or oxidative degradation occurs. Both amylose and amylopectin are then water soluble at elevated temperatures. Amylose chains tend to assume a helical arrangement giving it a compact structure. Each turn contains six glucose units.

Major modification efforts include the free-radical grafting of various styrenic, vinylic, and acrylic monomers onto cellulose, starch, dextran, and chitosan. The grafting has been achieved using a wide variety of approaches including ionizing and ultraviolet/visible radiation, charge-transfer agents, and various redox systems. Much of this effort is aimed at modifying the native properties such as tensile (abrasion resistence and strength) and care (crease resistence and increased soil and stain release) related properties, increased flame resistance, and modified water absorption. One area of emphasis has been the modification of cotton and starch in the production of super-absorbent material through grafting. These materials are competing with all synthetic crosslinked acrylate materials that are finding use in diapers, feminine hygiene products, wound dressings, and sanitary undergarments.

## 2.3 Other Polysaccharides

Polysaccharides can be divided by many means. Here we will look at the number of basic building units involved in the structure of the polysaccharide as a means of cataloguing some important polysaccharides.

### 2.3.1Homopolysaccharides

The best know homopolysaccharides are derived from D-glucose and known as glucans. Glucose has a number of reactive sites and a wide variety of polymers formed utilizing combinations of these reactive sites are found in nature. We have already visited the two most well known members of this group-cellulose and starch containing amylose and amylopectin. Here we will look at some other important members.

Glycogen is a very highly branched glucan or polysaccharide formed from glucose. It is structurally similar to amylopectin though more highly branched. This greater branching gives glycogen a greater water solubility. Glycogens are the principle carbohydrate food reserve materials in animals. They are found in both invertebrates and vertebrates and likely found in all animal cells.

Glycogen is an amorphous polymer of high molecular weight, generally $10^6$ to$10^9$ Daltons. In spite of its high molecular weight it has good water solubility because, as noted above, of its highly, but loosely branched character. It is polydisperse with respect to molecular weight as are other polysaccharides. The particular molecular weight and molecular weight distribution varies within and between cells and metabolic need. It stores D-glucose units until needed as an energy source. It also serves as a buffering agent helping control the amount of glucose in the blood. It is stored in tissues as spherical particles called β particles.

The average distance between branch points is only about 10 to 15 in comparison to amylopectin with about 20 to 30 units between branch points. Many glycogen particles contain small amounts of protein to which the polysaccharide chains are covalently bonded.

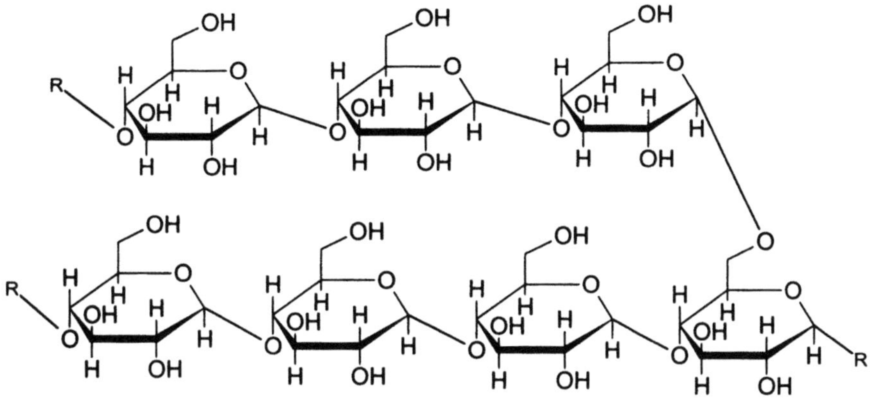


Glycogen

Dextrans are a high molecular weight branched extracellular polysaccharide synthesized by bacteria. These bacteria are found in many places including the human mouth where they flourish on sucrose-containing food which become trapped between our teeth. The generated dextrans become part of the dental plaque and thus are involved in tooth decay. Dextran-causing bacteria can also infect sugar cane and sugar beet after harvest and act to not only decrease the yield of sucrose but also interfere with sugar refining clogging filters and pipelines. These bacteria can also contaminate fruit juices and wines, in fact any ready source of glucose or sucrose.

On the positive side, dextran itself has been refined and employed as a therapeutic agent in restoring blood volume for mass casualties. Natural dextrans are very high molecular weight (on the order of $10^8$ to$10^9$ Daltons) and are found to be unsuitable as a blood-plasma substitute. Lower molecular weight (about $10^6$ Daltons) dextran is suitable and is often referred to as clinical dextran.

Dextran gels are commercially used. The gel formed from reaction with epichlorohydrin gives a crosslinked material used as a molecular sieve. Commercial crosslinked dextran is know as Sephadex(TM). Different series of Sephadex are used industrially and in research. Ionic groups are often incorporated to give anionic and cationic dextrans and ion-exchange molecular sieves. Sulfate esters of dextran are also used in separations.

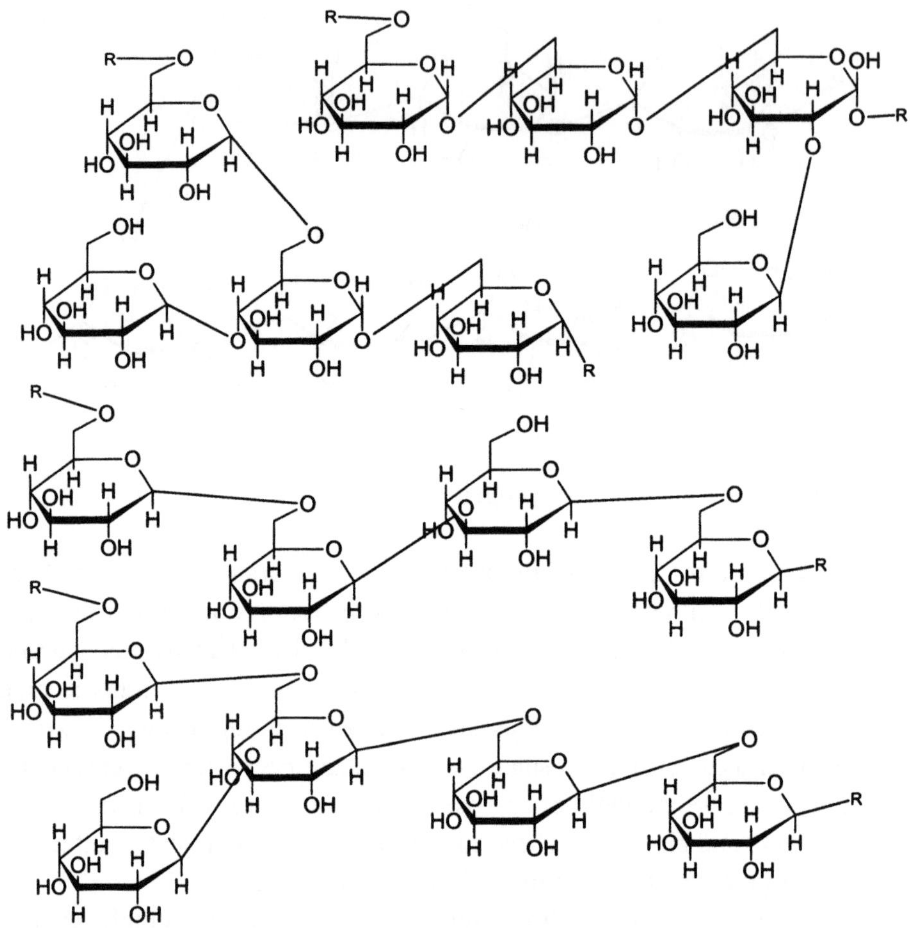

Above illustrates some typical units that compose dextrans. Representative Dextran Structures-top left- 1–>6 linked glucose units with a 1–>4 branch; top, right- linear 1–>6 linked glucose units with a 1–>2 branch; middle-linear chain with both 1–>6 and 1–>3 linkages; bottom- linear chain of 1–>6 linked glucose units with a 1–>3 branch. All links are alpha linkages.

### 2.3.2 Chitin and Chitosan

Chitin is generally a homopolymer of 2-acetamido-2-deoxy-D-glucose (N-acetylglucosamine) 1–>4 linked in a β configuration; it is thus an amino sugar analog of cellulose. While it is widely distributed in bacteria and fungi, the major source is crustaceans. In fact, chitin is the most abundant organic skeletal component of invertebrates. It is believed to be the most widely distributed polysaccharide with the Copepoda alone synthesizing on the order of $10^9$ tons each year. It is an important structural material often replacing cellulose in cell-walls of lower plants. It is generally found covalently bonded to protein. Invertebrate exoskeletons often contain chitin that provides strength with some flexibility along with inorganic salts such as calcium carbonate that provide strength. In a real sense this is a composite where the chitin holds together the calcium carbonate domains.

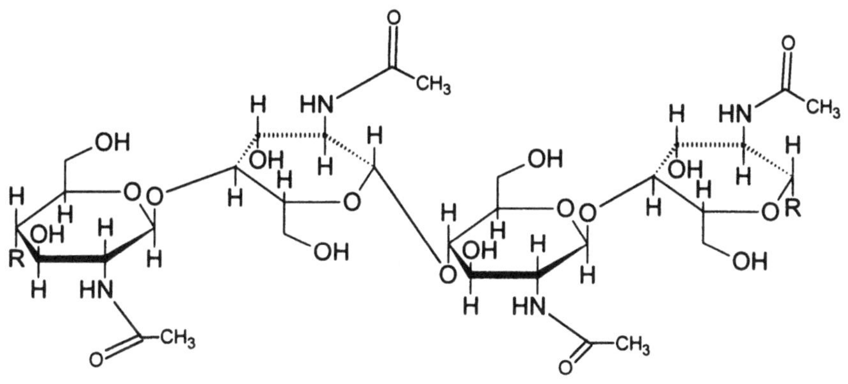

Chitin

Chitosan is produced from the deacetylation of chitin. Chitosan is employed in the food industry. It is a hemostatic from which blood anticoagulants and antithrombogenic have been formed. It is often sold as a body fat reducing agent or to be taken along with eating to encapsulate fat particles.

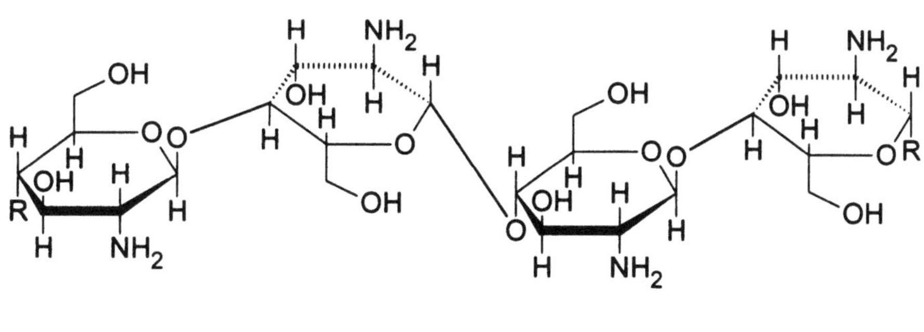

Chitosan

Both chitosan and chitin are greatly underused readily available abundant materials that deserve additional study as commercial materials and feedstocks. Chitin itself is not antigenic to human tissue and can be inserted under the skin or in contact with bodily fluids generally without harm. In the body chitin is slowly hydrolyzed by lysozyme and absorbed. Chitin and chitosan can be safely ingested by us and often we eat some since mushrooms, crabs, shrimp, many breads, and beer contain some chitin. Chitin and chitosan are believed to accelerate wound healing. Chitosan is also reported to exhibit bacteriacidal and fungicidal properties. Chitosan solutions were reported to be effective against topical fungal infections such as athlete's foot.

A continuing problem related to the introduction of bioengineering materials into our bodies is their incompatibility with blood. Many materials cause blood to clot (thrombosis) on the surfaces of the introduced material. Heparin, below, is an anticoagulant, non-toxic material that prevents clot formation when coated on vascular implants. While chitosan is a hemostatic material (stops bleeding by enhancing clotting), chitosan sulfate has the same anticoagulant behavior as heparin.

Cardiovascular disease is the leading cause of death in America. A contribution factor to cardiovascular disease is serum cholesterol. When ingested, chitosan exhibits hypocholesterolemic activity. Chitosan dissolves in the low pH found in the stomach and reprecipitates in the more alkaline intestinal fluid entrapping cholic acid as an ionic salt preventing its absorption by the liver. The cholic acid is then digested by bacteria in the large intestine. Chitosan may also act to increase the ratio of high density lipoprotein to total cholesterol. Chitosan has been studied in the formation of films including membrane-gels that immobilize enzymes and other materials because of the mild conditions under which they can be formed.

Chitosan has been used as a flocculate in wastewater treatment. The presence of the amine gives coacervation with negatively charged materials such as negatively charged proteins allowing removal of unwanted protein waste. The amine groups also capture metal ions, in particular polyvalent and heavy metal ions such as iron, lead, mercury, and uranium. The amine and hydroxyl groups can be modified through use of a wide range of reactions including formation of amides and esters.

Thus, there exists sufficient reason to consider these abundant materials in dietary, biomedical, cosmetic, etc. applications.

## 2.4 Heteropolysaccharides

Heteropolysaccharides contain two or more different monosaccharides. Glycosaminoglycans are polysaccharides that contain aminosugar units. Most are of animal origin.

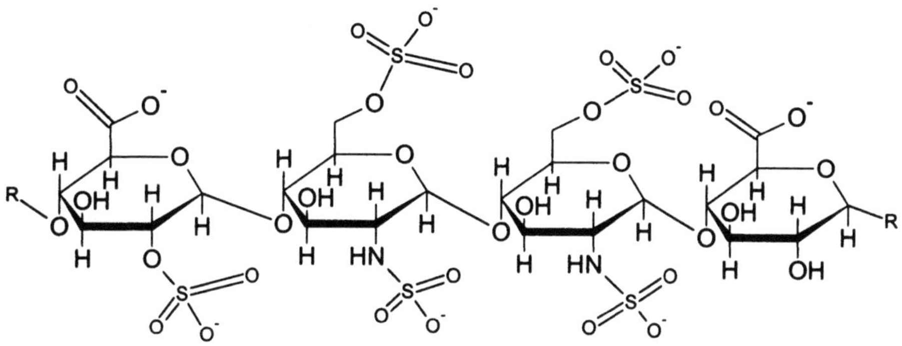

(Representative) Structure of heparin

Above is a representative structure of heparin that is complex containing D-glucuronic acid, L-iduronic acid, and D-glucosamine units. The glucosamine units may be N-acetylated or N-sulfonated. It is found in the lung, liver, and arterial walls of mammals. It is also found in intracellular granules of mast cells that line arterial walls and is released through injury. The glucuronic acid and iduronic acid units are not randomly present but occur in blocks. Heparin is found as the free polysaccharide and bonded to protein. Heparin acts as an anticoagulant, an inhibitor of blood clotting, and is widely used for this in medicine. In nature its purpose appears to be to prevent uncontrolled clotting.

Hyaluronic acid is found in connective tissues, umbilical cord, skin, and it is the synovial fluid of joints. It can have very large molecular weights, to $10^7$ Daltons making solutions of hyaluronic acid quite viscous. They can form gels. As a synovial fluid in joints it acts as a lubricant and in the cartilage it may also act, along with chondroitin sulfates, as a shock absorber. In some diseases such as osteoarthritis the hyaluronic acid of the joints is partially degraded resulting in a loss of elasticity of the area. The molecules can adopt a helical structure.

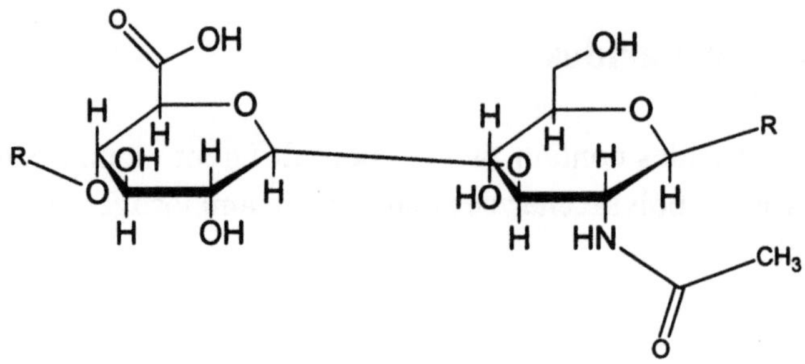


Hyaluronic Acid

Chondroitin sulfates are found in bone, skin, and cartilage but not as a free polysaccharide. Rather it exists as proteoglycan complexes where the polysaccharide is covalently bonded to a protein. The proteoglycan of cartilage contains about 10 % protein, keratan sulfate (below), and chondroitin sulfate, mainly the 4-sulfate in humans. The chondroitin sulfate chains have a weight average molecular weigh of about 50,000 Daltons but the complex has a molecular weight of several million. Again, chondroitin sulfates can adopt a helical conformation. The function of proteoglycan in cartilage is similar to that of non-cellulosic polysaccharides and protein in plant cell walls. In cartilage collagen fibers provide the necessary strength that is provided in plants by cellulose fibers. Thus, cartilage proteoglycan is an important part of the matrix that surrounds the collagen fibers giving it rigidity and incompressibility. This network can also act as a shock absorber since on compression the water is squeezed out to a nearby uncompressed region acting to "share the load" by distributing a shock or stress/strain.

Chondroitin sulfate is sold as a health aid to "maintain healthy mobile joints and cartilage".

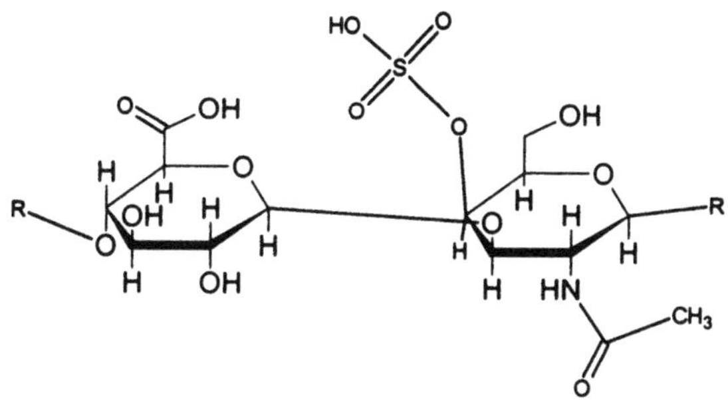


Chondroitin 4-sulfate

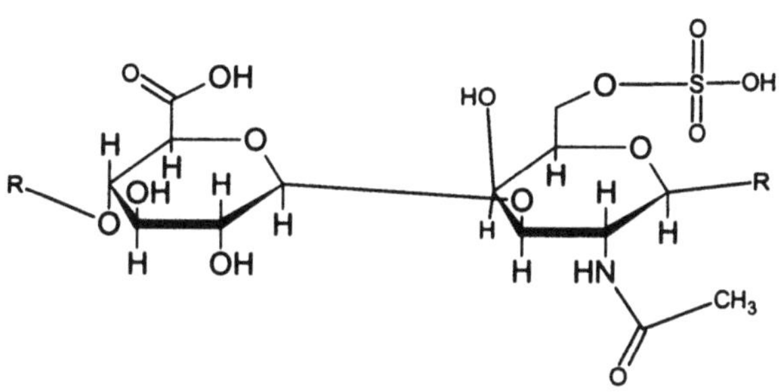


Chrondroitin 6-sulfate

Dermatan sulfate is found in the skin, arterial walls, and tendon where it is a part of another proteoglycan complex. It is about the same size as chondroitin sulfate and also able to form helical conformations.

There are two main divisions of polysaccharides that contain unmodified galactose groups-arabinogalactans that contain many plant gums and carrageenas and agar. Seaweeds represent a source of many polysaccharides including alginic acid, agar, and carrageenin. Alginic acid is a polymer of D-mannuronic acid and L-guluronic acid that may be arranged in a somewhat random fashion or in blocks. It is used as a stabilizer for ice cream, in paper coating, in the manufacture of explosives, and in latex emulsions.

The carrageenans and agar are generally linear galactans where the monomeric units are joined by alternating 1 –> 4 and 1 –> 3 bonds consisting then of disaccharide units. Carrageenan is the name given to a number of sulfated polysaccharides found in many red sea weeds where they play a structural role. The approximate repeat units for two industrially important carrageenans are given below. Both are able to form double helices containing two parallel staggered chains creating a gel. There are three disaccharide units per helix turn. The sulfate units are located on the outside of the helix with the helical structure stabilized by internal hydrogen bonds. In nature red seaweeds contain an enzyme that converts the galactose-6-sulfate of the k-carrageenan to 3,6-anhydrogalactose that causes a stiffening of the helix. It has been found that red seaweeds found where there is strong wave action contain a high proportion of anhydrogalactose. Thus, it appears that the seaweed is able to control its structure in response to external stimuli to minimize shredding by the increased wave action.

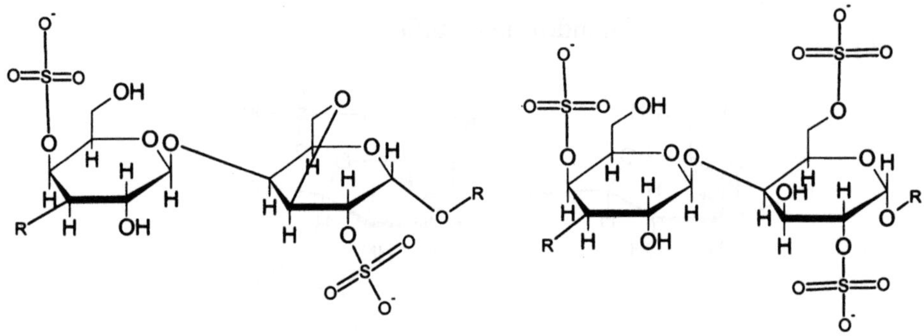

(Main repeat unit) (Less common repeat unit)

l-carrageenan

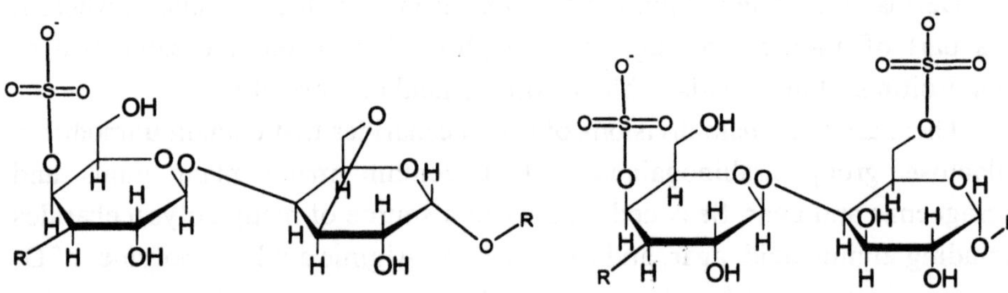

(Main repeat unit) (Less common repeat unit)

K-carrageenan

Because of its gelling ability, carrageenan is widely used as food thickeners and emulsion stabilizers in the food industry and is present in many dairy products including less expensive ice cream and other desert products providing a smooth, creamy texture. It is used as a stabilizer in foods such as chocolate milk.

The name agar refers to a family of polysaccharides that contain alternating β-D-galactopyranose and 3,6-anhydro-α-L-galactopyranose units and is thus similar to a carrangeenan where the anhydro-L-galactose is substituted for the anhydro-D-galactose. It is employed as the basis of many microbiological media and in canned food because it can be sterilized. The latter is an advantage over gelatin that is not able to withstand sterilization.

Agarose is the agar polysaccharide with the greatest gelling tendency. It contains no glucuronic acid units. It can form a compact double helix with the two chains being parallel and staggered, as in the case of carrageenan, forming a gel. Agarose gels are employed in gel-permeation chromatography, GPC, and gel electrophoresis.

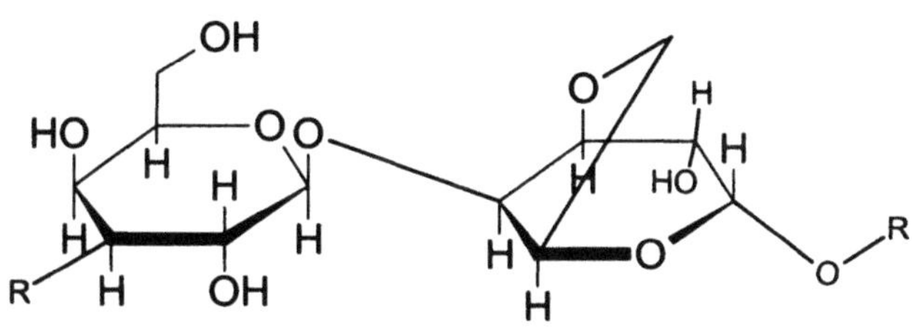


Agarose

Glycoproteins contain both saccharide and protein moieties with the protein being the major component, but both portions are involved in the overall biological activities.

## 3. NUCLEIC ACIDS

Nucleic acids comprise an amazing source of biomedical materials. On a molecular level, accomplishing so-called gene-splicing constitutes an example of DNA being used as a functionalized condensation polymer.

## 3.1 Primary Structure

The human genome is composed of natures most complex, exacting, and important macromolecule. It is composed of nucleic acids that appear complex in comparison to simpler molecules such as methane and ethylene, but simple in comparison to their result on the human body. Each unit is essentially the same containing a phosphate, and a deoxyribose sugar, below.

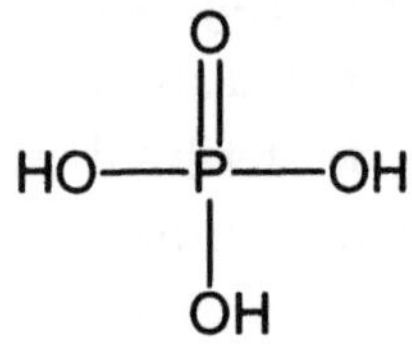

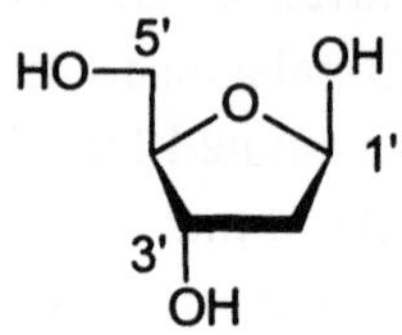

Deoxyribose

and one of four bases shown below with each base typically represented by the capital of the first letter of their name, from left to right G, C, A, and T.

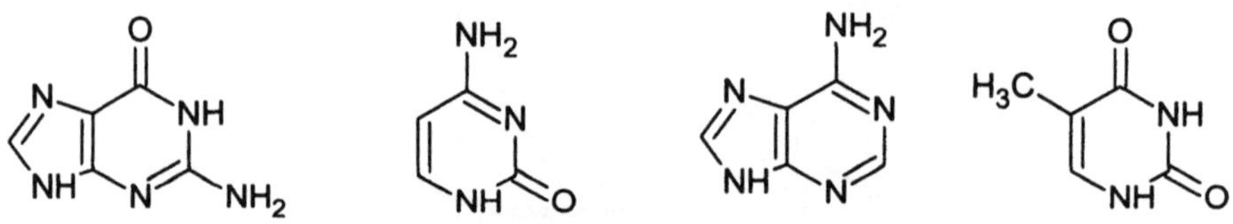

Guanine Cytosine Adenine Thymine

In fact, the complexity is less than having four separate and independent bases because the bases come in matched sets, they are paired. The mimetic Gee CAT allows an easy way to remember this pairing. The base, sugar, and phosphate combine forming nucleotides such as adenylic acid, adenosine-3'-phosphate shown below and represented by the symbols A, dA, and dAMP.

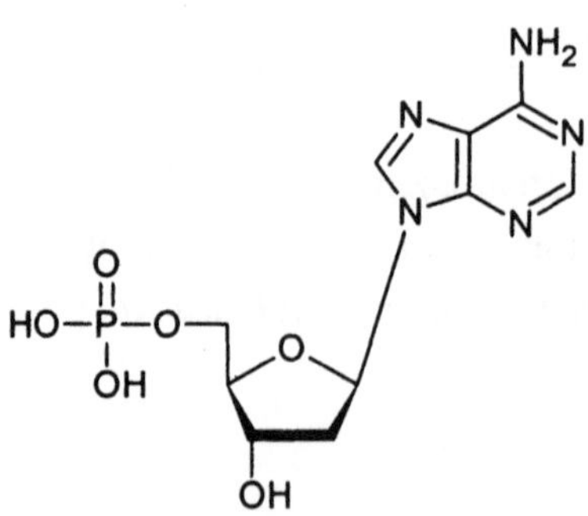

The backbone of nucleic acids is connected through the 3' and 5' sites on the sugar with the base residing at the 1' site. Because the sugar moiety is not symmetrical each unit can be connected differently but there is order (also called sense or directionality) in the sequence of this connection so that phosphodiester linkage between units is between the 3' carbon of one monomer and the 5' carbon of the next unit. Thus nucleic acids consist of units connected so that the repeat unit is a 3'-5' (by agreement we consider the start to occur at the 3' and end at the 5' though we could just as easily describe this repeat as being 5'-3') linkage. Thus, the two ends are not identical-one contains an unreacted 3' and the other an unreacted 5' hydroxyl.

Knowledge of other shorthand is also important in understanding descriptions being utilized to describe the particular genes and gene sequences. Following is a trimer containing in order the bases cytosine, adenine, and thymine.

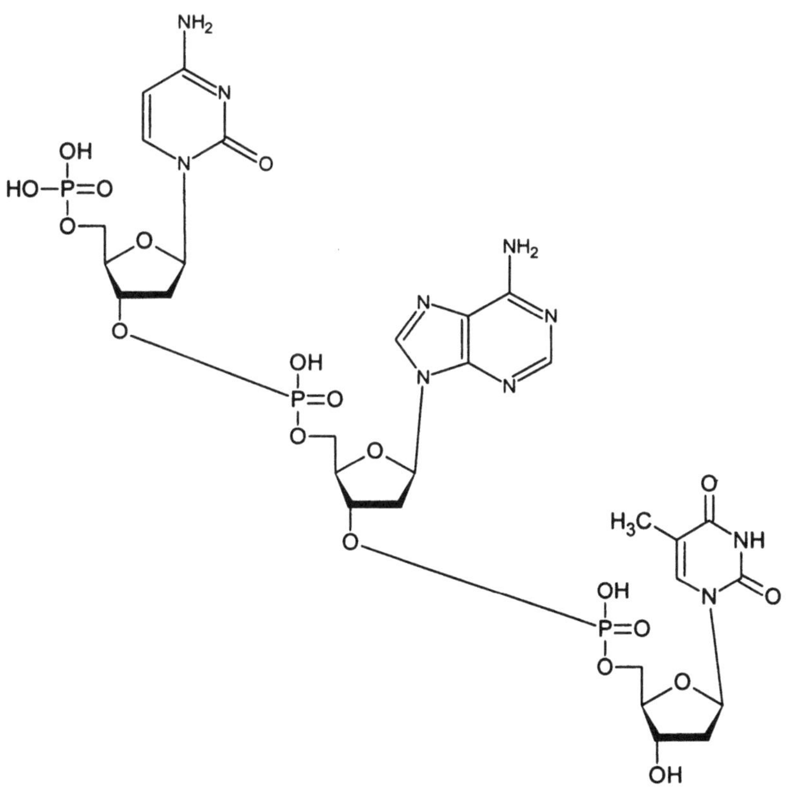

This sequence can be described as

p-5'-C-3'-p-5'-A-3'-p-5'-T-3' or pCpApT or usually as simply CAT.

## 3.2 Secondary Structure

Watson and Crick correctly deduced that DNA consists as a double-stranded helix in which a pyrimidine base on one chain or stand was hydrogen-bonded to a purine base on the other chain.

The combination AT has two hydrogen bonds while the combination GC has three double bonds contributing to making the GC a more compact structure as seen above. This results in a difference in the twisting resulting from the presence of the AT or GC units and combinations of these units result in structures that are unique to the particular combination. It is this twisting, and the particular base sequence, that eventually results in the varying chemical and subsequently biological activities of various combinations. The contributions to this twisting include other factors.

The stability of the DNA is due to both internal and external hydrogen bonding as well as ionic and other bonding. First, the internal hydrogen bonding is between the complementary purine-pyrimidine base pairs. Second, the external hydrogen bonding occurs between the polar sites along exterior sugar and phosphate moieties and water molecules. Third, ionic bonding occurs between the negatively charged phosphate groups situated on the exterior surface of the DNA and electrolyte cations such as $Mg^{+2}$. Fourth, the core consists of the base pairs, which, along with being hydrogen bonded, stack together through hydrophobic interactions and van der Waals forces. In order to take good advantage of pi-electron cloud interactions the bases stack with the flat "sides" over one another so that they are approximately perpendicular to the long axis.

The AT and CG base pairs are oriented in such a manner so that the sugar-phosphate backbones of the two twined chains are in opposite or antiparallel directions with one end starting at the 5' and ending at the 3' and the starting end of the other across from the 5' end being a 3' end and opposite the other 3' end is a 5' end. Thus, the two chains "run" in opposite directions.

The glucose bonds holding the bases onto the backbone are not directly across the helix from one another. Thus, the sugar-phosphate repeat units are not the same. This dislocation creates structures referred to as major and minor grooves.

In solution, DNA is a dynamic, flexible molecule. It undergoes elastic motions on a nanosecond time scale most closely related to changes in the rotational angles of the bonds within the DNA backbone. The net result of these bendings and twistings is that DNA assumes a compact shape. The overall structure of the DNA surface is not that of a reoccurring "barber pole" but rather because of the particular base sequence composition each sequence will have its own characteristic features of hills, valleys, bumps, etc.

## 3.3 Higher Structures

### 3.1.1. Supercoiling

Electron microscopy shows that individual DNA chains consist of two general structures: linear and circular. The chromosomal DNA in bacteria is a closed circle, a result of covalent joining of the two ends of the double helix but the DNA within eukaryotic cells, like our cells, is believed to be linear.

The most important secondary structure is supercoiling. Supercoiling simply is the coiling of a coil or in this case a coiling of the already helical DNA. The typical relaxed DNA structure is the thermally stable form. Two divergent mechanisms are believed responsible for supercoiling. The first, and less prevalent, is illustrated by a telephone cord. The telephone cord is typically coiled and represents the "at rest" or "unstressed" coupled DNA. As I answer the telephone I have a tendency to twist it in one direction and after answering and hanging up the telephone for awhile it begins forming additional coils. Thus, additional coiling tends to result in supercoiling. The second, and more common form, involves the presence of less than normal coiling. Thus, under winding occurs when there are fewer helical turns than would be expected. Purified DNA is rarely relaxed.

### 3.1.2. Compaction

Essentially all of human DNA is chromosomal with a small fraction found within the cells energy producing "plant", the mitochondria. The contour length, the stretched out helical length, of the human genome material in one cell is about 2 meters in comparison with about 1.7 meters for E. coli. An average human body has about $10^{14}$ cells giving a total length that is equivalent in length to traveling to and f rom the earth and sun about 500 times or 1000 one way trips.

Bacterial DNA appears as a loose, open arrangement of the closed loop DNA that exhibits supercoiling. These supercoiled DNA molecules are generally circular, right-handed in a negatively supercoiled DNA, tend to be extended and narrow rather than compacted, with multiple branches.

By comparison, our DNA is present in very compacted packages. One of the major compacting comes in the form of supercoiling. As noted before, our DNA is linear, but because of their large size they act as though they are looped forming coils about specific proteins. Subjection of chromosomes to treatments that partially unfold them show a structure where the DNA is tightly wound about "beads of proteins" forming a necklace-like arrangement where the protein beads represent precious stones imbedded within the necklace fabric. This combination forms the nucleosome, the fundamental unit of organization upon which higher-order packing or folding occurs. The bead of each nucleosome contains eight histone proteins. Histone proteins are small basic proteins with molecular weights between 11,000 to 21,000.

Wrapping of DNA about a nucleosome core compacts the DNA length about seven fold. The overall compacting though is about 10,000 fold. Additional compacting of about 100 fold is gained from formation of so called 30 nm fibers. These fibers contain one histone for each nucleosome core. The name "30 nm fibers" occurs because the overall shape is of a fiber with a 30 nm thickness. The additional modes of compaction are just beginning to be understood but may involve scaffold-assisting, that is DNA-containing segments wrapped about or within protein-containing units. Thus, certain DNA regions are separated by loops of DNA with about 20,000 to 100,000 base pairs with each loop possibly containing sets of related genes.

The scaffold contains several proteins, especially histone in the core and topoisomerase II. Both appear important to the compaction of the chromosome. In fact, the relationship between topoisomerase II and chromosome folding is so vital that inhibitors of this enzyme can kill rapidly dividing cells and several drugs used in the treatment of cancer are topoisomerase II inhibitors.

### 3.1.3. Replication

Replication occurs with a remarkably high degree of fidelity such that errors occurs only once per about 1,000 to 10,000 replications or an average single missed base for every $10^9$ to $10^{10}$ bases added. This highly accurate reproduction occurs because of a number of reasons including probably some that are as yet unknown. As noted before, the GC group has three hydrogen bonds while the AT has two. In vitro studies have found that DNA polymerases inserts one incorrect base for every $10^4$ to $10^5$ correct ones. Thus, other features are in place that assist in this process. Some mistakes are identified and then corrected. One mechanism intrinsic to virtually all DNA polymerases is a separate 3'-5' exonuclease activity that double-checks each nucleotide after it has been added. This process is very precise. If a wrong base has been added this enzyme prevents addition of the next nucleotide removing the mispaired nucleotide and then allowing the polymerization to continue. This activity is called proofreading and it is believed to increase the accuracy another $10^2$ to $10^3$ fold. Combining the accuracy factors results in one net error for every $10^6$ to $10^8$ base pairs, still short of what is found. Thus, other factors are at work.

# 4. PROTEINS

## 4.1. General Structures

In 1954 Linus Pauling received the Nobel Prize for his insights into the structure of materials - in particular, proteins. While the protein chain may assume an infinite number of shapes or conformations due to essentially free rotation about the various covalent bonds in the chain, Pauling showed that only certain conformations are preferred because of intramolecular and intermolecular hydrogen bonding. The simple sequence of amino acids is referred to as its primary structure (below). The preferred conformations are called secondary structures.

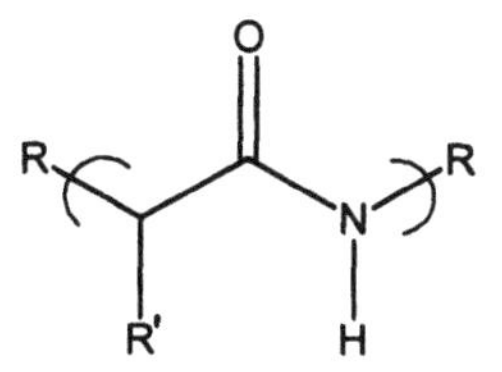

Two major secondary structures are found in synthetic and natural polymers- the helix and the sheet. The helix is a major structure for many polymers since it can take advantage of both intermolecular secondary bonding and relief of steric constraints. Some materials utilize a combination of helix and sheet structures. Thus wool consists of helical protein chains connected to give a "pleated" sheet. Further, many materials may offer only a limited range of one or the other structure that can be called short range order-long range disorder. The number of units within a "repeat helical unit" also varies.

The particular helix or sheet-like secondary structures generally also exist in preferred structures called tertiary structures . Tertiary structures focus on the overall chain folding, for instance, the gross shape of the protein chain. Secondary forces and primary forces in the form of crosslinks are important in determining and preserving the tertiary structure. It is important to remember that these tertiary structures still retain the primary and secondary structures. Thus, one of the myoglobin chains consists of amino acid units of a specific order that exist with about a 60-80% helical structure.

In turn, these "folded" chains can gather together giving larger, more intricate structural arrangements called quaternary structures. Thus hemoglobin is composed of four protein chains - each containing an iron site.

As noted before, proteins, and most polymeric materials, exist in solid and solution in one of two general secondary structures, helical or folded sheet. With respect to tertiary structures, proteins generally exist as fiberous or globular, with fiberous structures usually present when the protein is employed as a structural material, and globular with the protein is used as an enzyme.

While humans synthesize about a dozen of the twenty amino acids needed for good health, the other eight are obtained from outside our bodies, generally from eating foods that supply these essential amino acids.

Almost all of the sulfur needed for healthy bodies is found in amino acids as cysteine and methionine. Sulfur serves several important roles including as a crosslinking agent similar to that served by sulfur in the crosslinking, vulcanization of rubber. This crosslinking allows the various chains, that are connected by these crosslinks, to “remember’ here they are relative to one another. This crosslinking allows natural macromolecules to retain critical shapes to perform necessary roles. We will now briefly look at the secondary structures of some proteins. Remember that the primary secondary drives the secondary structure and the secondary structure, in turn, drives the tertiary structure.

## 4.2. Secondary Structure

The term secondary structure is used to describe the molecular shape or conformation of a molecule. The most important factor in determining the secondary structure of materials is its precise structure. For proteins, it is then the amino acid sequence. Hydrogen bonding is also an important factor in determining the secondary structures of natural materials and those synthetic materials that can hydrogen bond. In fact, for proteins, secondary structures are generally those that allow a maximum amount of hydrogen bonding. This hydrogen bonding also acts to stabilize the secondary structure while crosslinking acts to lock-in a structure.

In nature, extended helical conformations appear to be utilized in two major ways: to provide linear systems for the storage, duplication, and transmission of information (DNA, RNA), and to provide inelastic fibers for the generation and transmission of forces (F-actin, myosin, and collagen). Examples of the various helical forms found in nature are single helix (messenger and ribosomal DNA), double helix (DNA), triple helix (collagen fibrils), and complex multiple helices (myosin). Generally, these single and double helices are readily soluble in dilute aqueous solution. Often solubility is only achieved after the inter and intra-hydrogen bonding is broken.

As noted before, the structures of proteins generally fall into two groupings-fibers and globular. The structural proteins such as the keratines, collagen, and elastin are largely fiberous. A reoccurring theme with respect to conformation is that the preferential secondary structures of fiberous synthetic and natural polymers approximates that of a pleated sheet or skirt or helix. The pleated sheet structures in proteins are referred to as beta arrangements. In general, proteins with bulky groups take on a helical secondary structure while those with less bulky groups exist as beta sheets.

## 4.3. Keratines

As noted above, two basic "ordered" secondary structures predominate in synthetic and natural polymers. These are helices and the pleated sheet structures. These two structures are illustrated by the group of proteins called the keratines. It is important to remember that hydrogen bonding is critical in both structures. For helices, the hydrogen bonding occurs within a single strand, whereas in the sheets, the hydrogen bonding occurs between adjacent chains.

Hair and wool are composed of alpha-keratine. Both are available as functional condensation feedstocks for a variety of reactions. A single hair on our head is composed of many strands of keratine. Coiled, alpha-helices, chains of alpha-keratin intertwine to form protofibrils that in turn are clustered with other protofibrils forming a microfibril. Hundreds of these microfibrils, in turn, are embedded in a protein matrix giving a macrofibril that in turn combines giving a human hair.

While combing will align the various hairs in a desired shape, after a while, the hair will return to its "natural" shape through the action of the sulfur crosslinks pulling the hair back to its original shape.

The major secondary bonding is involved in forming the helical structures allowing the various bundles of alpha-keratine to be connected by weak secondary interactions that in turn allow them to readily slide past one another. This sliding or slippage along with the "unscrewing" of the helices allows our hair to be flexible.

Some coloring agents and most permanent waving of our hair involves breakage of the sulfur crosslinks and a reforming of the sulfur crosslinks at new sites to "lock in" the desired hair shape.

Fingernails are also composed of alpha-keratin, but keratin with a greater amount of sulfur crosslinks giving a more rigid material. In general, increased crosslinking leads to increased rigidity.

The other major structural feature is pleated sheets. Two kinds of pleated sheets are found. When the chains have their N—>C directions running parallel they are called parallel beta sheets. The N—>C directions can run opposite to one another giving what is called an antiparallel beta sheet. The beta keratin that occurs in silk produced by insects and spiders is of the antiparallel variety. While alpha-keratin is especially rich in glycine and leucine, beta-keratine is mostly composed of glycine and alanine with smaller amounts of other amino acids including serine and tyrosine. Size-wise, leucine offers a much larger grouping attached to the alpha carbon than does alanine. The larger size of the leucine causes the alpha-keratine to form a helical structure to minimize steric factors. By comparison, the smaller size of the alanine allows the beta-keratine to form sheets. This sheet structure is partially responsible for the "softness" felt when we touch silk. While silk is not easily elongated because the protein chains are almost fully extended, beta-keratin is flexible because of the low secondary bonding between sheets allowing the sheets to flow past one another.

In the silk fibroin structure almost every other residue is glycine with either alanine or serine between them allowing the sheets to fit closely together. While most of the fibroin exists as beta sheets, regions that contain more bulky amino acid residues interrupt the ordered beta structure. Such disordered regions allow some elongation of the silk.

The beta-keratin structure is also found in the feathers and scales of birds and reptiles.

Wool, while naturally existing in the helical form, forms a pleated skirt sheet-like structure when stretched. If subjected to tension in the direction of the helix axes, the hydrogen bonds parallel to the axes are broken and the structure can be irreversibly elongated to an extent of about 100%.

## 4.4. Collagen

Collagen is the most abundant single protein in vertebrates making up to one third of the total protein mass. Collagen fibers form the matrix or cement material in our bones where mineral materials precipitate. Collagen fibers constitute a major part of our tendons and act as a major part of our skin. Hence, it is collagen that is largely responsible for holding us together.

The basic building block of collagen is a triple helix of three polypeptide chains called the tropocollagen unit. Each chain is about 1000 residues long. The individual collagen chains form left-handed helices with about 3.3 residues per turn. In order to form this triple-stranded helix, every third residue must be glycine because glycine offers a minimum of bulk. Another interesting theme in collagen is the additional hydrogen bonding that occurs because of the presence of hydroxyproline derived from the conversion of proline to hydroxproline.

Collagen fibers are strong. In tendons, the collagen fibers have a strength similar to that of hard-drawn copper wire. Much of the toughness of collagen is the result of the crosslinking of the tropocollagen units to one another through a reaction involving lysine side chains. Lysine side chains are oxidized to aldehydes that react with either a lysine residue or with one another through an aldol condensation and dehydration resulting in a crosslink. This process continues throughout our life resulting in our bones and tendons becoming less elastic and more brittle. Again, a little crosslinking is essential, but more crosslinking leads to increased fracture and brittleness.

Collagen is a major ingredient in some "gelation" materials. Here, collagen forms a triple helix for some of its structure while other parts are more randomly flowing single collagen chain segments. The bundled triple helical structure acts as the rigid part of the polymer while the less ordered amorphous chains act as a soft part of the chain. The triple helix also acts as a non-covalently bonded crosslink.

## 4.5. Tertiary Structure

The term tertiary structure is used to describe the shaping or folding of macromolecules. These larger structures generally contain elements of the secondary structures. Often hydrogen bonding and crosslinking lock in such structures. As noted above, proteins can be divided into two broad groups-fiberous or fibrillar proteins and globular proteins that are generally soluble in acidic, basic, or neutral aqueous solutions. Fibrous proteins are long macromolecules that are attached through either inter or intra hydrogen bonding of the individual residues within the chain. Solubility, partial or total, occurs when these hydrogen bonds are broken.

## 4.6. Globular Proteins

There is a wide variety of so-called globular proteins. Many of these have varieties of alpha and beta-structures imbedded within the overall globular structure. Beta sheets are often twisted or wrapped into a "barrel-like" structure. They contain portions that are beta sheet structures and portions that are in an alpha conformation. Further, some portions of the globular protein may not be conveniently classified as either an alpha or beta structure.

These proteins are often globular in shape so as to offer a different "look" or polar nature to its outside than is present in its interior. Hydrophobic residues are generally found in the interior while hydrophilic residues are found on the surface interacting with the hydrophilic water-intense external environment. (This theme is often found for synthetic polymers that contain polar and not polar portions. Thus, when polymers are formed or reformed in a regular water-filled atmosphere, many polymers will favor the presence of polar moieties on their surface.)

Globular proteins act in maintenance and regulatory roles-functions that often require mobility and thus some solubility. Included within the globular grouping are enzymes, most hormones, hemoglobin, and fibrinogen that is changed into an insoluble fibrous protein fibrin that causes blood clotting.

Denaturation is the irreversible precipitation of proteins caused by heating, such as the coagulation of egg white as an egg is cooked, or by addition of strong acids, bases, or other chemicals. This denaturation causes permanent changes in the overall structure of the protein and because of the ease with which proteins are denatured, it makes it difficult to study protein structure. Nucleic acids also undergo denaturation.

## 5. LIGNIN

Lignin is the second most widely produced organic material, after the saccharides. It is found in essentially all living plants. It is produced at an annual rate of about $2 \times 10^{10}$ tons with the biosphere containing a total of about $3 \times 10^{11}$ tons. About $5 \times 10^{7}$ tons are produced yearly from pulp and paper production and so is available for use. It contains a variety of structural units including those pictured below.

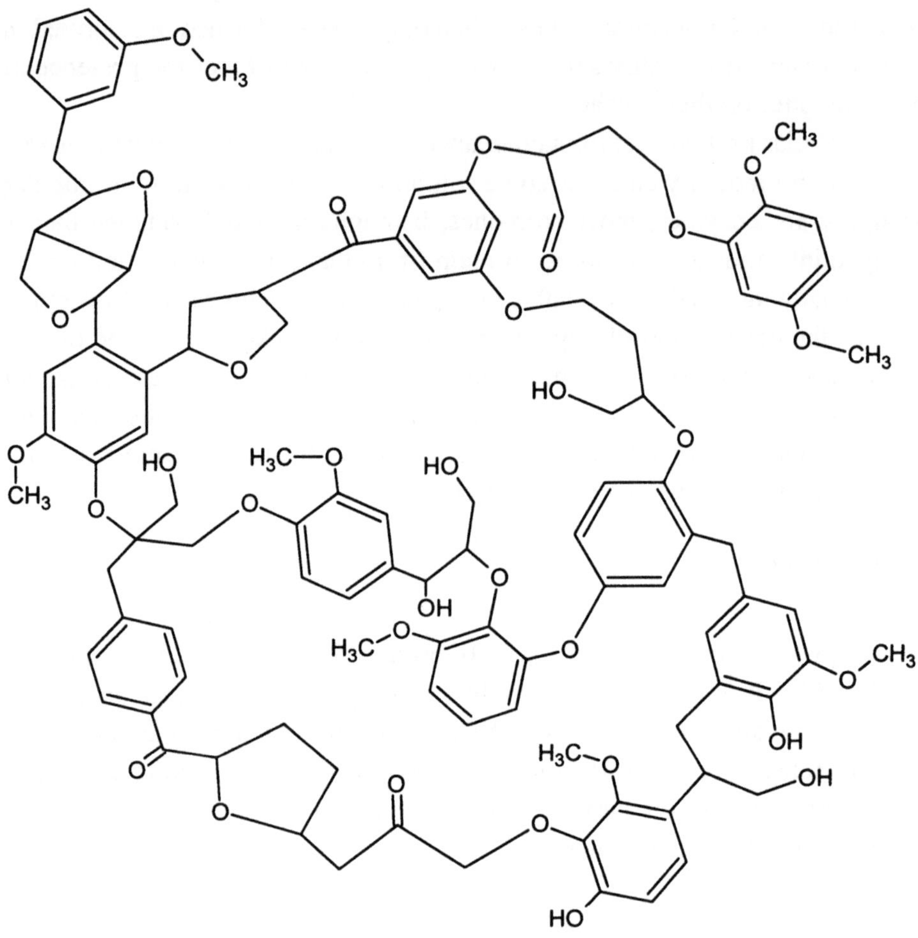


(Representative Structure) Lignin

Lignin contains an abundance and variety of functional groups that can be and have been modified. These include alcohols, aldehydes, ethers, and ketones. While some of these have been utilized commercially, lignin remains a major largely untaped storehouse of readily available, inexpensive, and renewable materials.

# 6. READINGS

1. Bloomfield, V., Crothers, D., Tinoco, I. (2000): Nucleic Acids: Structure, Properties and Functions, University Science Books, Sausalito, CA.
2. Carraher, C. (2000): Polymer Chemistry, Dekker, NY.
3. Carraher, C., Moore, J. (1983): Modification of Polymers, Plenum, NY.
4. Carraher, C., Sperling, L. (1983): Polymer Applications of Renewable-Resource Materials, Plenum, NY.
5. Carraher, C., Sperling, L. (1986): Renewable-Resource Materials, Plenum, NY.
6. Carraher, C., Tsuda, M.(1980): Modification of Polymers, ACS, Washington, DC.
7. Craver, C., Carraher, C. (2000): Applied Polymer Science, Elsevier, NY.
8. Gebelein, C., Carraher, C. (1995): Industrial Biotechnological Polymers, Technomic, Lancaster, PA.
9. Hecht, S. M. (1998): Bioorganic Chemistry:Carbohydrates, Oxford University Press, Cary, NC.
10. Moldave, K.(2000): Progress in Nucleic Acid Research and Molecular Biology, Academic Press, NY.
11. Paulsen, B. (2000): Bioactive Carbohydrate Polymers, Kluwer, NY.
12. Scholz, C., Gross, R. (2000): Polymers form Renewable Resources: Biopolyesters and Biocatalysis, ACS, Washington, DC.
13. Steinbuckel, A. (2001): Lignin, Humic, and Coal, Wiley, NY.
14. Vigo, T. (2001): Bioactive Fibers and Polymers, ACS, Washington, DC.
15. Woodings, C. (2001): Regenerated Cellulose Fibers, Woodhead Pubs., Cambridge, UK.

Chapter 13

# FUNCTIONAL HYDROPHILIC-HYDROPHOBIC HYDROGELS DERIVED FROM CONDENSATION OF POLYCAPROLACTONE DIOL AND POLY(ETHYLENE GLYCOL) WITH ITACONIC ANHYDRIDE

Monica Ramos and Samuel J. Huang
*Institue of Materials Science and Department of Chemistry, University of Connecticut, 97 North Eagleville Road, Storrs, CT, 06269.*

## 1. INTRODUCTION

Among some of the most important biorelated polymers are poly(ε-caprolactone), PCL, and poly(ethylene glycol), PEG. PCL, a fully biodegradable semi-crystalline polymer, has been used widely in the production of burn covers, elastomeric fibers, paints, and thermoplastics.[1-3] PEG, a water-soluble polymer, has been used extensively in the toiletry and cosmetic industry. Biomaterials based on polyethylene oxide (PEO), both crosslinked and linear, have shown low immunogenicity, reduced thrombogenicity and reduced protein and cell adherence.[4-7]

Water-containing gels derived from these biopolymers have been the subject of many studies in the field, as they are known to have low interfacial tension with the surrounding biological environment. While they are insoluble in water, their hydrophilicity enables them to swell in water without losing their structure. As a result, these polymers are able to mimic hydrated-based materials in the body. These gels have been used in prosthetic materials, soft lenses, catheters, membranes, and as bioactive releasing agents.[8-11] Moreover, they show promise in molecular imprinting as antibody- and receptor site-mimicking systems, shape memory devices, or enzyme-mimicking catalytic systems.[12]

Smart gels are known to have the ability to change their structures and/or function by the action of external signals such as light, temperature, electrical field, magnetic field, solvent, pressure, stress, ionic strength, and pH. These intelligent materials have been used to release bioactive agents in the body in a modulated way. Among some of these bioactive agents are cells for soft tissue engineering, genes, proteins, and pharmaceutical drugs [7, 13-18]

Developing materials derived from natural and renewable resources has become a highly regarded topic in many fields related to research in different areas.[19] This interest is due to concerns associated with health and the environment. Itaconic anhydride, ITA, is a compound synthesized from itaconic acid, which in turn is obtained from citric acid or fermentation of polysaccharides.[20] We have been interested in biomaterials and polymeric materials derived from renewable resources.[10, 21-28] Here we are reporting the condensation of ITA with different molecular weight PCLdiol and with different molecular weight PEG, resulting in modified macromonomers that after crosslinking, produced hydrogels with various structures and properties.

Previous work related to these types of hydrogels containing PEG-PCL can be found in the literature. Drug-releasing hydrogels for implantable delivery systems, PEG acrylate-terminated macromonomers and semi-IPNs polymer networks composed by PCL and PEG macromonomers have been synthesized. [29-33] In general, PCL-PEG block copolymers had shown increasing hydrophilicity as the content of PEG in the copolymers increases. It has been reported that the degradation rate decreases with reducing crystallinity and increasing hydrophilicity of the copolymer.[34]

We reported previously the synthesis of end-capped PCL diol with ITA.[35-38] The obtained PCLDIs were incorporated in PHEMA in order to synthesize semi-INPs and INPs, which resulted in hydrogels with outstanding mechanical properties. These networks are potentially suitable for artificial implants.

## 2. EXPERIMENTAL

### 2.1 Materials

Itaconic anhydride (ITA), polycaprolactone diol (PCL diol), poly(ethylene glycol) (PEG), 2,2-dimethoxy-2-phenylacetophenone (DMPA), ethylene glycol dimethacrylate (EGDMA), chloroform, and tetrahydrofurane (THF), were purchased from Aldrich. Stannous-2-ethyl hexanoate was purchased from Sigma. The reagents were used without further purification and the solvents were dried under standard procedures.

## 2.2. Instrumentation

IR spectra were obtained by using a Nicolet 560 Magna FTIR; single bounce -micro attenuated total reflectance (ATR), objective was used for films and resins. $^1$H-NMR were performed with a Bruker DMX 500 instrument at 25°C using deuterated chloroform; chemical shifts in parts per million (ppm), were referenced relative to tetramethylsilane (TMS), as an internal reference. UV lamp 100Watt UV 115V-60cps, long wave UV light for low temperature polymerization from Polysciences was used to carry out the crosslinking processes. TGA thermograms were recorded in a Perkin Elmer 7 instrument and a TA instruments Hi-Res TGA 2950 thermogravimetric analyzer. DSC thermograms were recorded in TA instrument DSC 29201; 1st and 2nd scans were recorded from −130 to 200°C at a heating rate of 15°C/min under nitrogen purge. About 10-15 mg of sample was sealed in aluminium pans manufactured by Rheometric Scientific. Glass transition temperature, (Tg's), was taken as the mid point of the heat capacity change, (inflection point). The melting temperature, (Tm), and enthalpy of fusion were determined from the endothermic melting peaks. Cross polar microscopy photographs were obtained by using a Nikon Labphot with crosspolars microscope equipped with a CCD camera.

## 2.3. Synthesis of Polycaprolactone Diitaconates, (PCLDIs)

In a typical procedure, PCL diol was end-capped with itaconic anhydride by reacting PCL diols in the molecular weight range of 530 to 2k, with 2.5 equivalents of ITA and 0.1% weight of catalyst, stannous-2-ethyl hexanoate. All the reactions were run under nitrogen gas, at 80°C for 8-12 hours. Reaction time was determined by monitoring the reactions by IR. The reaction products were purified by sublimation of the unreacted ITA under vacuum at 50°C for 24 hours. The presence of monoitaconate products were not detected by $^1$H-NMR. Macromonomers obtained by this method were completely soluble in chloroform and in THF; a brief scheme of this reaction is shown in Figure 1.

## 2.4. Synthesis of Poly(ethylene glycol) Diitaconates, (PEGDIs)

In a typical procedure, PEGs in the molecular weight range 300 to 4k, followed synthesis and purification as for PCLDIs described in section 2.3. The macromonomers were completely soluble in chloroform and THF; this reaction is described in Figure 2.

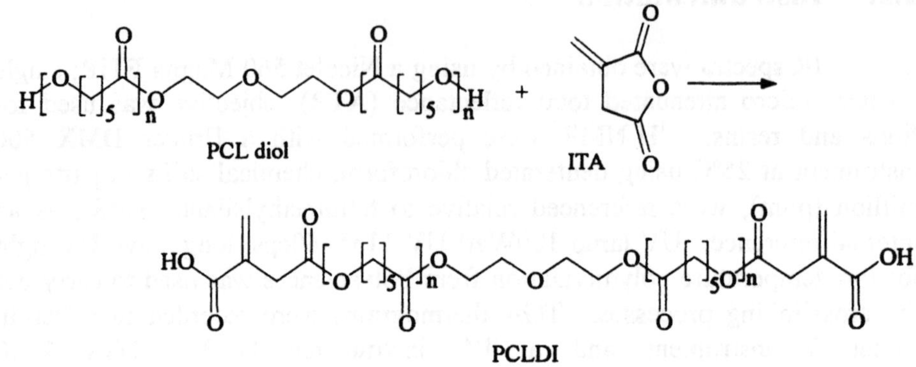


*Figure 1.* Synthesis of PCL diol end-capped with itaconic anhydride, (PCLDIs)

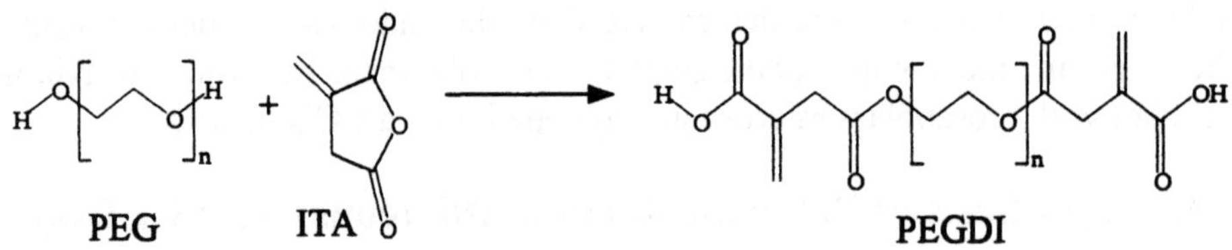


*Figure 2.* Synthesis of PEG end-capped with itaconic anhydride (PEGDIs)

## 2.5. Crosslinking Procedure

In a typical procedure, the obtained macromonomers for crosslinking, either PCLDIs or PEGDIs, or a combination of PCLDIs and PEGDIs in 1:1 (w/w) proportions were combined in THF with 0.3% weight of DMPA as photo initiator and 0.1% weight of EGDMA as a cross-linking agent. Once all the reagents were well mixed, the solutions were poured into a teflon dish and almost all of the solvent was allowed to evaporate after which, the systems were irradiated with UV light, (with λ of 365nm), for about 15 minutes at approximately 60°C. Films were peeled off and by using THF, remaining unreacted macromonomers were washed out from the gels. The resulting gels were dried in a vacuum oven at 40°C for 24 hours.

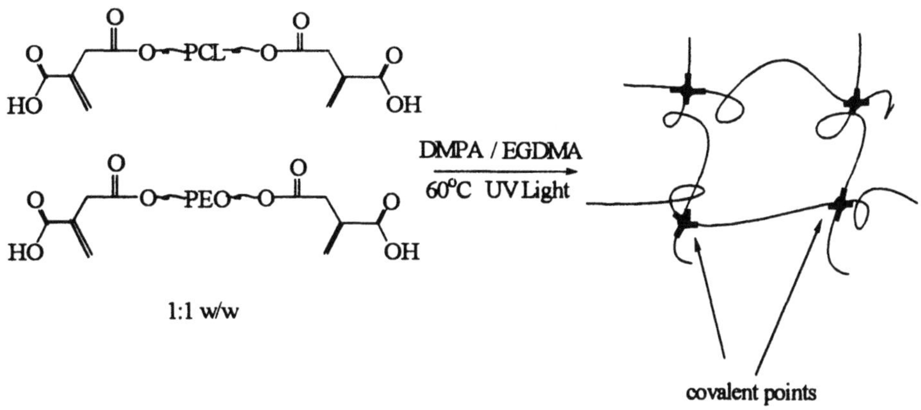


*Figure 3.* Crosslinking process to form hydrogels based on PCLDIs and PEGDIs, 1:1 (w/w)

## 2.6. Gel Swelling

Gel pieces were dried under vacuum at 40°C for about 24 hours. After reaching a constant weight, the pieces were immersed in an excess of buffered solutions at constant pH of 2, 7, and 10 at room temperature for 120 hours. The buffer solutions were changed every 24 hours. The swelling kinetics were followed by determining the weight until equilibrium was reached. After removing the swollen gels from the buffer solutions at regular intervals, they were dried superficially with filter paper, weighed and returned to the solutions.

# 3. RESULTS AND DISCUSSION

## 3.1. Spectral Characterization of PCLDIs and PEGDIs Macromonomers

IR spectra of the obtained macromonomers and hydrogels showed in all the cases, expected changes in the 3600 to 3000 $cm^{-1}$ range. Features associated to alcohols for the starting materials disappeared gradually when modified macromonomers, PCLDIs and PEGDIs, were formed. Associated with these changes, typical bands for carboxylic acid groups were observed. After crosslinking of the double bonds, the carboxylic groups, which are expected to be localized by the covalent points in the hydrogels, remained intact as it demonstrated by ATR-FTIR.

Results in the region associated with carbonyl group, C=C double bond and stretches for alkyl chains also showed expected transformations.

For instance, transformations of PCLdiol2k to PCLDI2k are depicted below in Figure 3. For comparison IRs obtained from gels derived from PCLDI2k, and from PCLDI2k and PEGDI300 mixture 1:1 (w/w), are also shown.

The $^{1}$H NMR of the PCLdiol530 as it was received from Aldrich, is shown on Figure 5. Changes in the $^{1}$H NMR after forming the itaconate derivatives can be clearly observed for PCLDI530. Hydroxyl group in the starting material, signal around 1.8 ppm, (top spectrum in figure 5), disappeared in the product, (bottom spectrum in figure 5). The product showed features corresponding to the double bond moieties that appeared at 5.8 and 6.4 ppm and to the carboxylic groups at around 9 ppm. Similar spectral characterization was carried out for all others PCLDIs and PEGDIs macromonomers. By $^{1}$H-NMR we were able to determine that condensation of the alcohol with ITA, happens at both carbonyl groups because of the complexity of the vinyl signals. We expected the mechanism of the reaction to be affected by electronic and steric factors producing more of the γ-vinyl ester.

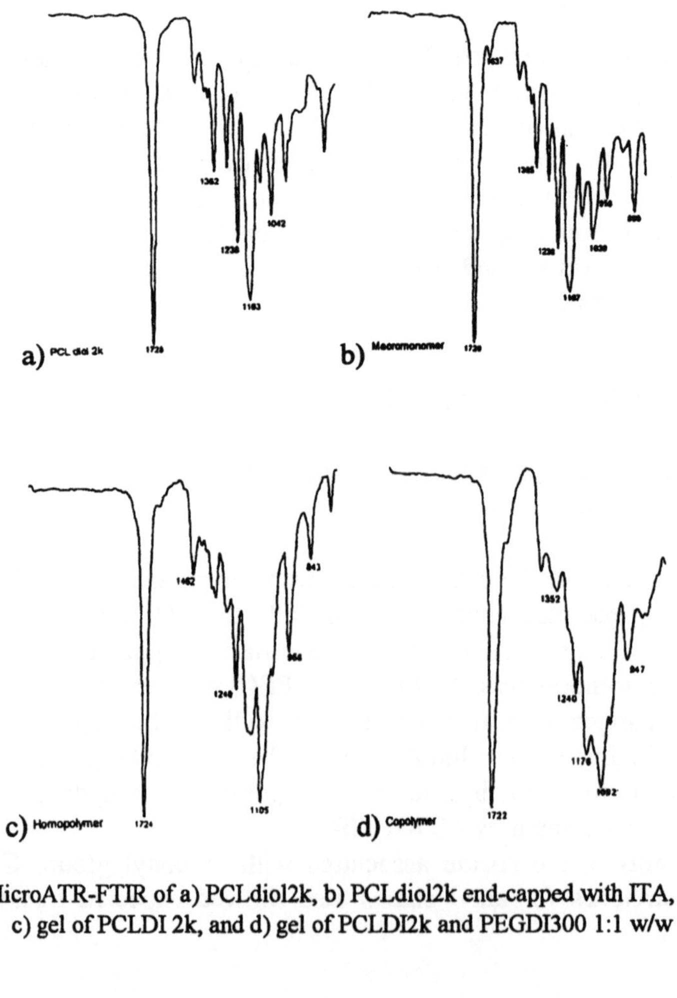


*Figure 4*. MicroATR-FTIR of a) PCLdiol2k, b) PCLdiol2k end-capped with ITA, (PCLDI2k), c) gel of PCLDI 2k, and d) gel of PCLDI2k and PEGDI300 1:1 w/w

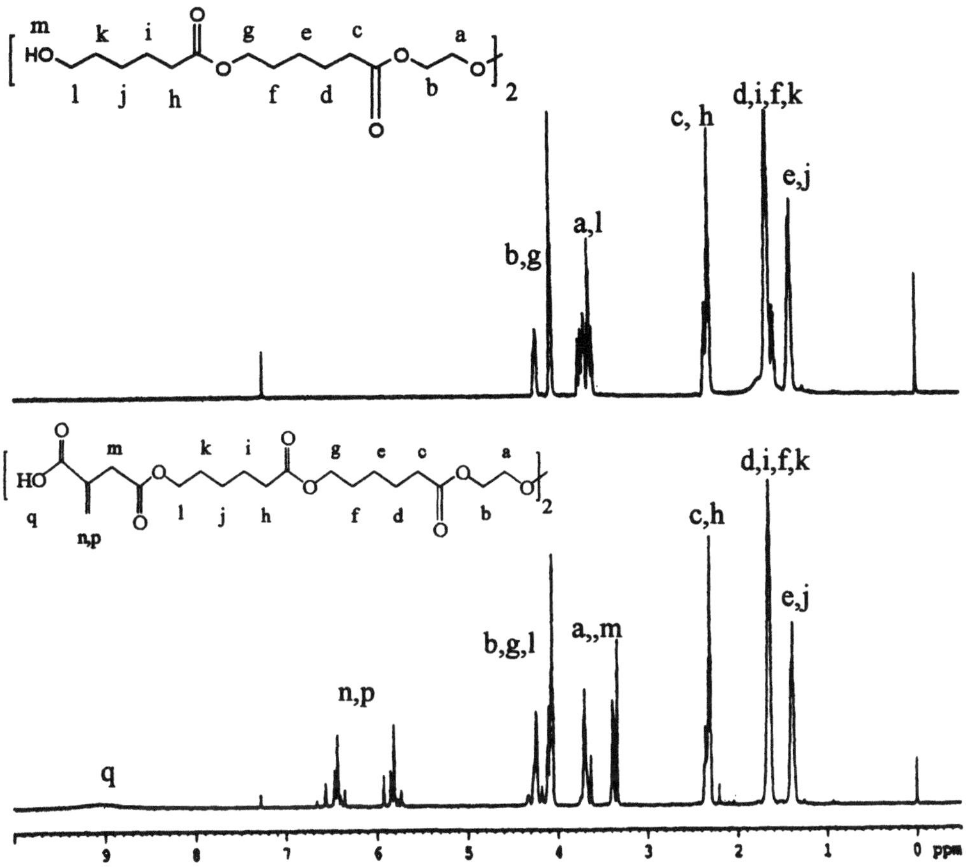


*Figure 1.* Top $^{1}$H NMR PCLdiol530, Bottom $^{1}$H-NMR PCLDI530. The signal for the hydroxyl groups in PCLdiol varies from 1.5 to 2.5 ppm

## 3.2. Characterization of the Hydrogels

Characterization of these gels included ATR-FTIR, as it was illustrated in Figure 4. Typically, some C=C double bond signals are apparent in the gels, indicating incomplete crosslinking of the material. Thermoanalysis of these new materials and swelling behaviour were also carried out. A typical TGA of these gels is depicted in Figure 6a. DSC analysis showed that Tg values are related to the amount of PEG in the original mixture as is shown in Figure 6b. Tg values were obtained from the second heat after quenching. The more PEG in the hydrogels, the lower the Tg obtained for the material. After crosslinking, ΔH values for melting transitions are smaller in magnitude than those for their corresponding macromonomers.

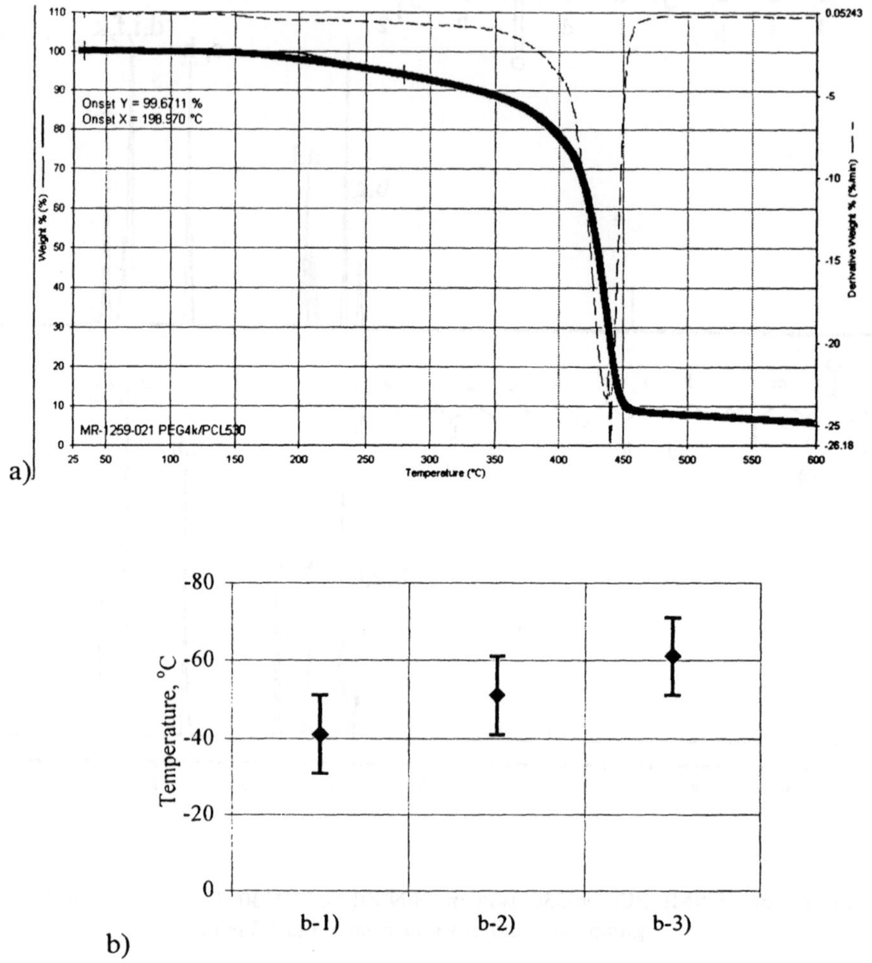


*Figure 6.* Thermoanalysis data a) TGA of PCLDI530-PEG4k hydrogel, b) Glass transition temperatures for gels obtained from PEGDI 2K and PCLDIs, approximate molar ratios, PEGD:PCLDI I b-1) 0.25:1, b-2) 0.50:1, and b-3) 1:1

Amorphous and crystalline phases are apparent in mostly all of these materials by thermal studies. Gels derived from macromonomers with higher molecular weight showed both glass and melting transitions. However, studying the gels by light microscopy, phase separation was observed in gels that have different molar ratios of macromonomers in the initial mixture. The greater the difference in molar ratios, the greater the phase separation detected by light microscopy; Figure 7 illustrates these findings.

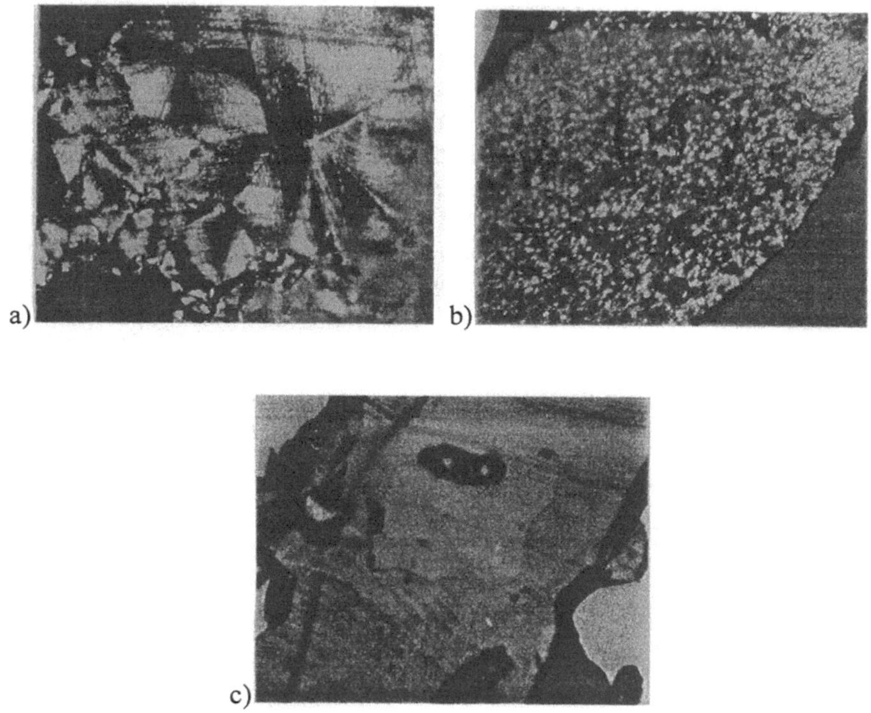


*Figure 7.* Transmitted Light Microscopy of a) PEGDI 4k-PCLDI 830 gel, approx molar ratio: 1:5; b) PEGDI 2k-PCLDI 830 gel, approx molar ratio: 1:2.5; c) PEGDI2k-PCLDI2k gel, molar ratio 1:1.

Presence of the carboxylic acid groups imparted pH sensitivity to the gels, as demonstrated by measuring the degree of swelling at different pH values. Results are shown in Figure 8. In this study degree of swelling (DS) is define by:

$$DS = \frac{W_{wet} - W_{dry}}{W_{dry}}$$

Where $W_{wet}$ is the weight of the gel after 120 hours in the buffered solution, $W_{dry}$ is the weight of the hydrogel at the beginning of the experiment.

Also, it was determined that the gels swelled by diffusion as depicted in Figure 10. The specific system illustrated corresponds to PEGDI300-PCLDI2k 1:1(w/w) hydrogel. Kinetics of all the gel-swelling process were examined and the results are illustrated in Figure 9. In the course of the swelling process, the solvent uptake behaviour could be described by the following equation obtained as the solution to Fick's second law for slob-shaped gels: [39]

$$\frac{M_{st}}{M_{s\infty}} = 4\left(\frac{Dt}{\pi l}\right)^{1/2}$$

Where $M_{st}$ is the total amount of water taken by the gel at the time t. $M_{s\alpha}$ is the total amount of water sorbed at the equilibrium state. D is the diffusion coefficient for a solvent in the polymer and l is the gel thickness.

Based on these studies we determined that the degree of swelling at room temperature was controlled by the molecular weight of the macromonomers involved. Hydrogels obtained from longer PCLDIs macromonomers underwent less swelling.

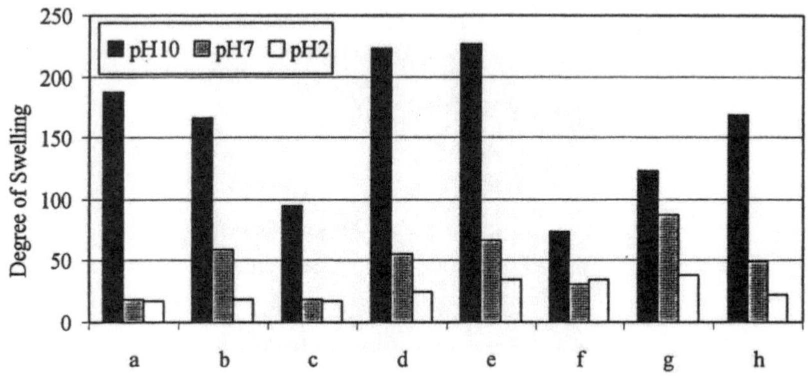


Figure 8. Degree of swelling for obtained hydrogels in water at different pH values at room temperature after 120hours; a)PEGDI300-PCLDI530, b)PEGDI300-PCLDI830, c)PEGDI300-PCLDI2k, d)PEGDI2k-PCLDI830, e)PEGDI2k-PCLDI830, f)PEGDI2k-PCLDI2k, g)PEGDI4k-PCLDI530, h) PEGDI4k-PCLDI830. All mixtures are 1:1(w/w).

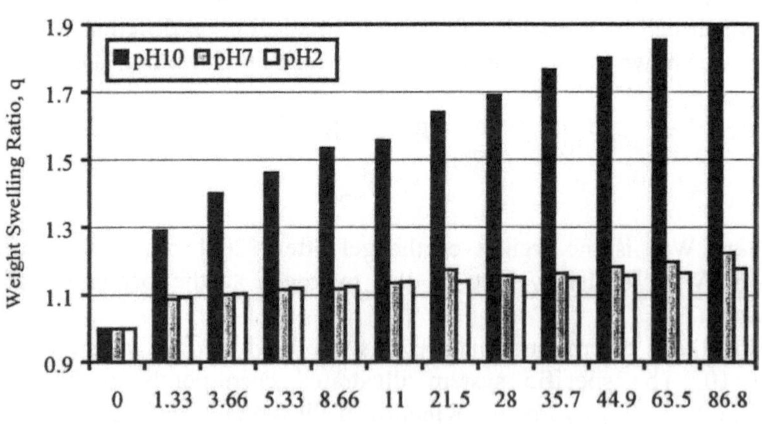


*Figure 9.* Kinetics of the weight swelling ratio in water at RT for PEGDI300-PCLDI2k hydrogel at different pH values.

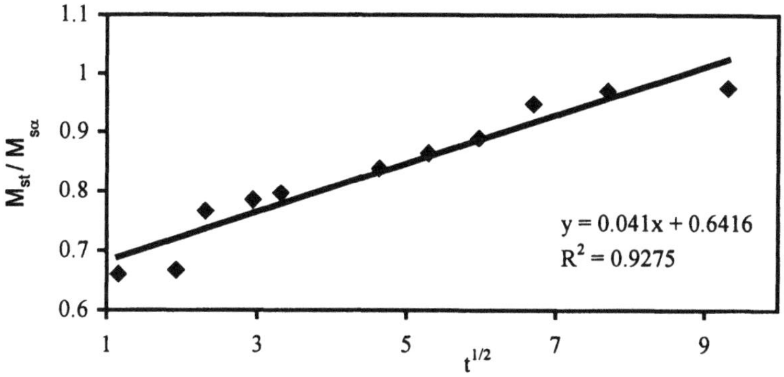


*Figure 10.* Plot of Fick's equation for swelling of PEGDI300-PCLDI2k 1:1 (w/w) hydrogel at basic pH

## 4. CONCLUSIONS AND EXTENSIONS

Condensation of itaconic anhydride with PCLdiol, (range molecular weight 530 to 2k), and PEG, (range molecular range 300 to 4k), in a one step reaction, has been demonstrated to be an efficient and versatile procedure in order to obtain PCL and PEG diitaconates. In general, it has been shown that anionic hydrophobic-hydrophilic hydrogels can be synthesized by using two different macromonomers, whose difference in molecular weights dictates its properties.

DSC analysis of the dry hydrogels suggested that in general, the more PEG in the hydrogels the lower the Tg obtained for the material. Also, that there coexists phase mixing and phase separation in the gels. These conclusions are based on the appearance of a glass transitions from amorphous domains, and melting transitions from crystalline domains of the PEG and PCL components.

By light microscopy phase separation was apparent but not in all the gels. The greater the difference in molar ratios, the greater the phase separation detected. Those hydrogels that have very close or even molar ratios in their components showed no phase separation at this level.

Based on the swelling studies, we have determined that the mechanism that best describes the solvent penetration into the gels as Fickian diffusion transport. Also, the degree of swelling at room temperature was controlled by the molecular weight of the macromonomers involved. Hydrogels obtained from longer PCLDIs macromonomers underwent less swelling.

Studies concerning the use of macromonomers with higher molecular weights could be expected to lead to different thermal stabilities and swelling rates. Furthermore, different chemical approaches are been explored in order to achieve different polymeric architectures keeping constant both, same biocompatible components and same condensation reaction with ITA. In any case, the presence of the carboxylic acid in these hydrogels will provide chemically modulated systems. These polymers are being explored as biomaterials.

## 5. ACKNOWLEDGMENTS

The authors thank Dr. Dawn Alison Smith for her valuable contributions to this project.

## 6. REFERENCES

[1]X. G. Zhang, Matteus, F. A., "Biodegradable Polymers," in *Polymeric Materials Encyclopedia*, J. C. Salamone, Ed.: CRC, 1996, pp. 593-600.

[2]P. Jarrett, C. Benedict, J. P. Bell, J. A. Cameron, and S. J. Huang, "Mechanism of the biodegradation of polycaprolactone," *Polym. Prepr. (Am. Chem. Soc., Div. Polym. Chem.)*, vol. 24, pp. 32-33, 1983.

[3]J. V. Koleske, "Poly(ε-caprolactone)," in *Polymeric Materials Encyclopedia*, J. C. Salamone, Ed.: CRC, 1996, pp. 5683-5891.

[4]L. Brannon-Peppas, "Poly(ethylene glycol): Chemistry and biological applications, edited by J. M. Harris and S. Zalipsky," in *J. Controlled Release*, vol. 66, 2000.

[5]N. B. Graham, "Poly(ethylene oxide)," in *Polymeric Materials Encyclopedia*, J. C. Salamone, Ed.: CRC, 1996, pp. 6042-6054.

[6]H. Otsuka, Y. Nagasaki, K. Kataoka, T. Okano, and Y. Sakurai, "Reactive-PEG-polylactide block copolymer for tissue engineering," *Polym. Prepr. (Am. Chem. Soc., Div. Polym. Chem.)*, vol. 39, pp. 128-129, 1998.

[7]N. A. Peppas and J. Klier, "Controlled release by using poly(methacrylic acid-g-ethylene glycol) hydrogels," *J. Controlled Release*, vol. 16, pp. 203-214, 1991.

[8]H. F. B. Mark, Norbert M.; Overberger, Charles G.; Menges, Georg, "Encyclopedia of Polymer Science and Engineering -gels," , vol. 7, 2nd ed: John Wiley & Sons, 1988, pp. 514-530.

[9]H. F. B. Mark, Norbert M.; Overberger, Charles G.; Menges, Georg, "Encyclopedia of Polymer Science and Engineering- hydrogels," , vol. 7, 2nd ed: John Wiley & Sons, 1988, pp. 783-807.

[10] R. M. Ottenbrite, S. J. Huang, K. Park, and Editors, *Hydrogels and Biodegradable Polymers for Bioapplications. (Symposium at the 208th National Meeting of the American Chemical Society, Washington, DC, August 21-26, 1994.) [In: ACS Symp. Ser., 1996; 627]*, 1996.

[11] S. W. H. Shalaby, Allan S.; Ratner, Buddy D.; and Horbett, Thomas A, "Polymers as Biomaterials," . New York: Plenum Press, 1984, pp. 323-385.

[12] R. A. Ansell, *Pharm News*, vol. 3, pp. 16, 1997.

[13] T. Aoyagi, M. Ebara, K. Sakai, Y. Sakurai, and T. Okano, "Novel bifunctional polymer with reactivity and temperature sensitivity," *J. Biomater. Sci., Polym. Ed.*, vol. 11, pp. 101-110, 2000.

[14] Y. H. Bae and I. C. Kwon, "Stimuli-sensitive polymers for modulated drug release," in *Biorelated Polymer Gels*, T. Okano, Ed.: Academic Press, 1998, pp. 93-134.

[15] N. B. Graham, "Hydrogels in controlled drug delivery," in *Polymeric Biomaterials*, E. H. Piskin, A. S., Ed.: Martinus Nijhoff Publishers, 1984, pp. 170-194.

[16] T. Okano, "Molecular design of stimuli-responsive hydrogels for temporal controlled drug delivery," *Proc. Int. Symp. Controlled Release Bioact. Mater.*, vol. 22nd, pp. 111-112, 1995.

[17] T. Okano and Editor, *Biorelated Polymers and Gels: Controlled Release and Applications in Biomedical Engineering*, 1998.

[18] M. Yokoyama, "Novel passive targetable drug delivery with polymeric micelles," in *Biorelated Polym. Gels*, 1998, pp. 193-229.

[19] C. Scholz, R. A. Gross, and Editors, "Polymers from Renewable Resources Biopolyesters and Biocatalysis. (Proceedings of the 216th American Chemical Society Meeting, held in 1998, Boston, Massachusetts.) [In: ACS Symp. Ser., 2000; 764]," , 2000.

[20] B. E. Tate, "Itaconic acid, itaconic esters and related compounds," in *Vinyl and Diene Monomers*, vol. 24, *High Polymers*, E. C. Leonard and Editor, Eds., 1970, pp. 206-261.

[21] A.-C. Albertsson, S. J. Huang, and Editors, "Degradable Polymers, Recycling, and Plastics Waste Management. (An International Workshop was organized in April 1994 in Stockholm, Sweden.) [In: Plast. Eng. (N. Y.), 1995; 29]," , 1995.

[22] A.-C. Albertsson, S. J. Huang, and Editors, "Biodegradable Polymers and Recycling. (Selected Papers presented at the International Workshop on Controlled Life-Cycle of Polymeric Materials, April 21-23, 1994, Stockholm, Sweden. [In: J. Macromol. Sci., Pure Appl. Chem., 1995; A32(4)]," , 1995.

[23] S. J. Huang, T. A. P. Seery, G. Swift, and Editors, *Selected Papers Presented at the Polymeric Materials: Science and Engineering Symposium at the 216th American Chemical Society Nation Meeting on Biomedical Applications of Water-Soluble Polymers and Hydrogels, held 23-27 August 1998, in Boston, Massachusetts. [In: J. Macromol. Sci., Pure Appl. Chem., 1999; A36(7 & 8)]*, 1999.

[24] S. J. Huang, "Biodegradable polyesters," presented at Book of Abstracts, 217th ACS National Meeting, Anaheim, Calif., March 21-25, 1999.

[25] J. A. Wallach and S. J. Huang, "Methacrylic group functionalized poly(lactic acid) macromonomers from chemical recycling of poly(lactic acid)," *ACS Symp. Ser.*, vol. 764, pp. 281-292, 2000.

[26] J. A. Wallach and S. J. Huang, "Cyclic anhydride containing copolymers derived from renewable resources," presented at Book of Abstracts, 219th ACS National Meeting, San Francisco, CA, March 26-30, 2000, 2000.

[27] J. A. Wallach and S. J. Huang, "Copolymers of Itaconic Anhydride and Methacrylate-Terminated Poly(lactic acid) Macromonomers," *Biomacromolecules*, vol. 1, pp. 174-179, 2000.

[28] J.-l. Jane, S. J. Huang, and Editors, *Biopolymers as Advanced Materials. (Selected Papers Presented at the ACS Symposium, April 2-7, 1995, in Anaheim, California.) [In: J. Macromol. Sci., Pure Appl. Chem., 1996; 33(5)]*, 1996.

[29] J.-Z. Bei, J.-M. Li, Z.-F. Wang, J.-C. Le, and S.-G. Wang, "Polycaprolactone-polyethyleneglycol block copolymer. IV: Biodegradation behavior in vitro and in vivo," *Polym. Adv. Technol.*, vol. 8, pp. 693-696, 1997.

[30] D.-r. Chen, H.-l. Chen, J.-z. Bei, and S.-g. Wang, "Preparation and degradation of polycaprolactone-polylactide-polyether tri-component copolymer microparticles," *Gongneng Gaofenzi Xuebao*, vol. 12, pp. 357-361, 366, 1999.

[31] S. Wang, B. Qiu, J. Gao, and Y. Duan, "Polycaprolactone-poly(ethylene glycol) block copolymer. II. Relationship among composition, crystallinity and degradability," *Gaofenzi Xuebao*, pp. 560-5, 1995.

[32] W. Wang, J. Bei, Z. Wang, and S. Wang, "Effect of surface properties of polycaprolactone-polyether copolymer and blended polymer on their drug release behavior," *Gaofenzi Xuebao*, pp. 253-6, 1995.

[33] J. Bei, W. Wang, Z. Wang, and S. Wang, "Surface properties and drug release behavior of polycaprolactone polyether blend and copolymer," *Polym. Adv. Technol.*, vol. 7, pp. 104-7, 1996.

[34] J. H. An, H. S. Kim, D. J. Chung, D. S. Lee, and S. Kim, "Thermal behavior of poly(.epsilon.-caprolactone)-poly(ethylene glycol)-poly(.epsilon.-caprolactone) tri-block copolymers," *J. Mater. Sci.*, vol. 36, pp. 715-722, 2001.

[35] F. O. Eschbach and S. J. Huang, "Hydrophobic-hydrophilic IPN and SIPN," *Polym. Mater. Sci. Eng.*, vol. 65, pp. 9-10, 1991.

[36] F. O. Eschbach and S. J. Huang, "Hydrophilic-hydrophobic binary systems: poly(2-hydroxyethyl methacrylate) and polycaprolactone," *Polym. Prepr. (Am. Chem. Soc., Div. Polym. Chem.)*, vol. 34, pp. 848-9, 1993.

[37].F. O. Eschbach and S. J. Huang, "Hydrophilic-hydrophobic binary systems of poly(2-hydroxyethyl methacrylate) and polycaprolactone. Part I: Synthesis and characterization," *J. Bioact. Compat. Polym.*, vol. 9, pp. 29-54, 1994.

[38] F. O. Eschbach, S. J. Huang, and J. A. Cameron, "Hydrophilic-hydrophobic binary systems of poly(2-hydroxyethyl methacrylate) and polycaprolactone. Part II: Degradation," *J. Bioact. Compat. Polym.*, vol. 9, pp. 210-21, 1994

[39] J. Crank, *The Mathematics of Diffusion*. London: Oxford University Press, 1975.

# Chapter 14

# ORGANOMETALLIC CONDENSATION POLYMERS AS ANTICANCER DRUGS

Deborah W. Siegmann-Louda(a), Charles E. Carraher, Jr.(a,b), Fred Pflueger(a), David Nagy(a), and John R. Ross(a)

(a) Florida Atlantic University, Department of Chemistry and Biochemistry, Boca Raton, FL 33431
(b) Florida Center for Environmental Studies, Palm Beach Gardens, FL 33410

## 1. INTRODUCTION

Cancer is one of the leading causes for adult death. Since the late 1970's we have worked towards the creation of more active and less toxic anti-cancer drugs, including more efficient drugs based on cisplatin, cis-dichlorodiaminoplatinum II. This parent platinum-containing drug is the most widely employed anti-cancer drug (for instance 1-4).

```
Cl Cl
\ /
-Pt-NH2-R-NH2-
```

platinum-containing polymer

We synthesized a variety of platinum-containing polymers as shown above from potassium tetrachloroplatinate and a diamine. The diamine component ranged from simple aromatic diamines to complicated compounds such as methotrexate. Many of these polymers showed significant biological activity, inhibiting the growth of several types of cells including cancer cells (5). Their effect upon cells resembled that of cisplatin, and their potential as anti-cancer drugs is currently being investigated.

More recently, we began testing a number of other metal-containing polymers to establish their biological activities. Activity against cancer cell lines was one of the biological screening techniques employed (for instance 6-8). Interestingly, some of these compounds have been found to inhibit the growth of a variety of cell types.

Our arguments for designing such polymeric drugs involve several features. First, because of their size, polymers are more limited with respect to their travel within the body. Passage though membranes depends upon a number of factors including size. Recent findings indicate, not surprisingly, that delivering the active drug to the specific site and limiting the mobility of the drug allow for reduced negative side effects and an enhanced activity of the drug at the delivery site. Second, coupling of two drug components allows synergetic effects to occur. Design of the components of the polymeric drug can allow the incorporation and subsequent activity of drugs that act differently but in a combined manner to eradicate the disease, or one might act to entice a cancer cell to accept part of the polymer combination with the second part carrying either a therapeutic or toxic "sting". Third, in the treatment of many diseases, including cancer, multiple attachments to the DNA or to a specific protein might increase the drug's chances of being successful, and the capacity for such multiple attachments increases for polymer drugs. Fourth, the polymeric drug can act in a controlled release mechanism that allows the drug to be present over an extended period of time. Fifth, most of these metal-containing condensation polymers exhibit poor stability to base but good stability to acid. If the drugs are to be taken orally, this allows the drug to bypass degradation in the stomach, which is acidic, and to move further along the digestive canal where the basic surroundings encourage polymer degradation. Through tailoring of the polymeric drug design, balances are sought between controlled release aspects and aspects related to advantages gained because of size.

These condensation polymers can be readily formed employing the interfacial polycondensation process. This process, under the right conditions and employing the right reactants, allows the rapid synthesis of the drugs (often within several seconds) under relatively mild conditions. Mild temperatures are particularly important when one of the reactants possesses thermally unstable sites. Below is a general representation for such a polymer made from the reaction of a dihalo or trihaloorganometallic reactant and a Lewis base-containing reactant, here an alcohol and an acid.

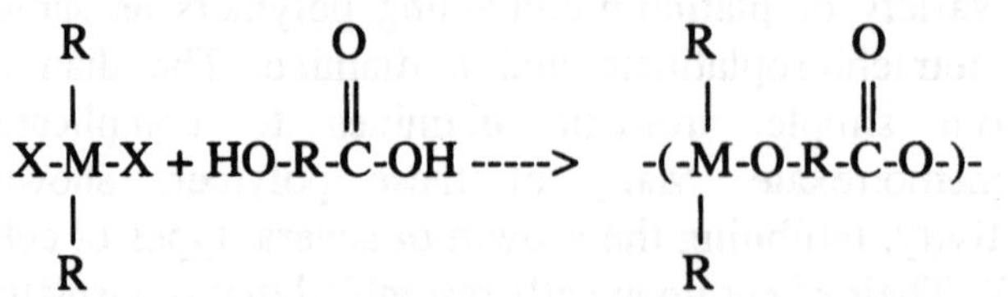

The Lewis bases can be amine-like groups (such as ureas, thioureas, hydrazines, hydrazides, and primary and secondary amines), thiols, hydroxyls, and salts of carboxylic groups. The reactant sites can be different or the same. For a linear polymer to occur there should be two reactive sites on each reactant. For some applications, crosslinking is permitted and reactants where functionalities are greater than two can be employed.

Recently we have emphasized the use of organotin-containing polymers. Tin-containing compounds of the form $R_2SnX_2L_2$ (where R = alkyl or aromatic group, X = halide, and L = oxygen or nitrogen donor ligand) have been synthesized as model compounds for cisplatin (9). Several show antitumor activity without the high nephrotoxicity often seen with platinum complexes. Structure-activity relationships indicate that butyl groups (R) in the compound produce a more effective drug than do ethyl or phenyl groups.

Here we investigate the biological activity of several organometallic condensation polymers, mainly tin-containing polymers, using cell lines. We have employed Lewis bases that are known to exhibit biological activity, as well as those that do not themselves exhibit biological activity, in an attempt to better understand the structure-activity window relationship. Our present efforts are aimed at designing polymers that show good activity against resistant cancer cells at low concentrations, and then studying the "best" candidates employing live animal studies.

## 2. EXPERIMENTAL

Normal Balb/3T3 cells (ATCC CCL163), which are contact-inhibited, non-tumorigenic cells, were employed. They were maintained in Dulbecco's Modified Eagle medium containing 10% calf serum at 35°C in a 5 % carbon dioxide atmosphere. Also used were human ovary adenocarcinoma cells (ATCC HTB161) grown in RMPI medium containing 10 μg/ml insulin and 20 % fetal bovine serum.

The testing of the compounds was done as previously described (6-8). Briefly, cells were harvested, counted, and plated into 35 mm dishes at a known concentration on day one. On day two, the cells are again counted. A stock solution of the compound was prepared in DMSO at a known concentration. On day two stock solutions, some containing the test compounds and other containing no added material (the controls), are diluted and added. The cells are again counted for the remaining days of study and inhibition/growth noted. Counting was done by removing the medium and adding 0.2 ml of 0.1 % trypsin in phosphate-buffered saline to each dish. The cells were incubated for 5 minutes until the cells detached. Then 0.8 ml of medium was added per dish, cells mixed, and aliquots counted using a Coulter counter.

## 3. RESULTS AND DISCUSSION

Testing often begins employing normal cell lines, here Balb/3T3 cells (mouse fibroblast cells). Compounds with anti-cancer activity generally show good ability to inhibit actively-growing normal cells. Compounds showing good activity are then tested against various cancer cell lines. Recently we have been employing so called "resistant" cell lines derived from patients that have already undergone various chemo treatments and where the cells have become resistant to the treatment(s). Generally, those compounds that show activity against non-resistant cell lines are also somewhat active against the resistant cell lines. For instance, for the polymer derived from reaction of diethyltin dichloride and ampicillin, good activity was found against the 3T3 cells down to a level of 5 micrograms/ml (Table 1), but a concentration of 50 μg/ml was required to partially inhibit the adenocarcinoma cells (Table 2).

*Table 1. Cell results employing Balb/3T3 cells for diethyltin dichloride, ampicillin, and the polymer derived from their reaction employing concentrations for 5(*) and 10 (**) micrograms/ml. All values are to be multiplied by 10,000 cells/dish. (from reference 8)*

| Day | DeSnDCl (*) | DeSnDCl (**) | AMP-POLY (*) | AMP-POLY (**) | Control |
|---|---|---|---|---|---|
| 1 | 5 | 5 | 5 | 5 | 5 |
| 2 | 12 | 12 | 12 | 12 | 12 |
| 3 | 19 | 12 | 23 | 20 | 28 |
| 4 | 29 | 9 | 45 | 35 | 69 |
| 5 | 40 | 9 | 87 | 62 | 163 |

*Table 2. Cell results employing HTB161 cells for the polymer derived from reaction of diethyltin dichloride and ampicillin. All values are to be multiplied by 10,000 cells/dish. (from reference 8)*

| Day | Control | Polymer 50 μg/ml | Ampicillin 50 μg/ml | DeSnCl 20 μg/ml |
|---|---|---|---|---|
| 1 | 20 | 20 | 20 | 20 |
| 3 | 36 | 36 | 36 | 36 |
| 5 | 78 | 72 | 76 | 16 |
| 7 | 110 | 96 | 103 | 12 |
| 9 | 188 | 152 | 194 | 12 |
| 11 | 264 | 205 | 268 | 11 |

We have also been looking at the effect of the various groups present on the tin. It appears that aliphatic groups are best (please compare results in Table 3). We have also found, by testing a series of polymers with different R-groups, that the use of the dibutyltin moiety can act to decrease the concentration needed for good activity by 10 fold or more, so that good activities can be found for agents at concentrations of less than 1.0 μg/ml. These findings are consistent with previous results involving monomeric tin compounds, where butyl groups were also found to confer superior biological activity compared to phenyl or ethyl groups (9).

*Table 3. Balb 3T3 cell results for the product of phenylmethyltin dichloride and adipic acid (left side) and diethyltin dichloride and adipic acid (right side). All values are to be multiplied by 10,000 cells/dish. (concentration of products tested was 25 μg/ml)*

| Day | Control | PhMeSn-Adipic Acid | Day | Control | DiEtSn-Adipic Acid |
|---|---|---|---|---|---|
| 1 | 5 | 5 | 1 | 5 | 5 |
| 2 | 11 | 11 | 2 | 14 | 14 |
| 3 | 37 | 26 | 3 | 31 | 22 |
| 4 | 81 | 74 | 4 | 60 | 34 |
| 5 | 145 | 147 | 5 | 129 | 60 |

The nature of the organometallic and Lewis base portions are both important in determining whether the materials will inhibit cell growth and to what extent. Tables 4 and 5 contain results for three different materials. Product 1 is derived from the reaction of titanocene dichloride and ampicillin. We have previously shown that selected organotin derivatives of ampicillin inhibit both Balb 3T3 cells and resistant cancer cell lines (8). Here, the polymer exhibited essentially no inhibition of cell growth.

The second compound is derived from triphenylantimony dichloride and thiopyrimidine. Here we were looking at the influence of having the Lewis base be a purine or pyrimidine containing an element, sulfur, that is needed, but which is not naturally widely found. These products are potent antibacterial agents (10). Whether the reasoning in designing the drug was correct or not, this product offers good inhibition of the cells.

The third product was derived from triphenylantimony dichloride and cobalticinium-1,1'-dicarboxylic acid (11, 12). It was synthesized in an attempt to vary the solubility and properties as a function of the associated anion. Here the associated anion is the nitrate ion. This product offered moderate inhibition of the cells.

Results in Tables 4 and 5 for the product of triphenylantimony dichloride and thiopyrimidine illustrate the typical concentration-inhibition pattern. Thus, as the polymer concentration decreased from 25 to 10 to 7.5 to

5 μg/ml, cell inhibition decreased from about 96 % inhibition to 40 % to 20 % and finally to only 8 % inhibition.

*Table 4. Balb 3T3 cell results for the products of titanocene dichloride and ampicillin, Ti-Am; triphenylantimony dichloride and thiopyrimidine, Sb-TPy; and triphenylantimony dichloride and cobalticinum-1,1'-dicarboxylic acid nitrate, Sb-Co. Results are to be multiplied by 10,000 cells/dish.*

| Day | Control | Ti-Am (12.5 μg/ml) | Sb-TPy (25 μg/ml) | Sb-Co (12.5 μg/ml) |
|---|---|---|---|---|
| 1 | 5 | 5 | 5 | 5 |
| 2 | 19 | 19 | 19 | 19 |
| 3 | 46 | 47 | 17 | 33 |
| 4 | 131 | 134 | 12 | 106 |
| 5 | 292 | 290 | 12 | 195 |

*Table 5. Balb 3T3 cell studies for the product of triphenylantimony dichloride and thiopyrimidine at various concentrations. Results are to be multiplied by 10,000 cells/dish.*

| Day | Control | Sb-TPy (5 μg/ml) | Sb-TPy (7.5 μg/ml) | Sb-TPy (10 μg/ml) |
|---|---|---|---|---|
| 1 | 5 | 5 | 5 | 5 |
| 2 | 10 | 10 | 10 | 10 |
| 3 | 27 | 25 | 24 | 20 |
| 4 | 118 | 115 | 91 | 80 |
| 5 | 201 | 185 | 164 | 120 |

These results show the importance of the metal, the R-groups, and the Lewis base portions of the polymer in influencing the biological activity of the compound. More experiments with additional polymers are needed to determine the broad principles governing the relationship between polymer structure and biological activity. Other factors also need to be considered, such as polymer size. Furthermore, the results of this and other studies are consistent with the polymers acting as more than simply controlled release agents. For instance, in cases where the Lewis base portion itself exhibits no activity, the polymer can be very active. There are also examples where the polymer exhibits greater biological activity than that of the organometallic halide itself. It is known that a major source of inhibition by organotin dichlorides is the hydrolysis and formation of hydrochloric acid that in turn actively inhibits cell growth on its own. To what extent this type of process may be occurring with the polymers remains to be determined. However, the polymers themselves appear to be involved in the inhibition of cell growth. This may be due to binding of the polymer to key biomolecules, the ability of the polymer to cross the cell membrane, the conversion of the polymer into smaller, active products, or some other route or combination of routes.

## 4. REFERENCES


1. Carraher, C.; Scott, W.; Schroeder, J.; Giron, D.; J. Macromol. Sci.-Chem., 1981, A15(4), 625.
2. Carraher, C.; Scott, W.; Giron, D.; Bioactive Polymeric Systems, Plenum, NY, Chpt. 20, 1985.
3. Siegmann, D.; Carraher, C.; Friend, A.; J. Polymer Materials, 1987, 4, 19.
4. Siegmann, D.; Carraher, C.; Brenner, D.; Progress in Biomedical Polymers, pages 371-388, Plenum, 1990.
5. Siegmann, D.; Brenner, D.; Colvin, A.; Polner, B.; Strother, R.; Carraher, C.; Inorganic and Metal-Containing Polymeric Materials, pages 335-661, Plenum, 1990.
6. Carraher, C.; Li, F.; Siegmann-Louda, D.; Butler, C.; Harless, S.; Pflueger, F.; PMSE, 1999, 80, 363.
7. Siegmann-Louda, D.; Carraher, C.; Pflueger, F.; Coleman, J.; Harless, S.; Luing, H.; PMSE, 2000, 82, 83.
8. Siegmann-Louda, D. Carraher, C., Ross, J.; Li, F.; Mannke, K.; Harless, S.; PMSE, 1999, 81, 151.
9. Crowe, A.; Smith, P.; Atassi, G.; Inorg. Chem. Acta, 1984, 93, 179.
10. Carraher, C.; Naas, M.; Giron, D.; Cerutis, D., J. Macromol. Sci.-Chem.., 1983, 8, 1101.
11. Sheats, J.; Blaxall, H.; Carraher, C.; Polymer P., 1975, 16(1), 655.
12. Carraher, C.; Venable, W.; Blaxall.; Sheats, J.;, J. Macromol. Sci.-Chem., 1980, A14(4), 571.

Chapter 15

# SYNTHESIS AND STRUCTURAL CHARACTERIZATION OF CHELATION PRODUCTS BETWEEN CHITOSAN AND TETRACHLOROPLATINATE TOWARDS THE SYNTHESIS OF WATER SOLUBLE CANCER DRUGS

Charles E. Carraher, Jr., Ann-Marie Francis, and Deborah W. Siegmann-Louda
*Florida Atlantic University, Department of Chemistry and Biochemistry, Boca Raton, FL 33431, and Florida Center for Environmental Studies, Palm Beach Gardens, FL 33410*

## 1. INTRODUCTION

### 1.1 Chitosan

Chitosan is a polysaccharide with a structure similar to that of cellulose. It is the deaceylated product of chitin, the main structural component in the cuticles of crustaceans, insects, mollusks, and in the cell wall of selected pathogens. Because it is a ready derivative of chitin it is potentially available in very large quantities inexpensively. The primary structure of chitosan contains a backbone of (1-4)β -D-glucose residues. Chitosan has no physiological toxicity with a LD50 greater than 16 g/kg of body weight, similar to that of sucrose or common salt It is certified for food application outside the USA. Drugs containing chitosan should offer a minimum toxicity associated with the chitosan itself. It is extensively used as a component in fat-reduction pills.

A number of derivatives of chitosan have been made. These include graft copolymers. While chitosan itself is an effective flocculating agent in only acid, derivatives having poly(acrylic acid) side chains exhibit high flocculation ability in both acid and base solutions because of the zwitterionic character of the graft copolymer (1). Chitosan-graft-polystyrene shows better adsorption of bromine than chitosan and films show less swelling and better elongation than chitosan films (2). Chitosan-graft-polyaniline products can form films and fibers (3). Other graft polymers have also been made that offer potentially useful properties (4). A number of other derivatives of chitosan have been made (for instance 5,6).

Chitosan can form a complex network when subjected to multivalent cations with the cations chelating largely through the amine groups. Interfacial gelation of chitosan with poly(acrylic acid) and heparin forms ultrathin membranes and fibers (7).

Chitosan and its derivatives are potentially useful in a number of areas including medical, cosmetics, biotechnology, food industry, agriculture, environmental protection, paper industry, textiles, etc. It is a widely underused material with good properties (8).

## 1.2. Platinum-Containing Anticancer Drugs

For more than 25 years, platinum-containing coordination complexes have been employed in the remedial treatment of cancer.

Cisplatin has firmly established itself as a prominent anticancer agent available to clinical oncologists (for instance 9,10). It is one of the most successful anti-tumor drugs with expenditures of about $500 million per year (11). It offers a wide spectrum of activity, being effective in 70-90 % cases of testicular cancer. It is also highly effective against ovarian cancer and contributes to the treatment of neck and head cancers (12-14). It shows good synergy with many other chemotherapeutic drugs such as 5-flurouracil, cytarabine, methotrexate, and bleomycin. Taken together with these drugs, cis-DDP can be active against refractive diseases including melanoma and breast cancer (15,16).

While thousands of cisplatin analogues have been synthesized and screened, only about 28 have entered the clinical trials as anticancer agents. Of these, four are currently approved-cisplatin, carboplatin, oxlinplatin, and nedaplatin.

The structure of cisplatin is well known. The platinum is present in a +2 oxidation state. It has a coordination number of four with a square planar arrangement of ligands with the amines (ammonias) present cis to one another. In general, a metal-nitrogen bond has a stronger sigma overlap than a metal-halogen bond making the ammonia ligands less easily displaced in comparison to the chloride ligands. It is believed that for platinum derivatives to be active against cancer the product must contain two labile ligands that are cis to one another as in the case of cisplain. It is reasonable to assume that the di-nitrogen containing sites of the present products have structures analogous to that of cisplatin.

Reaction of the tetrachloroplatinate with chitosan should occur through the amine groups, leaving the hydroxyls free to be aquated. The presence of the platinum moiety should lessen the tendency of the chitosan to form inter and intra chain bonds enhancing the ability of water to approach and dissolve the product. Both factors might allow the cis-dichloroplatinum II-containing chitosan to be water soluble.

In aqueous solution cis-DDP is known to undergo spontaneous hydrolysis. The reaction produces species such as monoaquo platinum and diaquo platinum complexes arising from nucleophilic substitution in water (17-19). In addition to the diaquo complex, the aquated species can also exist in the form of the dihydroxy complexes or the aquo hydroxy complex. The actual form is pH dependent (20,21). For instance, at a pH of 7.4, 85% of the material will exist in the dihydroxy form while at a pH of 6.0, 80% will be in the monohydroxy monoaquo form.

Thus, the possible number of aquated forms derived from cis-DDP is great and the proportion dependent on such variations as time, pH, temperature, and concentration of associated reactants (chloride ion and ammonia) (17).

While the active form within the cell is believed to be the monohydrated structure below, the "preferred" extracellular species contains two cis-oriented leaving groups which arc normally chloride ligands. Fortunately, due to the high chloride ion concentration in blood ($10^3$ mM) these leaving groups will remain in position causing the molecule to be electrically neutral until it enters the cell membrane (19).

$H_3N$ $OH_2^+$
\ /
Pt
/ \
$H_3N$ Cl

Cis-DDP enters cells by diffusion where it is converted to an active form. This is due to the lower intracellular chloride concentration (about 4 mM) which promotes ligand exchange of chloride for water, and thus formation of the active aquated complex (22). Thus, the platinum-containing complex should be neutral to enter the cell and labile chloride groups need to be present to form the active species within the cell. The antineoplastic activity of cis-DDP appears to be related to its interaction with DNA nucleotides as a monoaquo species (14). The monohydrated complex reacts with the DNA nucleotides forming intra and interstrand crosslinks. Of the four nucleic acid bases, cis-DDP has been shown to preferentially associate with guanine. The most common are intrastrand crosslinks between adjacent guanines (10).

There are several possible crosslinks with DNA. One favored option occurs between the 6-NH group of adenine on opposing strands in the A-T rich region. This is because these groups are approximately 3.5 A apart, close to the 3 A distance between the cis leaving groups on the platinum. The second favored option is crosslinking occurring between the amino groups of guanine and cytosine in opposing strands. This is favored because the platinum is at right angles to the bases that in turn are coplaner with one another. This implies that the bases will have to either "bend down" or "turn edge" to achieve the necessary configuration to bind to the platinum complex. This binding pattern is believe to lead to perturbation of the tertiary structure and minor disruption of the double helix. This is sufficient to cause inhibition of cell replication and transcription with eventual cell death, yet too small to cause a response by damage recognition proteins and consequent excision of the affected segment and repair of the stand (23).

Initially, it was believed that atom six of the guanine was the site of importance, being linked with both carcinogenesis and the mechanism of antitumor activity. This belief was amended when tests on guanine-like structures (such as 1,3,9-trimethylxanthine) showed that cis-DDP bound to the $N^7$ and not atom six of guanine. Further studies showed that cis-DDP preferentially bonded at $N^7$ of purines adenine and guanine, and at $N^3$ of pyrimidines cytosine and uracil (24).

Despite the unquestionable success story of cis-DDP, limitations remain including the powerful toxic side effects (24). These toxic side effects include gastrointestinal problems such as acute nausea, vomiting and diarrhea; occasional liver dysfunction; myelosuppression involving anemia, leukopenia and thrombocytopenia; nephrotoxicity, and less frequently cases of immunosuppression, hypomagnesia, hypocalcemia, and cardiotoxicity (25).

The most serious side effect is damage to the kidney. Much of the administered cis-DDP is filtered out of the body within a few hours, exposing the kidneys to bursts of high concentrations of platinum (26). The rapid rate at which the kidneys filter the platinum from the blood is believed to be responsible for the kidney problems. Another problem is the cumulative and irreversible hearing loss experienced first in the 4000-8000 Hz range and then later in the 1000-4000 Hz range. Complete deafness may occur just prior to death (27).

In general, for platinum-containing compounds, the best activities are found for complexes that contain two anionic leaving groups such as chloride, bromide, oxalate, or malonate. Complexes with more labile ligands, such as the nitrate ion, hydrolyze too quickly to permit them to be useful in vivo, and other ligands such as cyanide ion bind too tightly to platinum to be active (12,16,28). Further, the complexes should be of cis geometry and be neutral, containing relatively inert amine or amine donor groups (16). Some exceptions have been found (for instance 12,16).

We have been involved in the synthesis and biological characterization of polymeric analogues of cisplatin (cis-DDP; cis-diaminedichloroplatinum (II)) the most widely used drug employed in the treatment of cancer (for instance 29-32). Cis-DDP and the analogous platinum polyamines are made from the reaction between amine (or ammonium in the case of cis-DDP) and tetrachloroplatinate (II) giving exclusively the cis product illustrated below.

```
Cl   Cl
 \   /
  Pt
 /   \
      NH2-R-NH2-
```

Some of these products have shown equal to greater inhibition of certain cancers with lowered toxicities in cell lines and animal tests compared to cis-DDP (for instance 29). Unfortunately, these drugs are not water soluble but are delivered using DMSO and DMSO-water mixtures. The preferred method of delivery of chemo drugs is via aqueous solution employing IVs. The present project is aimed at creating water-soluble drugs.

Neuse and coworkers (33,34) and others have also been involved in the synthesis of water soluble analogues of cis-DDP. Recent work has involved designing carriers containing biodegradable polyamide backbone structures with attached hydrosolubilizing tert-amine side groups such as the use of presynthesized polyasparatamide carriers grafted with methoxy-terminated poly(ethylene-propylene) side groups. (33).

The goal of the current effort is the synthesis of chemical derivatives of cisplatin (cis-DDP; cis-diaminedichloroplatinum II) with lowered toxicity and (hopefully) enhanced activity (for instance 29). The synthesis of polymeric derivatives is based on the analogous reaction employed in the formation of cis-DDP except using diamines in place of ammonia.

```
 H   H                             H   H
 |   |                             |   |
H-N-R-N-H + K2PtCl4   ----->   -HN-R-NH
                                    \ /
                                    Pt
                                    / \
                                  Cl  Cl
```

Our approach involves inclusion of a cisplatin-like moiety into a polymer as shown above. It is hoped that inclusion of this moiety into a polymer will achieve the following:

1. Llimit movement of the biologically active drug. Because of their size polymers are not as apt to easily pass through membranes present in the body.

2. Enhance activity through an increased opportunity for multiple bonding at a given site.

3. Increase delivery of the bioactive moiety. Cisplatinum hydrolyzes in the body forming a wide variety of platinum-containing agents, none of which is as active as cisplatin itself, only one of which shows any activity against cancer

cells, and most of which exhibit increased toxicity to the body. It is believed that the polymeric nature of the drug will "protect" the active portion, both though steric constraints restricting the approach of water to the active site and because the polymer is not as hydrophillic as the cisplatin itself as shown by the lack of water solubility of most cisplatin-containing polymers.

## 2. EXPERIMENTAL

The following reagents were used without further purification: potassium tetrachloroplatinate (J and J Materials, Neptune City, NJ) and low molecular weight beta(1-4)-2-amino-2-deoxy-D-glucose (chitosan; Fluka Biochemika Chemicals, Switzerland).

The platinum-chitosan complexes were synthesized by mixing together solutions containing the two reactants. Chitosan solutions were prepared by dissolving the chitosan (1.00 mmole) in 30 ml of 0.100 M HCl aqueous solution (deionized water used throughout). The mixture was heated to about 50° C until a clear viscous solution formed. The mixture was cooled to room temperature and 0.10 M NaOH added to adjust the pH. An aqueous solution of potassium tetrachloroplatinate (1.00 mmole) in 15 ml water was added to the chitosan solution. The mixture was stirred for three days at room temperature. The product was collected employing suction filtration, washed repeatedly with water, transferred onto a glass petri dish, and air dried. The product is a light to medium brown-colored solid.

Solubility tests were conducted by placing 1 mg of product in 9 ml of liquid and observed over a period of several weeks. Elemental analysis of chloride was achieved using sodium fusion followed by precipitation of the chloride as silver chloride. Percentage platinum was determined using thermal analysis and energy dispersion analytical X-ray analysis (EDAX).

Molecular weights were determined employing a Brice-Phoenix Model BP-3000 Universal Light Scattering Photometer. Refractive index measurements were obtained using a Bausch and Lomb Abbie Refractometer Model 3-L. Infrared spectra were obtained employing potassium bromide pellets using a Mattson Instrument Galaxy 4020 FT-IR. Mass spectroscopy was carried out at the Washington University Center for Biomedical and Bioorganic Mass Spectroscopy (St. Louis). A VG-Micromass ZAB-SE 2 sector instrument was employed. Operating conditions included 1000 resolving power over a mass range of 35-1000 Daltons with continuous scan acquisition using a water-cooled solids probe with a heating rate of 120 C/min to 450 C until evolution of sample vapor decreased below 10 % of the maximum. The solids probe is fitted with a special copper tip and thin wall melting point capillary sample cups to optimize thermal contact between sample and probe heater.

## 3. RESULTS AND DISCUSSION

### 3.1. Synthesis

In solution, the structure of the chitosan chain is dependent upon a number of factors including ionic strength, concentration, time , nature of the medium, and pH. Here we will focus on changes as a function of pH. A major factor is the presence, or absence, of charges on the amine. The pK of the chitosan amine is 6.3. Below this pH the amine on the chitosan becomes increasingly protonated and it might be assumed the chain becomes extended, approaching a stiff helix. In truth, the precise structure is more complex. For the present study, solubility is achieved through addition of acid and then base, sodium hydroxide, is added to neutralize the amine allowing it to act as a Lewis base chelating agent. As excess acid is neutralized, the ionic strength of the solution increases and the size of the chitosan chain aggregates decrease. Deprotonation of the amine becomes significant above a pH of 5.2 where upon free amino groups form intermolecular hydrogen bonds with the hydroxyl groups of adjacent chains. Aggregate size increases up to a pH of 6.3 where precipitation normally occurs.

The reaction between tetrachloroplatinate II and chitosan was carried out as a function of chitosan solution pH and ratio of reactants. Results appear in Table 1.

*Table 1. Product yield from the chelation reaction between chitosan and potassium tetrachloroplatinate II.*

| *Sample pH** | *Tetrachloro-platinate (mmol)* | *Chitosan (mmol)* | *Yield (g)* | *Yield (%)* | *Product Number* |
|---|---|---|---|---|---|
| *5-6* | *1.00* | *1.00* | *0.026* | *44* | *1* |
| *High*** | *1.00* | *1.00* | *0.029* | *49* | *2* |
| *Low**** | *1.00* | *1.00* | *0.56* | *56* | *3* |
| *High*** | *2.00* | *1.00* | *0.47* | *49* | *4* |
| *Low**** | *1.00* | *2.00* | *0.46* | *53* | *5* |

**pH of chitosan solution, **pH >6, ***pH <5*

Below is an abbreviated structure of some of the possible unit structures that include inter and intra-molecularly "di-bonded" platinum products that are similar to the cis-DDP structure (structures A and B below), the mono-bonded product (structure C), and finally unreated units (structure D). Depending upon the particular reaction conditions, the resulting product probably contains varying ratios of all four units.

```
       A             B         C          D
     C-O-C    C-O-C   C-O-C   C-O-C   C-O-C
 \   /     \  /    \  /    \  /    \  /   \
  C-C       C-C      C-C     C-C     C-C
   |         |        |       |       |
  HNH       HNH      HNH     HNH     HNH
    \       /         |       |
       Pt          Cl-Pt-Cl Cl-Pt-Pt
      /  \            |       |
     Cl  Cl          HNH      Cl
                      |
                     C-C
                    /   \
                C-O-C   C-O-C
```

Product yields were approximately constant being about 50 % based on a di-chelated product and percentage of platinum found. Structure A represents the preferred structure containing two labile Pt-N sites. Structure B also contains two labile sites in the cis position, but represents a crosslink or internally cyclized structure, both giving a product with decreased solubility. The amount of non-adjacent same-chain-chelation should be low since the chitosan chain is a fairly rigid helix in solution, limiting the close association of different parts of the same chitosan chain with itself.

## 3.2 Structural Characterization

Solubility was studied in a number of liquids. Table 2 contains results for DMSO and water. Lack of complete solubility is consistent with the product containing at least some crosslinks. The molecular weight for some of the samples was determined and is given in Table 3. Chitosan has a reported molecular weight (for the sample employed for these studies) of 70,000. By light scattering photometry the molecular weight of the employed chitosan was determined to be about 60,000 in near agreement with that reported by the manufacturer. This corresponds to an average chain containing about 380 hexose units.

*Table 2. Solubility of chitosan-tetrachloroplatinate II product after two days.*

| *Product Number** | *Solubility in Water-%* | *Solubility in DMSO-%* |
|---|---|---|
| 1 | 12 | 34 |
| 2 | 4 | 17 |
| 3 | 2 | 7 |
| 4 | 25 | 48 |
| 5 | 85 | 87 |

**please see Table 1.*

*Table 3. Molecular weight for products determined in DMSO.*

| *Product Number** | *%-Soluble in DMSO* | *Molecular Weight* |
|---|---|---|
| Chitosan | | $6.0 \times 10^4$ |
| 1 | 34 | $4.6 \times 10^7$ |
| 2 | 17 | $1.0 \times 10^6$ |
| 3 | 7 | $1.6 \times 10^5$ |
| 4 | 48 | $2.5 \times 10^9$ |
| 5 | 87 | $8.9 \times 10^5$ |

**Please refer to Table 1.*

Product 1 gives a molecular weight of 1 x $10^6$ after being left in DMSO for four weeks. These results are consistent with the following calculations. A chain fully substituted with Cl-Pt-Cl would have a molecular weight of about 111,000. The number of chains that would have to be connected thorugh crosslinking is calculated by dividing the molecular weight found by the molecular weight per chain and is about 9.

Since the elemental analysis indicated that the chain is not fully substituted, but only about 35 % substituted, then the new chain would have a calculated molecular weight of 78,000 and the number of chains connected would be about 13.

The amount of crosslinking needed to make a polymer insoluble varies but is generally in the range of 1 to 5 % crosslinking per unit or 1 to 5 per 100 units. We can make further calculations that allow us to get an impression of the number of crosslinks and number of Cl-Pt-Cl's per chain. Again, assuming only a 35 % reaction (substitution) we would have 66 Cl-Pt-Cl's per uncrosslinked or original chain of which only about 2 to 3 of the platinum units need to be involved in crosslinking assuming that only 1 % crosslinking is present. This indicates that the vast majority, about 95 %, of the platinum sites are internally bonded rather than acting as a crosslink (assuming only units of A, B, and D). Infrared spectra of the chelation products prepared using different molar ratios and different pHs are similar. The spectra contain bands characteristic of the presence of both reactants and the presence of new bands consistent with the formation of the Pt-N moiety. The number of Pt-N stretching bands is often taken as evidence of the geometry of the product. Trans-platinum diamines exhibit one Pt-N band, while cis-products exhibit two bands. The second band is weak and sometimes missed, and while the presence of two bands indicates the cis geometry, presence of only one band is not firm evidence for the trans geometry. The products contain two Pt-N associated stretching bands, one at 564 (all bands given in 1/cm) and one at 497, consistent with the products being in the cis conformation.

A strong band centered around 3430 is assigned to O-H stretching. It remains unchanged in shape, location, and intensity from chitosan itself. The band around 3000 is assigned to N-H deformation. The presence of bands characteristic of O-H stretching and the presence of new Pt-N bands are consistent with reaction occurring through the amines and not the hydroxyl groups. The bands about 1071 are assigned to the C-O stretch in chitosan (both within the hexose ring and the connective ether or acetal linkage).

Mass Spectroscopy was carried out on the products. The results for the products are similar. Here we will focus on the product produced at high pH (product 1 in Table 1).

Chlorine has two isotopes present at 35 (all masses are given at m/e = 1 and in amu or Daltons) and 37 in about a 3:1 ratio (natural abundance 35 = 75 %; 37 = 25 %). Since the mass spectroscopy unit used identifies positive ions the intensity for the HCl (36, 38) ions are greater than those associated with Cl itself (35, 37) because of the difficulty of forming positive Cl ions compared with HCl. This tendency is found in the present spectra with the relative intensities for Cl (35), HCl (36), Cl (37) and HCl (38) being 2.68, 14.77, 1.07, and 4.26 with the percentage Cl from Cl 35 being 71.5 % and from HCl (35) being 77.6 % in rough agreement with the natural abundance of Cl (35). Thus, the data is consistent with the presence of Cl.

Platinum has six natural isotopes, four present in abundances of 1 % and greater. Compounds containing both Pt and Cl will contain isotopic abundance distributions dictated by the relative abundances of each isotope. Tables 1-3 contain the expected isotope abundances for ion fragments that indicate the presence of the formation of the amine-complexes with platinum containing one, two, and three chlorides, that is N-PtCl, N-$PtCl_2$ , and N-$PtCl_3$ . (The product will also contain unreacted chitosan units.) The matches are in reasonable agreement with those predicted from the relative isotopic abundances for each ion fragment grouping considering the low intensities found for the higher molecular weight ion fragments and the fact that the intensity of each ion fragment will be divided among the various isotopic species.

*Table 1. Relative percentage isotopic abundance match for PtCl for the ion fragment derived from one half chitosan unit minus Me-OH, OH, and plus PtCl.*

| *m/e* | *325* | *326* | *327* | *328* | *329* | *330* |
|---|---|---|---|---|---|---|
| *Rel % Found* | 23 | 21 | 35 | 7 | 10 | 4 |
| *Rel % Calc.* | 25 | 25 | 27 | 8 | 12 | 2 |

*Table 2. Relative percentage isotopic abundance match for $PtCl_3$ NH.*

| *m/e* | *315* | *316* | *317* | *318* | *319* | *320* | *321* |
|---|---|---|---|---|---|---|---|
| *Rel % Fd.* | *16* | *18* | *29* | *10* | *13* | *5* | *8* |
| *Rel % Cal.* | *14* | *14* | *25* | *14* | *18* | *5* | *7* |

Special effort was made to identify higher (>500 Daltons) molecular ion fragments. Here, only intact chitosan rings were considered. Of the 75 ion fragments between m/e 500 to 917 (highest ion fragment recorded) 71 % (or 53) were associated with the presence of Pt (Table 4). This represents 77 % of the ion fragmentation abundance for ion fragments above 500 Daltons. Identified non-Pt containing ion fragments accounted for 17 % of the ion fragmentation abundance. The (higher molecular weight) ion fragmentation abundance values are consistent with a high "loading" of platinum onto the chitosan units with an ion fragmentation abundance ratio of platinum-containing chitosan units/unsubstituted chitosan units = 4.5.

*Table 3. Relative percentage isotopic abundance matches for selected dichloro-platinum compounds.*

| | | | | | |
|---|---|---|---|---|---|
| *$PtCl_2(NH_2)_2$* | | | | | |
| *m/e* | *295* | *296* | *297* | *298* | *299* |
| *Rel % Found* | *20* | *16* | *34* | *11* | *19* |
| *Rel % Calc.* | *18* | *19* | *27* | *13* | *6* |
| *$PtCl_2NH_4$* | | | | | |
| *m/e* | *282* | *283* | *284* | *285* | *286* |
| *Rel % Found* | *17* | *21* | *27* | *14* | *20* |
| *Rel % Calc.* | *18* | *19* | *27* | *13* | *16* |

Elemental analysis was carried out on the various products. EDAX analysis was carried out on product 1 (Table 1). EDAX analysis gave a % Cl of 28.14 and Pt = 71.86 which is consistent with an approximate 2:1 Cl:Pt ratio. While this is what we would predict from our proposed structure, EDAX looks at only the surface and the "surface" structure may be different from the "bulk" structure. Even so, it does show that both Pt and Cl are present in the sample.

Wet and thermal analysis for Cl and Pt results are given in Table 5. In general, the results are consistent with the products containing a majority of the dichelated repeat units. The extent of reaction was calculated using the elemental analysis results for Pt. Extent of chelation varied from 9 to 35 %. As expected, the percentage chelation was less for the water soluble portions because the water soluble portion would be expected to be less crosslinked and (probably) lower molecular weight because of a lower extent of chelation by the tetrachloroplatinate.

The percentage chelations ranged from 7 to 43 %. Partial chelation is expected for several reasons. First, the product precipitates from the reaction mixture, isolating unreacted sites from a ready source of unreacted tetrachloroplatinate II. Second, steric factors generally restrict complete substitution on polymers.

*Table 4. Identification of platinum-containing higher (>500 Daltons) molecular weight ion fragments.*

| *m/e* | *Ion Fragment* | *m/e* | *Ion Fragment* |
|---|---|---|---|
| 504 | 1 C+PtCl-OH | 654 | 1.5C+ClPtCl-OHMeOH |
| 508 | 1C+ClPt$_2$-OHMeOH | 667 | 1.5C+PtCl |
| 521 | 1C+PtCl | 689 | 1.5C+PtCl$_3$ -OHMeOH |
| 525 | !CPt$_2$ -MeOH | 702 | 1.5C+Pt$_2$ |
| 539 | 1C+Pt$_2$-OH | 706 | 1.5C+PtCl$_3$-MeOH |
| 543 | 1C+PtCl$_3$-HOMeOH | 720 | 1.5C+PtCl$_3$ -OH |
| 556 | 1C+Pt$_2$ | 730 | 2C+Pt-HOMeOH |
| 560 | 1C+PtCl$_3$-MeOH | 765 | 2C+PtCl-HOMeOH |
| 574 | 1C+PtCl$_3$-OH | 778 | 2C+Pt |
| 591 | 1C+PtCl$_3$ | 782 | 2C+PtCl-MeOH |
| 601 | 1.5C+Pt-MeOH | 817 | 2C+Pt$_2$ -MeOH |
| 615 | 1.5C+Pt-OH | 852 | 2C+PtCl$_3$-MeOH |
| 632 | 1.5+Pt | 866 | 2C+PtCl$_3$-OH |
| 636 | 1.5C+PtCl-MeOH | 924 | 2.5C+Pt |
| 650 | 1.5C-OH | | |

**1C indicates two hexose rings; 1.5C represents three hexose rings; 2C represents four hexose rings; etc.*

In summary, the mass spectral results are consistent with the coordination of platinum, through the amines, onto the chitosan. The results are also consistent with the presence of at least the $PtCl_2$ and $PtCl_3$ being coordinated to the chitosan. IR data indicates the formation of Pt-N bonds with a cis geometry. Elemental analysis results are consistent with the presence of an approximate 2:1 ratio of Cl:Pt and a low to moderate extent of chelation.

*Table 5. Elemental analysis results for the products of chitosan andtetrachloroplatinate II formed at different pHs.*

| *pH** | *%Pt* | *%Cl* | *Ratio Cl/Pt* | *ChelationProduct %* | *Number*** |
|---|---|---|---|---|---|
| *Results for total sample* | | | | | |
| *5-6* | *16.4* | *7.1* | *2.3/1* | *35* | *1* |
| *High* | *16.4* | *5.7* | *2/1* | *35* | *2* |
| *Low* | *5.2* | *2.1* | *2/1* | *9* | *3* |
| *High**** | *4.9* | – | – | *9* | *4* |
| *High***** | *19.1* | *6.9* | *2/1* | *43* | *5* |
| *Results for water soluble portions only* | | | | | |
| *5-6* | *12.7* | *5.1* | *2/1* | *23* | *1* |
| *High**** | *4.2* | – | – | *7* | *4* |
| *High***** | *12.4* | *3.5* | *2/1* | *22* | *5* |

**pH of chitosan solution, **From Table 1 , ***Reaction carried out using a 2:1 ration of tetrachloroplatinate to chitosan, ****Reaction carried out using a 1:2 ratio of tetrachloroplatinate to chitosan*

# 4. REFERENCES


1. Kin, Y. B.; Jung, Y. B.; Kang, Y.; Kim, K. S.; Kim, J.; Kim, K. H., Pollino 13, 126 (1989).
2. Shigeno, Y.; Kindo, K.; Takemoto, K., J. Macromol. Sci. Chem A17, 571 (1982).
3. Berkovich, L.;Tsyurupa, M.; Davankov, V., J. Polymer Sci. Polym. Chem Ed. 21, 1281 (1983).
4. Kurita, K., **Polymeric Materials Encyclopedia**, page 1205, CRC, Boca Raton, FL 1996.
5. Van Luyen, D.;Huong, D., **Polymer Materials Encyclopedia**, J. Salamone, Ed., 1208, CRC, Boca Raton, FL 1996.
6. Domard, A.; Rinado,M.; Terrassin, C., Int. J. Biol. Macroml. 8, 105 (1986).
7. Dutkiewicz,J.; Tuora, M.; Jugkiewicz, M.; Ciszewski,R. in **Advances in Chitin and Chitosan**, C. J. Brime, et al. Eds., Elsevier, NY 1992.
8. Pavlath, A.; Wong, D.; Robertson, G., **Polymer Materials Encyclopedia**, 1230, CRC, Boca Raton, FL 1996.
9. Cadwell, G.; Neuse, E., J. Inorganic and Organometallic. Polymers, 7, 117 (1997).
10. Murry, V.; Whittaker, J., J. Biochimica et Biophysica Acta, 1354, 261 (1997).
11. Lebwohl D.; Canetta, R., European J. Cancer, 34, 1522 (1998).
12. Hambley, W., Coordination Chem. Revs. 166, 181 (1997).
13. Y. Mika and M. Yokoyama, Inorganic Chim. Aceta, 51 (1998).
14. Heudi, O.; Cailleus, A., J. Inorganic Biochem. 71, 61 (1998).
15. Dalla Via, L.; Noto, C., Chemico-Biological Interaction, 110, 203 (1998).
16. Neuse, E.., to be published.

17. Carraher, C.; Ademu-John, C.; Fortman, J.; Giron, D.; Turner, C.; Linville, R., **Polymeric Materials in Medication**, Plenum, NY, 1985.
18. Carraher, C.; Ademu-John, C.; Fortman, J; Giron, D., **Metal-Containing Polymeric Systems**, Plenum, NY, 1985.
19. Carraher, C; Gasper, A., Polymeric Science and Technology, 25, 149 (1983).
20. Siegmann,D.; Carraher, C.; Brenner, D., **Progress in Biomedical Polymers**, Plenum, NY, 1990.
21. Yachnin, J., Cancer Letters 132, 175 (1998).
22. Carraher, C.; Deremo-Reese, C., J. Polymer Sci. 16, 133 (1983).
23. Gebelein, C.; Carraher, C., **Bioactive Polymeric Systems**, Plenum, NY, 1985.
24. Cadwell, G.; Neuse, E., J. Inorganic and Organometallic Polymers, 7, 217 (1997).
25. Carraher, C.; Lopez, I., Polymer science and Technology, 35, 311 (1987).
26. Gebelein, C.; Carraher, C., **Polymeric Materials in Medication**, Plenum, NY. 1986.
27. Lipperd, B., Coordination Chem. Revs. 182, 263 (1999).
28. Carraher, C.; Sheats, J.; Pittman, C., **Organometallic Polymers**, Academic Press, NY, 1978.
29. Carraher, C.; Scott, W.; Giron, D., **Bioactive Polymeric Systems**, Plenum, NY, Chapter 20, 1985.
30. Carraher, C.; Scott, W.; Schroeder, J., J. Macromol. Sci.-Chem., A15(4), 625 (1981).
31. Siegmann, D. Brenner, D.; Carraher, C.; Strother, R., PMSE 61, 214 (1989).
32. Siegmann, D.; Carraher, C.; Friend, A., J. Polymer Mater., 4, 19 (1987).
33. Chiba, U.;Neuse, E.;Swarts, J.;Lamprecht, G., Angew. Makromol. Chem.214, 137 (1994) and Schechter, B.; Caldwell, G.; Meirim, M.; Neuse, E. Applied Organometallic Chem., in press.
34. Neuse, E.; Mbonyana, , C., **Inorganic and Metal-Containing Polymeric Materials**, Sheats, J.; Carraher, C.; Pittman, C.; Zeldin, M.; Currell, B., Eds., Plenum Press, NY, 1990.

Chapter 16

# CONDENSATION POLYMERS AS CONTROLLED RELEASE MATERIALS FOR ENHANCED PLANT AND FOOD PRODUCTION: INFLUENCE OF GIBBERELLIC ACID AND GIBBERELLIC ACID-CONTAINING POLYMERS ON FOOD CROP SEED

Charles E. Carraher, Jr., Herbert Stewart, Shawn M. Carraher, Donna M. Chamely, Wesley W. Learned, James Helmy, Kumudi Abey, and Alicia R. Salamone

*Florida Atlantic University, Departments of Chemistry and Biochemistry and Biological Sciences, Boca Raton, FL 33231; Florida Center for Environmental Studies, Palm Beach Gardens, FL 33410; Texas A&M Commerce, Department of Marketing and Management, Commerce, TX TX 75429; and Flying Circle L. Ranch, OK 74630-2018*

## 1. INTRODUCTION

### 1.1 General

Increased plant and food production has been an aim since humankind began raising their own food (for instance 1-3). These have been accomplished in many ways through history. Plant antagonists are also a problem. Recent research has focused on the genetic modification of plant seeds towards the production of plants that give fruit of increased yield in some selected manner and that are able to withstand the onslaught of certain antagonists. Some of the results, while giving plants with the desired outcome, also produced plants that had other, unforseen, affects to an extent that much of this research is being rethought. Another major recent effort has been aimed at the production of selected chemical agents, insecticides and others, that combat these unwanted plant enemies. This approach has been largely successful but again sometimes with unwanted side results to the extent that even here a rethinking is underway with some of this rethinking involving the genetic modifications noted above.

The worm has in some sense turned so that there is a revival in the idea that we should be investigating ways to assist nature itself to combat antagonists and to enhance plant and food production. Our research with plant growth hormones is aimed at accomplishing this goal.

A living green plant is a stunning result of a whole series of events with chemical agents, referred to as plant growth hormones, PGHs, acting to regulate germination, seedling and plant development, flowering and fruiting, and finally death (4).

Plant growth hormones, PGHs, generally are effective at very low concentrations over the life of the plant in many and varied ways that are plant specific (4). Thus, our initial research is based largely on offering plants various materials and then observing the outcome. This "offering" is done by presenting the PGH either itself or as part of a more complex molecule, here as an integral part of a polymer (for instance 5-10). As part of a polymer, the introduction of the PGH occurs via a sustained manner

We selected incorporation into condensation polymers for several reasons. The major reason was that condensation polymers offered a better possibility that release would occur at a reasonable, but not too slow or too rapid, rate. Because of the polarity of the bonds connecting the various components of the polymer backbone, the chain is susceptible to hydrolysis allowing release of the PGH..

PGHs effects are often plant specific. Specific factors that are often measured to determine the effect and extent of effect include rooting time and incidence, flowering, greening, rate of maturing, growth rate, fruit retention and maturing.

Plants have need of many metals including a number of so-called trace elements. As our study of plants increases we are finding that the number of needed trace elements is increasing and may be different for different plants. Currently, the list of trace metals includes iron, boron, copper, nickel, cobalt, potassium, magnesium, calcium, manganese, zinc, and molybdenum.

While some metals are essential for good plant health, other metals such as cadmium, lead, chromium, mercury, silver, and gold are not essential for plant functioning and are toxic at low concentrations. Even too high concentrations of essential metals can be toxic to plants. It appears that the toxic role of metals is varied but in cases involves the chelation of certain sites in essential materials in the plant. Thus, it is important that the metal chosen to be included is carefully chosen as to not act in an unwanted manner. Also, of consideration is the low concentrations of PGHs needed to affect desired behavior and the corresponding low concentrations of associated metals employed is so low as to not act as a threat to plants or to us even when cumulated ten fold.

Currently, we have included tin, manganese, titanium, zirconium, and hafnium along with specific PGHs into polymers. This group includes one metal, tin, that can act to deter unwanted fungi and bacterial growth; includes one essential metal, manganese, that may assist in the plant growth and development; and three members that are generally believed to be benign with respect to plant activity and human response at the levels employed.

Our efforts have focused on several aspects of PGH activity including seed germination, seedling development and rooting of plant parts. Because results are plant species specific, results from studying one plant are at best indicative of the behavior towards another plant and where application is desired, the specific plant species should be studied.

The PGHs employed by us are naturally occurring. This is intentional since their presence offers nothing "new" to the plant kingdom except maybe the particular metal-containing moiety that can be tailored to plan an active role in the ongoing enterprise or to be benign.

In general polymers containing PGHs can act as control release agent where these formulations can offer

*greater shelf life

*sustained release

*greater retention of the active agent due to lowered solubility of the polymer and

*(in some cases) the "co-reactant" can offer additional properties such as antifungal activity.

There are a number of PGHs. The three major groupings tested by us thus far are the auxins, gibberllins, and cytokinetins.

## 1.2 Gibberellins

Gibberellins are cyclic diterenes with the ability to induce a number of plant responses including cell elongation and cell division. Widespread in nature, the major commercially employed gibberellin is call gibberellic acid, GA3. While widespread in nature, most are not biologically active. The formation of inactive species from active species may represent a mechanism by which excess biologically active gibberellins are removed.

Gibberellins act primarily by inducing the expression of the gene for a-amylase. The transcription of information from DNA to RNA which in turn translates into making the protein a-amylase is believed to be mediated by DNA-binding proteins similar to regulatory proteins in operons of prokaryotic cells. Using this model of an operon in a prokaryotic cell, it is believed that active gibberellins increase the level of protein that switches on the production of a-amylase mRNA by binding to an upstream regulatory sequence of the a-amylase gene. While this is a proposed model, it is not known how active gibberellins actually increase the level of this protein.

The most active of the naturally occurring gibberellins is GA3 and it is GA3 that is the most widely utilized in research and to increase plant and food production. GA3, 2,4,7-trihydroxy-1-8-methylene gibb-3-ene-1,10-carboxylic acid-1-4 lactone, is sold under a number of names including Gib-tabs, Ceku-Gib, Gibrel, Brellin, Gib-Sol, Pro-Gibb, Berfelex, Activol, Grocel, Cekugib, Regulex, and floraltone. It has a LD50 of 6300 mg/kg so is relatively non toxic. It is registered in the USA for use on grapes, cotton, spinach, turf, artichokes, hops, rye, peas, oats, soybeans, beans, rhubarb, citrus, cherries, ornamentals, potatoes, sugar cane, celery, blueberries, strawberries, etc. Uses are varied but include increase in fruit size, fruit loosening, increased fruit set, prevents color change, helps against cold stress and light frost, helps break dormancy, stimulates sprouting, increases fruit yield, etc. Application rates are generally on the order of 1 ppm.

We found with common pole beans (Phaseolus vulgaris L.; Kentucky Wonder, rust resistant pole; Ferry-Morse) that use of 1000 ppm of GA3 alone caused too rapid growth so that the pole beans outgrew bean stalk strength and after two weeks they were so thin and watery in appearance that they could no longer support their own weight and died (5,6). But with the use of titanocene and manganese-GA3 containing polymers that growth, while greatly accelerated, was controllable so that the resulting plants were robust enough to be considered viable food producing plants (5,6 ). In fact, the use of 10 through 1000 ppm GA3 alone and the same concentrations for the polymers gave more rapid growth over untreated beans. Similar results were found for Red Kidney pole beans.

Of interest, is the intermediate results for the polymer-containing systems so that the magnitude of growth rate was 1000 ppm GA3 > 1000 ppm GA3-Polymer > 100 ppm GA3 > 100 ppm GA3-Polymer > 10 ppm GA3 > 10 ppm GA3-Polymer > Untreated. This trend is consistent with the polymer acting as a controlled release agent. These results are also consistent with the polymer degrading to give the active GA3 since other, even very similar gibberellin structures are inactive.

Another study involved the rooting and flowering of Hibiscus rosa-sinensis, variety Albo Lacinatus (Anderson's Crepe Pink) (6). Here addition of GA3 or GA3-containing polymer inhibited the normal rooting of the hibiscus stocks. Of interest were the large number of stock that formed roots along the stem, above the treated portion at sites where small branches had been removed. Thus, the hibiscus "wanted" to root but the presence of the GA3 discouraged this activity. As an extension to this study , the rate and extent of flowering of hibiscus bushes that formed from treated stock generally decreased as the concentration of GA3 or GA3-containing polymer increased. The rate of flowering is most rapid for bushes formed from stocks treated with low levels of GA3 or GA3-containing polymer.

In a study of organotin-GA3 containing polymers related to some of our Everglades restoration efforts, it was found that the use of both polymers containing GA3 and GA3 itself gave positive results (9,10). Sawgrass and water are the two major signatures of the Everglades. Water is seasonal and its control is under the influence of the South Florida Water Management District, SFWMD. By comparison, the coverage of the Everglades by sawgrass is decreasing due to a number of factors including encroachment by the Southern Cattail. Reclaiming the Everglades by simple hand-planting of green house grown sawgrass is time consuming and not needed. Our efforts are aimed at allowing treated seeds of sawgrass to be delivered by air boat or air plane. Thus, initial efforts were aimed at increasing the germination rate of sawgrass seeds from a previously published rate of 0 to 2 %. (We have now increased this germination rate to above 50%.) With the use of GA3 and GA3-containing polymers we were able to increase the germination rate to a high of 13 % for GA3 and 25 % for the polymers. We were also able to decrease the time for germination from about a month to several weeks.

By comparison, the use of GA3 decreased the rate of germination from about 25 % to a low of 6 % but for the polymers the rate of germination increased to a high of about 40 %. A related study, except employing Group IV metallocenes in place of the organotin moieties, gave similar results (7,8).

## 1.3 Auxins

Auxin is a generic name for compounds with the ability to induce elongation in shoot cells. The most rapidly growing parts of plants typically contain the highest concentrations of indole-3-acetic acid, IAA-in the tip of the coleoptile, in buds, and in the tips of of leaves and roots. The concentration of auxin drops from the tip of the coleoptile to the base. From the base of the coleoptile the auxin concentration increases until it reaches a peak at the root tip. The amount of auxin in the shoot tip, however, is usually much greater than in the tip of the root. In fact, the auxin concentration that promotes cellular elongation in the shoot appears to be too great for the root, and somehow results in actual inhibition of cell elongation in the root.

While the application of relatively high concentrations of IAA retard elongation, there may be a considerable increase in the development of branch roots and root hairs due to this high concentration of auxin.

The known biological effects of auxin are numerous. In addition to cellular elongation, apical dominance and root initiation, the more important effects include callus formation, parthenocarpy, phototropism, and geotropism. While auxins play an important role in cell elongation, they also are active in cell division.

The various PGHs interact with one another. For instance IAA interacts with kinetin such that a low ratio of cytokinin to auxin gives a mass of loosely arranged, undifferentiated cells or callus, but a high cytokinin to auxin ratio results in the growth of cultured plantlets with stems and leaves.

The major commercially employed auxin is indole-3-butyric acid, IBA. It is sold under a number of tradenames and is a member of almost all commercial rooting mixtures. It is sold under such names as Indole Butyric, Hormodin, Jiffy Grow, Hormex Rooting Powder, Seradix, Rhizopon AA, Chryzopon, Chryzosan, and Chryzotek. It has an LD50 of about 100 mg/kg in mice. Formulations generally contain on the order of 100 ppm IBA. Commercially it is mainly employed to promote and accelerate the rooting of plant cuttings.

Studies were carried out employing dipyridine manganese and titanocene derivatives of IBA (11). In general, the polymer treatments produced greater and more rapid rooting in comparison to IBA alone and both were better than the control. Further, the average number of roots were greater for the polymers in comparison to IBA itself which in turn was greater than no treatment with the least rooting activity found for the monomers titanocene dichloride and dipyridine manganese dichloride. Thus, for this study, incorporation of the IBA into a polymer structure had an overall favorable affect on rooting and production of side roots.

## 1.4 Cytokinetins

Cytokinetins are most influential on bud and shoot initiation from leaf cuttings and in tissue culture systems. Natural and synthetic cytokinins include zeatin, thidiazuron, benzyladenine, zeatin riboside, and kinetin. They are commercially available under a number of tradenames including Burst, Yield Booster, Jump, Arise, and Cytogen for use on fruits and vegetables, cotton, corn, rice, sorghum, soybeans, and wheat to increase plant vigor and yield. While they are sold for promoting bud initiation and development, they are said to also interact with the plant hormonal system affecting cell division and reproductive activity, promoting female vigor, promoting root initiation and development, and increasing tolerance to stress.

Our work involves the cytokinetin kinetin ( 6-furfurylaminopurine) itself. It is the most widely studied cytokinetin and is naturally found in plants. Past work emphasized the use of organotin-kinetin containing polymers and their effect on the germination and seedling development of sawgrass and cattail as an extension of our Everglades restoration work noted above.

In general, treatment with kinetin and polymers containing kinetin resulted in an increase in the germination rate of both sawgrass and cattail. In this study there was no real advantage to using the polymer-containing kinetin in comparison to kinetin itself (12).

## 1.5 Current Study

Today, most commercial seeds have high rates of germination. Yet, in developing countries where seeds have become damaged through time (age), insect infestation, excessive heat/cold/water/etc., crop production may be severely limited by the lack of high germination because of these and other factors. The current study is aimed at the evaluation of a number of older seeds to study the effect of adding kinetin and GA3, PGHs, as well as polymers containing these PGHs, on the germination of these seeds.

As noted before Gibberellins are cyclic diterpenes with the ability to induce a number of plant responses including cell elongation and cell division. GA3 is also known to play a key role in the germination of most seeds.
Gibberellins are found in fungi, algae, ferns, angiosperms, bacteria, gymnosperms, and mosses and appear to be ubiquitous, generally present in minute amounts typically less than 1 microgram per kilogram of organism. They can be active at very low concentrations. For instance, in stem elongation, GA3 has been found to be active to the nano and less levels for rice and lettuce seedlings. Lower levels have been found for activity to be apparent, down to 3 femtograms(4).

Gibberellin concentration appears to be very important. Thus, the synthesis of gibberellin-containing condensation polymers that act through hydrolysis, or other means, to release minute amounts of GA3 might be useful in contributing to the world's supply of food and fiber.

We have previously reported on the synthesis and structural characterization of metal-containing polymers based on the reaction between metal dihalides and GA3 (for instance 5-7). The metal acts as a bridge connecting the GA3 moieties through M-O- and M-O-CO- linkages. Because GA3 is trifunctional, with two alcohol and one acid group, the products are crosslinked and insoluble presumably allowing the GA3 to be present over a longer lifetime than simply GA3 itself which is water soluble.

## 2. EXPERIMENTAL

Synthesis of the polymer was achieved using the interfacial polycondensation process as noted elsewhere (for instance 5-12).
Germination experiments were carried out in the Florida Atlantic University greenhouse in January 2000. Regular tap water was used. The water level was maintained to the level of the seeds and represented what is called a "saturation" system. Watering was performed so that water came through the bottom of the trays minimizing the movement of the treatment.

Rectangular four inch plastic pots with hole in the bottom were employed. The pots were filled with potting soil (Earthgro All-Purpose Potting Soil, Earthgro Inc., Marysville, OH.). Food crop seeds were mixed with the treatment and cast over the top of the soil. For larger seeds such as peas, the seeds, 25 in each pot, were pressed into the soil about 1/4 inch below the seed surface. Three replicates were planted. The seeds were monitored for two months. Viability was determined using a standard staining technique.

## 3. RESULTS AND DISCUSSION

The present study is aimed at the use of these materials to increase the seed germination for food crop seeds.

Seed sources and description are given in Table 1. Seed viability is given in Table 2.

*Table 1. Seed source and description.*

| *Seed/Variety/Type* | *Source* | *Year* |
|---|---|---|
| *Soybeans* | *Johnstons Seed Company* | *1996* |
| *Turnip/Purple Top* | *Johnstons Seed Comapny* | *1994/95* |
| *Peas/Little Marvel* | *Johnstons Seed Company* | *1995* |
| *Jagger Wheat* | *Flying Circle L Ranch* | *1997* |
| *Mustard/Florida Broad Leaf* | *Johnstons Seed Company* | *1993* |
| *Bad Wheat** | *Flying Circle L Ranch* | *1993* |
| *Broccoli/Calabrese* | *Johnstons Seed Company* | *1993* |

**Wheat that contained mold and which was exposed to animals and insects.*

*Table 2. Viability results for seeds.*

| *Seed Variety* | *% Viability* |
|---|---|
| *Peas* | *91* |
| *Wheat, Jagger* | *32* |
| *Wheat, Bad* | *61* |
| *Mustard* | *14* |

Table 3 contains total average percentage germinations for seeds where the germination averaged above 5 %. Table 4 contains the statistical treatment of the data given in Table 3 treating the three replications.

While eight different seeds were tested for results for only four are reported since the germination rates for the other seeds were less than five percent. Statistically important (significant values >0.1) results were found for many of the seeds but the results for mustard and turnips were only marginally significant.

*Table 3. Gemination results for seeds treated with GA3 and GA3-containing polymers. Figures given are averages of three studies and are the average germination percentages.*

| *Conc.,ppm* | *Wheat* | *Turnip* | *Mustard* | *Pea* |
|---|---|---|---|---|
| *Control* | *9* | *23* | *24* | *3* |
| *GA3,10* | *43* | *20* | *23* | *17* |
| *GA3,1000* | *17* | *29* | *12* | *12* |
| *MeSn/GA3,10* | *21* | *32* | *16* | *17* |
| *MeSn/GA3,1000* | *27* | *16* | *29* | *13* |
| *EtSn/GA3,10* | *21* | *32* | *11* | *4* |
| *EtSn/GA3,1000* | *20* | *31* | *16* | *11* |
| *BuSn/GA3,10* | *23* | *28* | *15* | *4* |
| *BuSn/GA3,1000* | *19* | *25* | *33* | *20* |
| *CpTi/GA3,10* | *17* | *24* | *31* | *17* |
| *CpTi/GA3,1000* | *13* | *23* | *29* | *23* |
| *CpZr/GA3,10* | *15* | *9* | *29* | *8* |
| *CpZr/GA3,1000* | *12* | *16* | *29* | *20* |
| *CpHf/GA3,10* | *29* | *29* | *20* | *24* |
| *CpHf/GA3,1000* | *23* | *33* | *16* | *1* |

*where MeSn/GA3 represents the polymeric product from the reaction of GA3 and dimethyltin dichloride; EtSn/GA3 represents the product from diethyltin dichloride; BuSn/GA3 represents the product from dibutyltin dichloride; CpTi the product from titanocene dichloride; CpZr/GA3 the product from zirconocene dichloride; and CpHf/GA3 the product form hafnocene dichloride.*

For Jagger wheat the largest significant levels were for GA3,10 ppm itself, dimethyltin/GA3, 1000 ppm, and dibutyltin/GA3, 10 ppm. These also represented most of the highest percentage germinations (control 9 % verses 43 % for GA3, 10 ppm; 27 % for dimethyltin/GA3, 1000 ppm; and 23 % dibutyltin/GA3, 10 ppm).

In all cases, seedling development appeared to be independent of the particular seed treatment.

*Table 4 Statistical data for data given in Table 3.*

| | *Wheat* | | *Turnips* | | *Mustard* | | *Peas* | |
|---|---|---|---|---|---|---|---|---|
| *Conc.,ppm* | *T* | *Sig* | *T* | *Sig* | *T* | *Sig* | *T* | *Sig* |
| *GA3,10* | 2.8 | 053 | 1.0 | .21 | .10 | .46 | 1.5 | .12 |
| *GA3,1000* | .65 | .29 | 1.3 | .15 | 1.5 | .13 | 1.0 | .21 |
| *MeSn/G,10* | 1.7 | .11 | 1.3 | .15 | 2.0 | .092 | 1.5 | .13 |
| *MeSn/GA3,1000* | 2.9 | .048 | .64 | .29 | 1.5 | .13 | 1.0 | .21 |
| *EtSn/GA3,10* | 1.5 | .13 | 1.9 | .096 | 1.4 | .14 | .38 | .37 |
| *EtSn/GA3,1000* | .84 | .24 | 1.0 | .21 | .65 | .29 | 2.0 | .092 |
| *BuSn/GA3,10* | 3.7 | .031 | .92 | .27 | .82 | .25 | 1.0 | .21 |
| *BuSn/GA3,1000* | .82 | .25 | 2.0 | .092 | 1.7 | .11 | 1.5 | .13 |
| *CpTi/GA3,10* | .79 | .25 | 1.0 | .21 | .48 | .33 | 2.0 | .086 |
| *CpTi/GA3,1000* | .38 | .37 | 00 | 1.0 | .72 | .27 | 2.8 | .051 |
| *CpZr/GA3,10* | .66 | .29 | 1.5 | .13 | .76 | .26 | 1.1 | .19 |
| *CpZr/GA3,1000* | .46 | .35 | .52 | .37 | 1.1 | .19 | 1.9 | .083 |
| *CpHf/GA3,10* | 1.5 | .13 | 1.3 | .15 | .33 | .38 | 1.8 | .10 |
| *CpHf/GA3,1000* | 1.2 | .17 | .98 | .21 | 1.3 | .16 | 1.0 | .21 |

By comparison, treatment with GA3 and GA3-containing polymers gave generally equivalent results. Binding the GA3 with polymers will result in additional cost so that use of GA3 alone might be recommend on this account. Counter, the GA3-containing polymers can be stored in the open under room conditions for over two years without any noticeable change (by IR examination) while GA3 itself must be stored under refrigeration.

## 4. REFERENCES


1. Guerinot, M.; Salt, D., Plant Physiology, 125, 164 (2001).
2. Somerville, C.; Bonetta, D., Plant Physiology, 125, 168 (2001).
3. Trewavas, A., Plant Physiology, 125, 174 (2001).

4. Metraus, J. P., **Plant Hormones and Their Role in Plant Growth and Development**, Davies, P. J., Ed., pp 296-317, Kluwer, Boston, 1987.
5. Carraher, C. E.; Stewart, H.; Soldani, W.; Reckleben, L.; Pandya, B. PMSE, 67, 270 (1992).
6. Stewart, H.; Carraher, C.; Soldani, W.; Reckleben, L.; de la Torre, J.; Miao, S-L, **Metal-Containing Polymeric Materials**, pp 93-107, Dekker, NY, 1996.
7. Salamone, A.; Carraher, C.; Stewart, H.; Miao, S-L.; Peterson, J.;Francis, A. PMSE 81, 147 (1999).
8. Salamone, A.; Carraher, C.; Carraher, S.; Stewart, H.; Miao, S-L.;Cowan, C., PMSE 82, 79 (2000).
9. Carraher, C.; Gaonkar, A.; Stewart, H.; Miao, S-L.; Carraher, S., **Tailored Polymeric Materials for Controlled Delivery Systems**, (Mculloch, I., Shalaby, S. Eds.) pp 295-308, ACS, Washington, DC, 1998.
10. Carraher, C; Gaonkar, A.; Stewart, H.; Miao, S-L.; Carraher, S.; PMSE, 80, 367 (1999).
11. Stewart, H.; Soldani, W.; Carraher, C.; Reckleben, L, **Inorganic and Metal-Containing Polymeric Materials**, Plenum, NY, 1990.
12. Carraher, C.; Nagata, M.; Stewart, H.; Miao, S. L.; Carraher, S. M.; Gaonkar, A.; Highland, C.; Li, F., PMSE, 79, 52 (1998).

# D. Enhanced Physical Properties

Chapter 17

# 2,6-ANTHRACENEDICARBOXYLATE-CONTAINING POLYESTERS AND COPOLYESTERS

David M. Collard[1] and David A. Schiraldi[2]
[1]*School of Chemistry abd Biochemistry, and the Polymer Education and Research Center, Georgia Institute of Technology, Atlanta Georgia 30332-0400, and* [2]*Next Generation Polymer Research, KoSa, P.O. Box 5750, Spartanburg, South Carolina 29304.*

## 1. INTRODUCTION

Polyesters prepared from aromatic acids and aliphatic diols, such as poly(ethylene terephthalate) (PET) and poly(butylene terephthalate) (PBT), are widely used in film, fiber, and packaging applications (1,2). The development of monomer-grade 2,6-naphthalenedicarboxylic acid has led to the introduction of poly(ethylene 2,6-naphthalate), PEN, which produces stronger fibers and films, retains the optical clarity of PET, and can be formed into food and beverage containers with superior gas barrier properties (1).

The desirable thermomechanical properties of PET ($T_g$=78 °C; $T_m$=256 °C) arise from the combination of rigid 1,4-phenylene units, short ethylene units, and dipolar esters. The additional flexibility of the 1,4-butanediyl unit of PBT allows for faster crystallization and imparts a lower glass transition temperature ($T_g$=40 °C) and a lower melting point ($T_m$=230 °C). PBT also has lower modulus and tensile strength than PET, and is often reinforced with fibers or fillers. On the other hand, the rigidity of the naphthalene units of PEN imparts a higher glass transition temperature (123 °C) that facilitates its use for food packaging applications that require hot-fill and heat sterilization. Current impediments to the more widespread use of PEN include the high cost of 2,6-naphthalenedicarboxylic acid and the high melt viscosity of the polymer.

Small amounts of comonomers (e.g., isophthalic acid, diethylene glycol, 1,4-cyclohexane dimethanol) are added to PET, PBT, and PEN to improve their physical properties. However, other poly(alkylene arenedicarboxylate) homopolymers have not been commercialized. The vast choice of flexible and rigid structural units available for incorporation in polyesters has led us to investigate new copolymers prepared from other aromatic diacids, in particular 2,6-anthracenedicarboxylic acid. As a homolog of terephthalic and napthoic acids, we expected the anthracenedicarboxylic acid unit to provide further enhancement of the physical properties of polyesters. In addition, while the anthracenedicarboxylic acid is aromatic, and therefore stable to the harsh conditions required for polyesterification, we also hoped to exploit the reactivity of the anthracene unit to chemically modify these new polymers and copolymers. Our work has been extended to a number of other fused arenes (3-7).

Previous reports of poly(alkylene 2,6-anthracenedicarboxylate)s (P*x*A) have been limited to characterization of the polymer prepared by condensation of dimethyl 2,6-anthracenedicarboxylate with 1,6-hexanediol (8), and solution phase polymerization of ethylene glycol and a Diels-Alder adduct of anthracenedicarboxylate followed by heating to effect the retro Diels-Alder reaction to give poly(ethylene 2,6-anthracenedicarboxylate), P2A (9). In this chapter we provide an overview of our work to incorporate 2,6-anthracenedicarboxylate into homopolymers and copolymers, how this unit effects the thermal properties of the polymers, and how it presents new opportunities to modify polymers by addition reactions.

## 2. MONOMER SYNTHESIS AND POLYMERIZATION

### 2.1 Monomer Synthesis

Our synthesis of dimethyl 2,6-anthracenedicarboxylate starts with the $AlCl_3$-catalyzed Friedel-Crafts acylation of *p*-xylene with 4-methylbenzoyl chloride to give 2,5,4'-trimethylbenzophenone, Figure 1 (10). Heating the neat product to reflux results in ring closure by the Elbs reaction to give a mixture of 2,6-dimethylanthracene and 2,6-dimethylanthrone. Both components of the mixture obtained from this cyclization undergo oxidation upon treatment with chromium trioxide to give a single product: 9,10-anththraquinone-2,6-dicarboxylic acid. The anthraquinone is reduced to 2,6-anthracene-dicarboxylic acid upon treatment with zinc metal in aqueous ammonium hydroxide. Our initial attempts to perform this reduction made use of a catalytic amount of copper(II) sulfate. However these reactions led to the formation of a large amount of the over-reduced product, 9,10-dihydroanthracene-2,6-dicarboxylic acid. Omission of the copper sulfate leads to a high yield of the desired product with no over-reduction. The diacid is treated with methyl iodide and lithium carbonate in DMF to give the methyl ester. This method has proved successful for the synthesis of diesters from diacids that do not undergo Fischer esterification because of their insolubility in acidic refluxing methanol. The monomer is isolated and purified by recrystallization from benzene as a yellow crystalline solid (mp=274-275 °C)

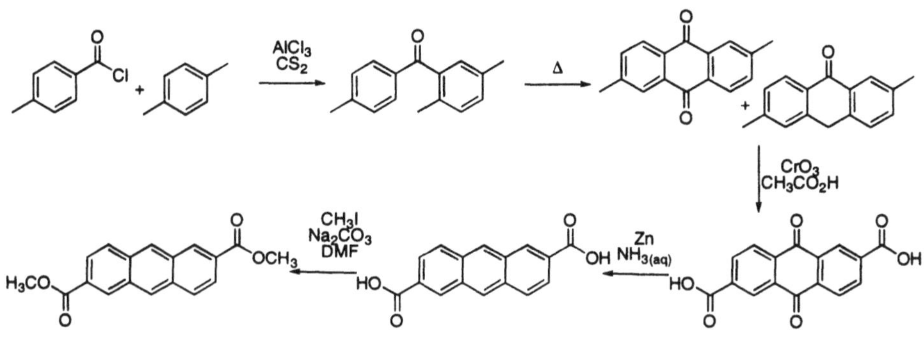


*Figure 1. Synthesis of dimethyl 2,6-anthracenedicarboxylate.*

An alternative route was explored to avoid the low conversion of the Elbs cyclization in the previous procedure and to allow access to the 2,7-isomer. Diels-Alder reaction between isoprene and benzoquinone, Figure 2, gives a mixture of 2,6- and 2,7-dimethyltetrahydroanthraquinones. Treatment of the crude mixture with oxygen in basic ethanol gives a mixture of the corresponding dimethylanthraquinones that can be separated by repeated recrystallization from ethanol (11). Oxidation of the separate dimethylanthraquinones with chromium trioxide yields the anthraquinonedicarboxylic acids that can be converted to the dimethyl anthracenedicarboxylates by the methods described above (reduction of the anthraquinone followed by esterification). While this route allows access to both the 2,6- and 2,7-isomers, the tedious separation of the dimethylanthraquinone presents a severe limitation.

Another attempt to synthesize the 2,6-substituted anthracene core made use of a Diels-Alder reaction between benzoquinone and 2-carboxy-1,3-butadiene. The latter compound can be generated by *in situ* extrusion of $SO_2$ from 3-carboxysulfolene, which itself is readily prepared from butadiene, sulfur dioxide and carbon dioxide, Figure 2 (12).

2-Anthracenecarboxylic acid was prepared by reduction of 2-carboxy-9,10-anthtraquinone, and converted to its 2-hydroxyethylester to prepare anthracene end-capped macromers (Section 7), Figure 3.

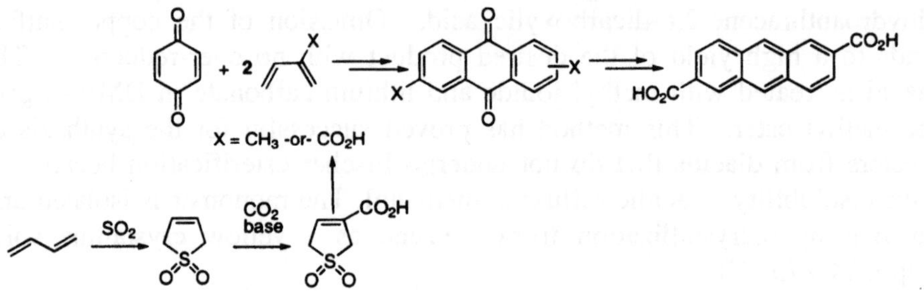


Figure 2. Alternate syntheses of dimethyl 2,6-anthracenedicarboxylate.

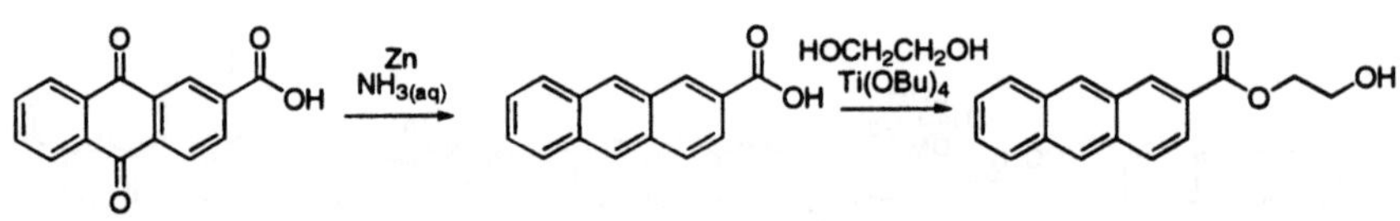


*Figure 3. Synthesis of 2-anthracenecarboxylic acid and its 2-hydroxyethyl ester.*

## 2.2 Polymerization

Dimethyl 2,6-anthracenedicarboylate is amenable to polyesterification under a variety of conditions appropriate for synthesis of polyesters. In general, it was incorporated into PET-copolymers, PET-A (Sections 4-6), by standard a two-step polymerization process. A mixture of ethylene glycol (2.2 equivalents), dimethyl terephthalate and dimethyl 2,6-anthracenedicarboylate (one equivalent of dimethyl esters in a ratio to define the composition of the copolymer), together with antimony trioxide and manganese acetate is heated to 190-230 °C for 2-3 hours to prepare a mixture of corresponding bis(2-hydroxyethyl) esters. This is followed by addition of polyphosphoric acid to deactivate the manganese catalyst and heating to 250-290 °C under vacuum for 2-3 hours with the removal of excess ethylene glycol to afford polymer.

Poly(alkylene 2,6-anthracenedicarboxylate)s, P*x*A (Section 3), were prepared from dimethyl 2,6-anthracenedicarboxylate and the appropriate diol, $HO(CH_2)_xOH$, in the presence of tetra(butoxy)titanium(IV) at 250-290 °C.

# 3. POLY(ALKYLENE ANTHRACENE 2,6-DICARBOXYLATE)S, P*n*A

Poly(alkylene 2,6-anthracenedicarboxylate)s (P*x*A) were prepared from dimethyl 2,6-anthracenedicarboxylate and α,ω-diols, $HO(CH_2)_xOH$, to explore the effect of extending the aromatic diacid in polyesters from 1,4-phenylene (i.e., terephalate) to 2,6-naphthalene to 2,6-anthracene (11). Poly(ethylene 2,6-anthracenedicarboxylate), P2A, the anthracene analog of PET and PEN, is an insoluble, intractable solid. The polymer solidifies as it is formed in the stirred polymerization reactor at 290 °C. An alternate synthesis of P2A entailed monitoring mass lost during thermal trans-esterification of bis(hydroxyethyl) 2,6-anthracenedicarboxylate by thermal gravimetric analysis (TGA). The thermogram indicates no loss of mass up to 225 °C and then a mass loss of 17.8% by 300 °C, which correlates well with the removal of one equivalent of ethylene glycol (18 mass%) from the monomer to form the polymer P2A. Further heating led to decomposition at temperatures above 400 °C. The mass loss was confirmed by isothermal gravimetric analysis at 310 °C. The evolution of ethylene glycol was confirmed by Fourier transform infrared spectroscopy of the gas evolved from the TGA. The material resulting from this thermolysis is insoluble in all common organic solvents including those typically used for polyesters (trifluoroacetic acid, 1,1,1,3,3,3-hexafluoroisopropanol, 2-chlorophenol). Differential scanning calorimetry (DSC) analysis of the material shows no thermal transitions up to 400 °C.

The homopolymer derived from dimethyl 2,6-anthracenedicarboxylate and 1,4-butanediol is also insoluble and infusible, and displays no thermal transitions below 400 °C by DSC. The higher homologs are more soluble and display melting points. Homologs from P6A to P12A display two endotherms upon melting, and a single exotherm upon cooling. As expected, the melting points of the polymers decrease as the alkyl spacer length increases. The temperature difference between the two endotherms also decreases and the relative peak areas of the two endotherms change as the rate of cooling is varied: The enthalpy associated with the lower-temperature peak increases upon more rapid cooling. This is consistent the with formation of a polymorphic sample whereby the polymer crystallizes into two different crystal forms, and it is analogous to the behavior of other aliphatic-aromatic polyesters, including poly(hexamethylene terephthalate), P6T, the terephthalate analog of P6A. Examination of the series of homologous anthracene-containing homopolymers by polarized light microscopy and x-ray diffraction provides no evidence for the formation of liquid crystalline phases.

In general, the melting points of the anthracene-containing polymers, P*n*A, are 80-90 °C higher than the corresponding terephthalates and 50-60

°C higher than the 2,6-naphthalate polymers. There is a pronounced odd-even effect of the number of carbons in the aliphatic diol on the melting point of the anthracene-containing homopolymers, similar to that found for the corresponding homologs in the P*x*T and P*x*N series.

## 4. POLY(ETHYLENE 2,6-ANTHRACENEDICARBOXYLATE-*co*-TEREPHTHALATE)S, PET-A

Incorporation of small amounts of the 2,6-anthracenedicarboxylate structural unit into copolymers of PET results in an increase in glass transition temperature and a decrease in melting temperature (10). The intensity of the melting transition gradually decreases upon increasing the amount of the anthracene comonomer in the polymer, and copolymers with greater than 15% of the anthracenedicarboxylate structural unit are completely amorphous. The copolymers with greater than 20% of the anthracene comonomer are insoluble in common solvents. We have explored the effect of the anthracenedicarboxylate in copolyesters on the post-polymerization chemical modification of polymer structure (Sections 4-6), on the fluorescent properties of PET fiber (3,4) and on gas barrier properties of PET films (13).

## 5. DIELS-ALDER CROSSLINKING AND GRAFTING REACTIONS OF PET-A

Anthracene undergoes rapid [4+2] cycloaddition with electron deficient alkenes by the Diels-Alder reaction. This reaction is reversible upon proceeding to higher temperatures. Reaction of copolyesters, PET-A, with various maleimides leads to rapid Diels-Alder reactions, Figure 4 (14). While initial experiments were performed over long times (a number of hours) in the melt (i.e., >270 °C), we have also shown that this reaction proceeds at considerably lower temperatures in the solid state (i.e., 150 °C).

Differential scanning calorimetry (DSC) of mixtures of PET-A and *N*-octadecylmaleimide shows a melting endotherm at 60-70 °C ($\Delta H$ = 150 J/per gram of maleimide in the mixture) corresponding to the melting of phase-separated maleimide. This is followed by a broad exotherm with a peak at approximately 150 °C corresponding to the heat of the Diels-Alder reaction (approximately -90 kJ per mole Diels-Alder adduct). Subsequent heating and cooling cycles show only a glass transition at 45 °C, ($\Delta Cp$=0.4 J/g). Further exidence for the Diels-Alder reaction included changes in the 1HNMR and

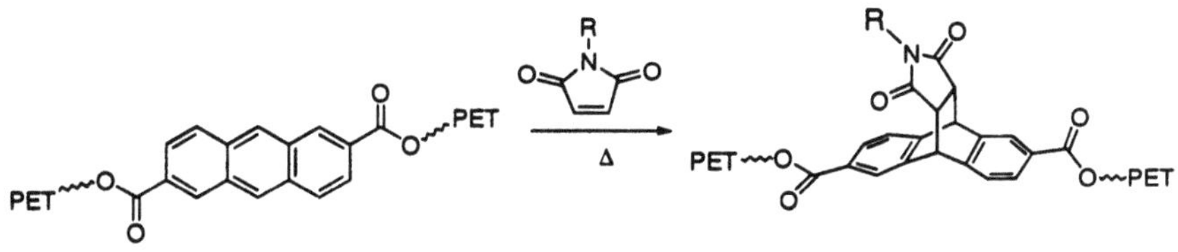


*Figure 4. Diels-Alder addition of N-substituted maleimide to the anthracenate structural unit of PET-A.*

ultraviolet-visible (UV-vis) spectra which were consistent with the conversion of the anthracene unit and maleimide to the bicyclic adduct. Having established that PET-A is amenable to grafting reactions by Diels-Alder addition of maleimides we have used this procedure for crosslinking with bismaleimides and for the selective modification of the surfaces of polyesters with hydrophobic and hydrophobic groups.

Spin-coated films of PET-A were treated with solutions of PEG-5000 with maleimide end groups, PEG-5000M (Shearwater Polymers), or *n*-octadecylmaleimide, and then heated to 170 °C. Unreacted PEG-5000M was removed from the film by exhaustive washing with water in a soxhlet extractor. The contact angles of water droplets on control samples (copolymers not treated with maleimide, and PET homopolymer treated with maleimide) remained the same after the heating and rinsing procedure. Treatment of PET-15A (the copolymer containing 15mol% of the anthracenecomonomer) films with PEG-5000M showed a decrease in the contact angle of water droplets from 75° to 55°, while grafting with *N*-octadecylmaleimide resulted in an increase in contact angle to 83°. Thus, the reaction of PET-A copolymers can be used to selectively modify the surface properties of a film without effecting the bulk properties of the material.

## 6. PHOTOCROSSLINKING OF PET-A

Anthracene undergoes rapid [4+4] face-to-face dimerization through the 9- and 10- positions upon irradiation with visible light. This reaction is reversed upon irradiation with ultraviolet light or upon heating. Model studies of the photodimerization of dioctyl 2,6-anthracenedicarboxylate indicate efficient dimerization upon irradiation of a solution at 350 nm and reversal (50% conversion in 1.5 hours upon irradiation at 254 nm in $CHCl_3$ solution, or 95% conversion upon heating at 145 °C for 12 hours).

Irradiation of spin-coated films of PET-4A and PET-18A at 350 nm in air for one hour affords a polymer that gels in trifluoroacetic acid but which does not dissolve, suggesting the formation of a crosslinked material, Figure 5 (14). Ultraviolet-visible (UV-vis) spectroscopy of the irradiated films shows a decrease in the absorbance of anthracene. The photolysis is rapid: irradiation of PET-4A for 5 minutes results in a 25% decrease in the anthracene absorbance at 420 nm. Dilute solution viscometry (single point intrinsic viscosity) indicates an increase in molecular weight prior to gelation.

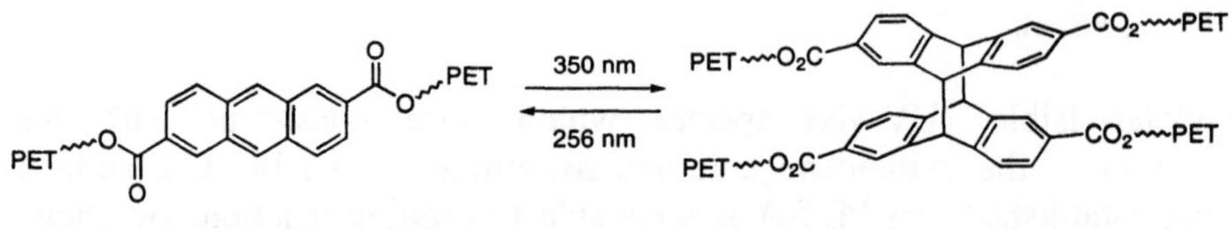


*Figure 5. Photochemical crosslinking of anthracenate structural unit of PET-A by irradiation at 350 nm and reversal upon irradiation at 256 nm.*

Crosslinking disrupts the crystallinity of the polymers. Whereas PET-4A has a melting point of 236 °C, the material obtained by irradiation at 350 nm in air for one hour has a melting point of 219 °C. No crystallization exotherm appears upon cooling from the melt and the melting transition does not appear in the second heating cycle. This can be attributed to the crosslinks impeding crystallization, and is consistent with crosslinking in the amorphous regions of the semicrystalline polymer.

Although we set out to demonstrate that crosslinking of PET-A could be attributed to anthracene photodimerization, we have been unable to confirm the structure of the crosslink. Cleavage of the crosslinks by irradiation at 254 nm or heating at 145 °C would provide strong evidence for the anthracene photodimer as the crosslink. However, films exposed to these conditions did not return the polymers to their original form. A series of experiments under both air and nitrogen suggest a competition between [4+4] photodimerization and the photochemical reaction between the anthracene structural units in the polymer and molecular oxygen to form anthracene-9,10-endoperoxide units. This endoperoxide undergoes subsequent irreversible thermal and photochemical decomposition, which lead to crosslinked polymer chains and a number of oxidized products. Thus, the crosslinking of PET-A could be a result of the irreversible radical reactions other than photodimerization.

Our search for photocrosslinkable copolyesters also led us to investigate the use of phenylene-1,4-bisacrylic acid as a monomer. This monomer is stable to the conditions for polymerization of PET. Copolymers containing a small amount of the bisacrylate monomer undergo rapid crosslinking upon irradiation with ultraviolet light as a result of an irreversible [2+2] cycloaddition of the cinnamate-type units (15).

## 7. CHAIN EXTENSION OF ANTHRACENE-TERMINATED PET

The thermal grafting reaction between the 2,6-anthracenedicarboxylate units of PET-A copolymers and dienophiles by the Diels-Alder reaction (Section 4) led us to investigate the possibility of chain extending anthracene end-capped macromers with bisdienophiles. Such an approach could be used to rapidly increase the molecular weight of the precursor polyester ([16]).

Anthracene-terminated macromers were prepared by condenation polymerization of bis(hydroxyethyl) terephthalate in the presence of monofunctional anthracene derivatives (acid, methyl ester, and hydroxyethyl ester) which serve as end capping reagents to limit the molecular weight, Figure 6, and by alcoholysis of high molecular weight PET with 2-hydroxyethyl 2-anthracencarboxylate. The determination of the average number of anthracene end groups per chain, $f_{AN}$, was made by the independent quantification of hydroxyl end groups (by $^1$H NMR spectroscopy, integrating the peak for the hydroxyl-substituted methylene at the chain end), carboxyl end groups (by titration), and 2-anthracenecarboxylate end groups (by $^1$H NMR spectroscopy). The amount of anthracene in a sample was determined by ultraviolet-visible spectroscopy.

Several bis(maleimide)s were investigated for their suitability as bisdienophiles in this study. 1,2-Ethylene bis(maleimide) and 1,6-hexamethylene bis(maleimide) are not stable at temperatures approaching 300 °C. Heating a neat, dry sample of bis(4-phenylmaleimido)methane (MDBM) to 290 °C for 3 minutes shows no mass loss by TGA, but gives an insoluble gel that swells in chloroform. Addition of 2,6-di-*t*-butylphenol prior to heating suppresses this gelation, suggesting that radical reactions of MDBM are potential side reactions in the absence of an antioxidant.

Mixtures of anthracene-terminated macromers and MDBM chain extender were initially prepared by mixing the components in a 20% v/v mixture of hexafluoroisopropanol and chloroform, followed by rapid removal of the solvent on a rotatory evaporator. This procedure led to mixtures in which approximately 50-75% of the anthracene end groups and maleimide had already reacted to form the Diels-Alder adduct, Figure 6. This initial conversion takes place during the removal of the solvent, and it is accelerated by the increasing concentrations of the anthracene end groups and chain extender. A dilute solution of anthracene-endcapped macromer

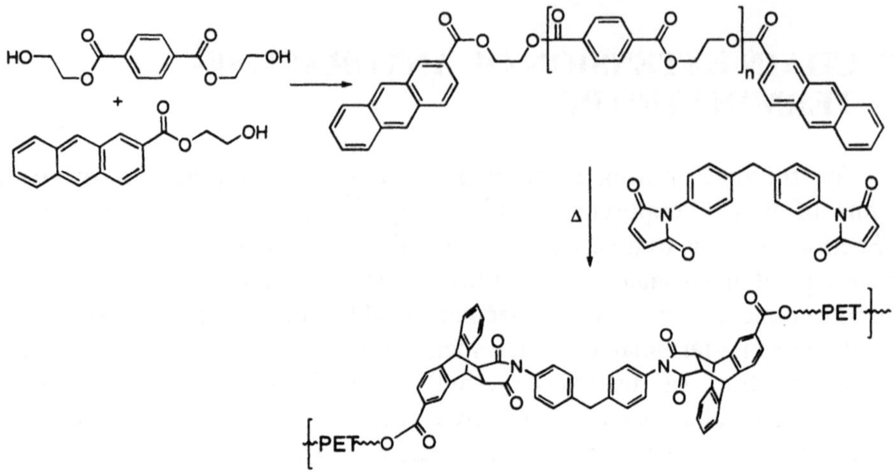


*Figure 6. Synthesis of anthracene-terminated macromers and their chain extension with bis(4-phenylmaleimido)methane.*

(ca. 40 mg/mL) with a stoichiometric amount of MDBM in HFIP/$CHCl_3$ resulted in only 1-2% conversion of the maleimide and anthracene to the Diels-Alder chain extension adduct after 24 hours (as shown by $^1H$ NMR spectroscopy).

After removal of the solvent, heating the mixtures of anthracene-terminated PET macromers and bismaleimides results in conversion of the remaining anthracene and maleimide to the Diels-Alder adduct. For example, heating a sample of anthracene-terminated PET ($f_{AN}$=1.83; $M_v$=6,600 g/mol) and MDBM at 260 °C for 30 minutes gives a polymer with no anthracene chain-ends or unreacted maleimide, as shown by $^1H$ NMR and UV-vis spectroscopies. Chain extension raises the intrinsic viscosity of the polymer from 0.25 dL/g to 0.64 dL/g ($M_v$=20,300 g/mol). Similarly, chain extension of end-capped polyester with lower functionality and higher initial molecular weight ($f_{AN}$=1.83; $M_v$=10,200 g/mol) with MDBM results in an increase in molecular weight to 25,500. While this crosslinking is slower at lower temperatures, the reaction still takes place at 150 °C in the solid-state.

Whereas the anthracene-terminated macromers are highly crystalline ($T_m$=249-251 °C), the chain extended materials derived from low molecular weight macromers are amorphous by DSC analysis. Thus, the relatively high density of bulky Diels-Alder adduct branch points appear to impede crystallization. Chain extended materials derived from higher molecular weight macromers, with a correspondingly lower density of crosslinks, are semicrystalline, displaying both a melting transition and glass transition. The enthalpy change of the melting endotherm of the chain extended polyester is smaller than that of the initial polymer and the supercooling of the melt, $\Delta T$ ($\Delta T = T_m - T_c$), is also larger, consistent with the formation of a less crystalline material with a higher molecular weight and higher melt viscosity.

Anthracene-terminated macromers were also chain extended with MDBM by reactive extrusion at 260-270 °C. A macromer with $f_{AN}$=1.83 and $M_v$=6,600 gave a chain extended material with $M_v$=29,800 g/mol. Preliminary analysis of mechanical properties on compression molded films of chain extended polymers indicate that these polymers are softer than semicrystalline PET, but considerably tougher, with greater extension prior to failure.

To investigate the possibility of reversible control of molecular weight in our polymers, we determined the equilibrium constant for the reaction of model compounds *N*-phenylmaleimide and methyl 2-anthracenecarboxylate at 260 °C. A neat sample of the Diels Alder adduct was heated to 260 °C for one hour, with samples being withdrawn and quenched into ice water to stop the reaction and freeze-in the equilibrium mixture. The ratio of diene, dienophile and adduct were determined by $^1$H NMR spectroscopy. This analysis indicated that the equilibrium is established rapidly, with an equilibrium constant of 9.6 ± 2.1 $M^{-1}$. Heating the chain extended polymer to 350 °C in an attempt to shift the equilibrium results in degradation of the polymer with the sublimation of 2-anthracenecarboxylic acid and small anthracenate ester-terminated oligomers. The observation that 2-anthracene-carboxylic acid is produced shows that the chain extension reaction is reversible, but only at temperatures where polymer degradation is significant.

The rapidity of the chain extension of anthracene-terminated macromers with bismaleimides by a Diels-Alder addition reaction contrasts dramatically with the slow rate of increase in molecular weight by polycondensation reactions. Whereas polycondensation relies on removal of a condensate, the Diels-Alder chain extension proceeds by an addition reaction and does not rely removal of a byproduct. The production of high molecular weight PET relies on driving the polycondensation to completion with removal of ethylene glycol, requiring thermal energy, vacuum, agitation, and time. Thus, the rapid thermal Diels-Alder chain extension presents the opportunity to convert low molecular weight precursors to high molecular weight polymer with potential savings, or in a reactive extrusion process.

## 8. CONCLUSIONS

Dimethyl 2,6-anthracenedicarboxylate is a thermally stable monomer that can be incorporated into polyesters under standard polymerization conditions. The rigid anthracenecarboxylate unit increases the glass transition and melting points of polymers and copolymers relative to the terephthalate and 2,6-naphthoate analogs.

While the monomer is stable to the harsh conditions required for the polymerizaton by virtue of its aromaticity, it possesses two modes of reactivity that allow for post-polymerization modification of the polymer structure. The anthracene unit undergoes addition reactions with electron deficient alkenes via a Diels alder reaction, and it undergoes photochemically-promoted [4+4] cycloadditon to form the face-to-face anthracene dimer. Both modes of reactivity have been used to modify the properties of copolyesters containing the 2,6-anthracenedicarboxylate unit. The Diels-Alder reaction between anthracene and maleimides also allowed us to demonstrate the rapid increase in molecular weight of anthracene-terminated macromers by reactive extrusion with a bimaleimide.[17]

## 9. REFERENCES


1. Silvis, H.C. Trends Polym Sci., 5, 75 (1997).
2. Callander, D.D. Polym. Eng. Sci., 25, 453 (1985).
3. Connor, D.M.; Kriegel, R.M.; Collard, D.M.; Liotta, C.L.; Schiraldi, D.A. J. Polym. Sci., 38, 1291 (2000).
4. Connor, D.M., Collard, D.M., Liotta, C.L., Schiraldi, D.A., Dyes Pigments, 43, 203 (1999).
5. Connor, D.M.; Collard, D.M.; Liotta, C.L.; Schiraldi, D.A. J. Appl. Polym. Sci., 81, 1675 (2001).
6. Connor, D.M.; Allen, S.D.; Collard, D.M.; Liotta, C.L.; Schiraldi, D.A., J. Org. Chem,. 64, 6888 (1999).
7. Connor, D.M.; Allen, S.D.; Collard, D.M.; Liotta, C.L.; Schiraldi, D.A., J. Appl. Polym. Sci., 80, 2696 (2001).
8. Inada, H. Japanese Patent #745545, to Teijin, Inc. (1973).
9. Anderson, B.C.; Frazier, A.H. US Patent 4371690, to Du Pont (1981).
10. Jones, J.R.; Liotta, C.L.; Collard, D.M.; Schiraldi, D.A., Macromolecules, 32, 5786 (1999).
11. Kriegel, R.M.; Collard, D.M.; Liotta, C.L.; Schiraldi, D.A., Macromol. Chem. Phys., 202, 1776 (2001).
12. Berkner, J.E., PhD thesis, Georgia Institute of Technology, 1996. Andrade, G.S., Berkner, J.E., Collard, D.M., Liotta, C.L., Schiraldi, D.A., *Synth. Commun.*, in press.
13. Polyakova, A.; Connor, D.M.; Collard, D.M.; Schiraldi, D.A.; Hiltner, A.; Baer, E. Journal of Polymer Science, Part B: Polymer Physics, 39, 1900 (2001).
14. Jones, J.R., Liotta, C.L.; Collard, D.M.; Schiraldi, D.A., Macromolecules, 33, 1640 (2000).
15. Vargas, M.; Collard, D.M.; Liotta; C.L.; Schiraldi, D.A., J.Polym. Sci., 38, 2167 (2000).
16. Kriegel, R.M.; Collard, D.M.; Liotta, C.L.; Schiraldi, D.A., Polym. Prepr. (Am. Chem. Soc., Div. Polym. Chem.), 2001.


17. We gratefully acknowledge a grant from KoSa which supported this research, and the contributions of coauthors of papers appearing here as references 3-7 and 10-16.

Chapter 18

# SYNTHESIS AND CHARACTERIZATION OF IONIC AND NON-IONIC TERMINATED AMORPHOUS POLY(ETHYLENE ISOPHTHALATE)

Huaiying Kang, R. Scott Armentrout*, Jianli Wang, and Timothy E. Long
*Department of Chemistry and the Center for Adhesive and Sealant Science, Virginia Polytechnic Institute and State University, Blacksburg, Va 24061-0212*

* *Eastman Chemical Company, Polymer Research Division, B-150B Research Laboratories, Kingsport, TN 37662*

## 1. INTRODUCTION

The presence of low concentrations of covalently bonded ionic substituients in organic polymers is known to exert a profound effect on their mechanical and rheological properties.(1-4) In fact, ionomers (polymers containing less than 20 mol % of ionic groups) have been shown to exhibit considerably higher moduli, and higher glass transition temperatures compared to their non-ionic analogues. Improvements in mechanical and thermal performance are generally attributed to the formation of ionic aggregates, which act as thermo-reversible cross-links (1) and effectively retard translational mobility of the polymeric chains. The precise form and size of these aggregates continues to receive significant attention, but their existence, as evidenced by small angle X-ray scattering (SAXS), neutron scattering, and other techniques, is firmly established for many compositions.(4) Significant attention has been devoted to ionomers derived from chain-growth polymers, such as polyethylene, polystyrene and polyisobutylene, and only limited attention has been directed towards step-growth polymers containing ionic functionalities.(5) Most investigations dealing with polyester ionomers have focused on random polyester ionomers, in which ionic groups were randomly distributed on the polymer main chains as pendent groups. However, there are significantly fewer studies concerning telechelic polyester ionomers, where the ionic groups are located at the polymer chain ends.(6-9) Telechelic ionomers are generally recognized as model systems for random ionomers since the molecular weight between ionic endgroups is controlled and the ionic groups are located exclusively at the polymer chain ends.(10) In addition, telechelic association, coupled with adjacent ordered sequences, serves to strengthen subsequent non-covalent associations.(11-15) Metal sulfonates are known to strongly associate in the solid state as ionic clusters and subsequently disassociate at elevated temperatures. Sulfonated isophthalate monomers have received significant attention in the patent literature as co-monomers in polyester fibers to improve the adhesion of various organic dyes.(16,17)

In this investigation, amorphous polyester ionomers were initially investigated due to their improved solubility compared to semicrystalline polyesters such as poly(ethylene terephthalate) and improved solubility facilitated subsequent molecular weight and solution viscosity characterization.(18, 19) These features will facilitate the initial establishment of structure-property relationships for telechelic polyester ionomers, and subsequent efforts will involve the investigation of semicrystalline telechelic polyester ionomers. In our current research, a series of novel amorphous poly(ethylene isophthalate) (PEI) ionomers were synthesized using sodio 3-sulfobenzoic acid (SSBA) as a monofunctional, ionic, endcapping reagent. In addition, this manuscript will describe the preparation of dodecane terminated poly(ethylene isophthalate) as non-ionic telechelic oligomers for comparison to ionic analogs.

## 2. EXPERIMENTAL

### 2.1. Materials

All reagents were used without further purification. Ethylene glycol (EG) was generously donated by Eastman Chemical Co. Dimethyl isophthalate (DMI), 3-sulfobenzoic acid, sodium salt (SSBA, 97 %), sodium acetate (NaOAc, ACS reagent grade, 99.5+ %), and 1-dodecanol were purchased from Aldrich. In addition, phosphoric acid (crystals, 98 %), cobalt acetate (99 %), antimony (III) oxide (99 %), manganese acetate (99 %) and titanium isopropoxide were all purchased from Aldrich and used without purification.

## 2.2. Preparation of Catalyst Solutions

Sb catalyst: $Sb_2O_3$ (3.00 g) solid was dissolved in 250 mL ethylene glycol (EG). The mixture was heated at 100 °C and stirred for 24 hours under nitrogen purge. The mixture was then filtered and a clear solution was obtained at a concentration of 0.012 g / mL based on Sb.

Mn catalyst: $Mn(OAc)_2 \cdot 4H_2O$ (2.685 g) and acetic acid (1.319 g) were added to 125 mL of EG and heated to produce the catalyst solution at a concentration of 0.0215 g / mL based on Mn.

Ti catalyst: The catalyst solution was obtained by mixing titanium isopropoxide (3.8 mL, 3.65 g) with 62.5 mL of n-BuOH in a dry bottle under nitrogen at a concentration of 0.055 g / mL based on Ti.

## 2.3. Synthesis of non-terminated high molecular weight poly(ethylene isophthalate) (PEI)

To a mixture of 48.5 g (0.25 mol) of DMI and 31 g (0.50 mol) of EG (100 % molar excess), manganese catalyst (2.31 mL, 60 ppm), antimony catalyst (3.8 mL, 200 ppm) and titanium catalyst (0.51 mL, 25 ppm) were added under nitrogen. A multi-step temperature procedure was used for the reaction, i.e., the reaction mixture was heated and stirred at 190 °C for 2 hours, 220 °C for 2 hours and 275 °C for 0.5 hour. At the final stage, vacuum (0.5 mm Hg) was applied for an additional 2 hours. The final product was obtained by breaking the reaction flasks. Since no solvent was utilized in the reaction, no further purification was performed.

## 2.4. Synthesis of sulfonate terminated PEI ionomers (PEI-SSBA)

To a mixture of DMI (48.5 g, 0.25 mol) and EG (31 g, 0.50 mol, 100 % excess), various amounts of end capping reagent, SSBA (PEI-xSSBA, x = 1 mol %, 3 mol %, 5 mol % compared to DMI) were added to obtain a series of PEI ionomers with different molecular weights (Scheme 1). The identical catalysts and reaction procedures were used as described above. The final product was obtained by breaking the reaction flasks and no further purification was needed. NaOAc (0.1:1 mol ratio with SSBA) was also used to react with any non-neutralized impurity in SSBA to maintain SSBA in the sodium salt form.

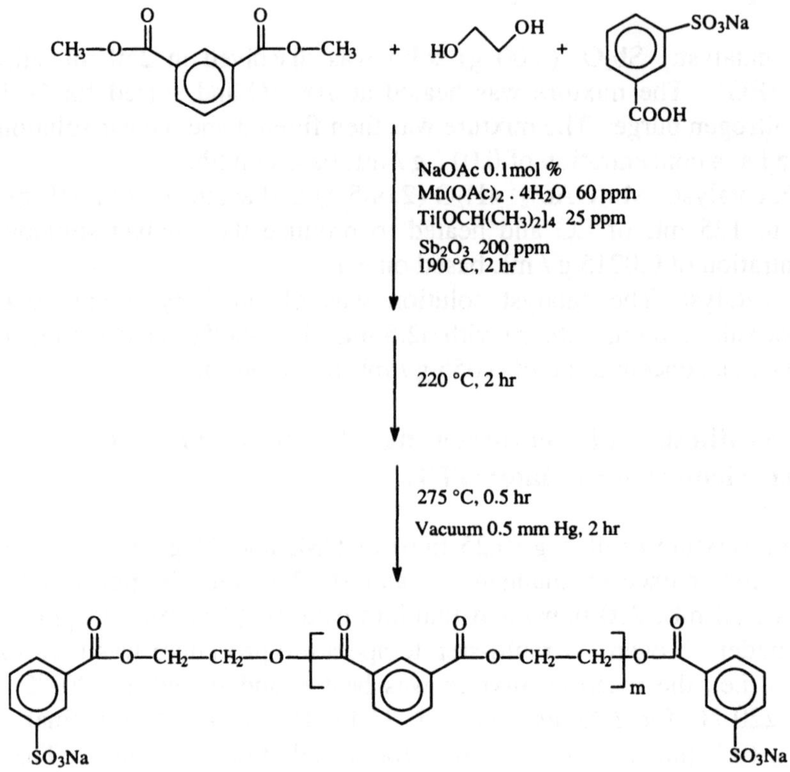


**Scheme 1.** Synthesis of sulfonate terminated PEI ionomers by melt polymerization

## 2.5. Synthesis of dodecanol terminated poly(ethylene isophthalate) (PEI-Dode-OH)

In order to obtain PEI with controlled molecular weights, 1-dodecanol (b.p. = 262 °C) was utilized. To the mixture of dimethyl isophthalate (48.5 g, 0.25 mol), ethylene glycol (31 g, 0. 5 mol, 100 % excess) and dodecanol (PEI-yDode-OH, y = 5 mol %, 10 mol %, 15 mol %, 20 mol %, 30 mol % and 40 mol %) were added (Scheme 2). The same catalysts and reaction procedure were used as above. The final product was obtained by breaking the reaction flasks and no further purification was needed.

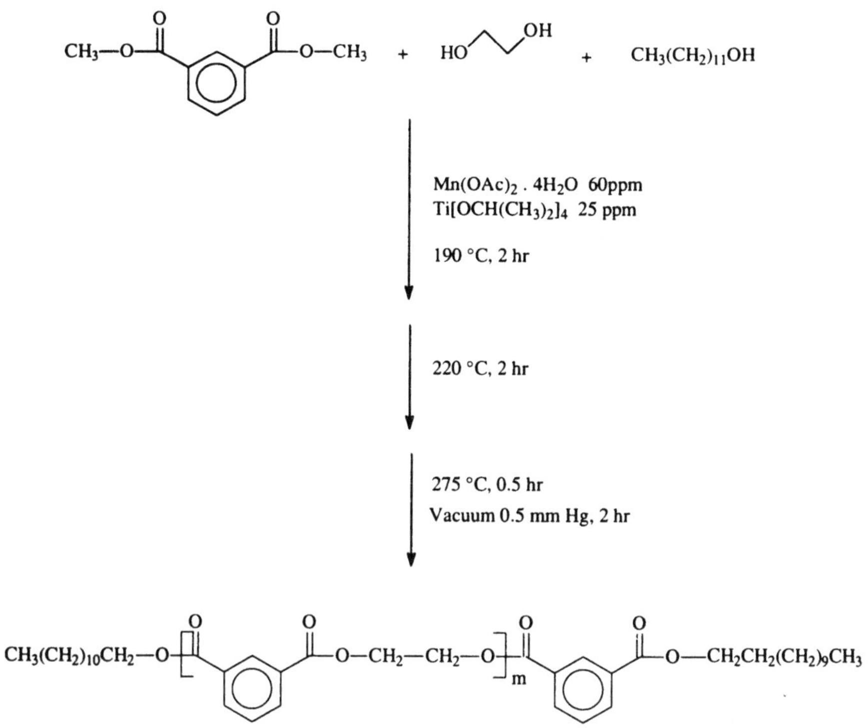


**Scheme 2.** Synthesis of dodecanol terminated poly(ethylene isophthalate) (PEI-Dode-OH)

## 2.6. Polymer Characterization

A 400 MHz NMR (Varian-400) was used to characterize all PEI samples in $CDCl_3$. FTIR spectra were collected using an infrared spectrometer (Perkin Elmer, 283B) to detect the sulfonate end groups. Gel permeation chromatography (GPC) was performed to obtain molecular weights and molecular weight distributions using a Waters 2690 chromatograph. Chloroform was utilized as the mobile phase and the data were recorded at 25 °C at a flow rate of 1.0 mL / min. Melt rheology behavior was analyzed using a melt rheometer (TA Instruments, Advanced Rheometer AR1000). DSC (Perkin Elmer, Pyris 1) was used to study glass transition temperatures. All the samples were maintained at 290 °C for 3 min to eliminate any thermal history, then quenched from 290 °C to room temperature at a rate of 200 °C/min, and finally ramped to 290 °C at a rate of 10 °C/min. All DSC experiments were performed under nitrogen and all the data are reported using the second heat. TGA (Thermal Analyst 2100, Du Pont Instruments) was used to study the thermal stability of the polymers under a nitrogen atmosphere. The solution behavior of the PEI samples was studied using an Ubbelohde viscometer at 25 ± 0.1°C in chloroform.

## 3. RESULTS AND DISCUSSIONS

### 3.1. GPC and NMR Analysis

Amorphous polyester systems were selected as a model for our initial studies because of their high solubility in $CHCl_3$, which facilitated subsequent molecular weight and compositional analysis. PEI-SSBA ionomers with various concentrations of sodium sulfonate end groups (1 mol %, 3 mol %, 5 mol % of SSBA) were synthesized to study the effect of telechelic ionic interactions on the polymer properties. Dodecanol terminated PEI polymers with similar molecular weights were also synthesized for comparison. NMR spectroscopy of PEI-5%SSBA indicated the quantitative disappearance of the hydroxyl ethyl groups at $\delta = 3.9$ and 4.2 ppm, suggesting that the $CH_2CH_2$-OH end groups in the oligomer were replaced with SSBA. As expected, the non-terminated polyester displayed hydroxyl ethyl end groups. In the NMR spectrum of dodecanol terminated PEI-12%Dode-OH, dodecanol resonances were observed at $\delta = 4.3$ ppm (2 H, *t*), $\delta = 1.7$ ppm (2 H, *m*), $\delta = 1.3$ ppm (18 H, *m*) and $\delta = 0.8$ ppm (3 H, *t*). The absence of free dodecanol was also confirmed using $^1H$ NMR spectroscopy. The amount of dodecanol reacted at the polymer chain ends was 4.8 mol % which is significantly less than the original feed (12 mol % compared to DMI). This is presumably due to the distillation of dodecanol (b.p. = 260-262 °C), and a significant fraction was lost during the conventional high temperature reaction process. GPC was also employed for the characterization of the dodecanol terminated polyesters and non-terminated polyesters (Table 1). The empirical relationship of number average molecular weight to dodecanol amount was used as a guideline to determine the amount of dodecanol that was necessary to prepare a desired molecular weight. Unfortunately, the exact molecular weights of PEI-SSBA ionomers were not be determined using GPC due to the strong interaction between the ionic endgroups and the GPC columns. Table 2 also includes the calculated number average molecular weights of the ionomers as determined, using Carothers' Equations:

where

$m_e$ = the molar mass of the combined end groups

$m_u$ = the molar mass of an internal repeat unit

N(A) = moles of monofunctional end capping reagent

x = the number of internal repeat units

The number-average molar weight of the reaction product is then:

$$\langle Mn \rangle = \text{(total mass of product molecules)} / \text{(moles of product molecules)}$$
$$= [\Sigma (m_e + x * m_u)] / (N(A)/2)$$

Table 1. DSC and TGA results of non-terminated and dodecanol terminated (PEI-Dode-OH) polymers

| PEI-Dode-OH | $M_n$[a] (g / mol) | PDI[b] | % DEG[c] | $T_g$ (°C)[d] | 5% wt loss T (°C)[e], TGA, $N_2$ |
|---|---|---|---|---|---|
| PEI (non-terminated) | 37,800 | 2.11 | 2.1 | 59 | 350 |
| PEI-5%Dode-OH | 22,300 | 2.12 | 2.7 | 54 | 356 |
| PEI-12%Dode-OH | 8,400 | 2.33 | 2.5 | 49 | 360 |
| PEI-24%Dode-OH | 5,200 | 2.53 | 2.4 | 42 | 355 |
| PEI-36%Dode-OH | 4,400 | 2.44 | 2.3 | 33 | 353 |
| PEI-48%Dode-OH | 2,600 | 2.18 | 2.6 | 24 | 355 |

[a] Number average molecular weight was determined using GPC at room temperature in $CDCl_3$.
[b] PDI was obtained by GPC at room temperature in $CDCl_3$.
[c] Percentage of diethylene glycol (DEG) was calculated using $^1H$ NMR
[d] Glass transition temperature was measured using DSC under nitrogen; second scan was used for the data collection, 10 °C/min
[e] 5 % weight loss temperature was detected by TGA under nitrogen.

Table 2. DSC and TGA results for sulfonate terminated PEI-SSBA, $N_2$, second heat; heating rate is 10 °C/min for DSC measurement

| PEI-SSBA | $M_n$[a] (g / mol) | $T_g$ (°C) DSC | 5% wt loss T (°C)[b] TGA, $N_2$ |
|---|---|---|---|
| PEI (non-terminated) | 37,800 | 59 | 350 |
| PEI-1%SSBA | 38,100 | 58 | 378 |
| PEI-3%SSBA | 13,300 | 58 | 374 |
| PEI-5%SSBA | 8,400 | 60 | 365 |

[a] Number average molecular weight was calculated using Carothers' equation.
[b] 5 % weight loss temperature was detected using TGA under nitrogen.

### 3.2. FTIR Analysis

In the FTIR spectra, the peak for the S-O stretching mode, which normally appears between 600-700 $cm^{-1}$ for organic sulfonates,(20) was used to identify the sulfonate endgroups of telechelic oligomers. The FTIR spectra of the PEI and PEI-SSBA are shown in Figure 1. As expected, a S-O stretching mode was not observed for the non-terminated polyester. For PEI-SSBA, however, the peak observed at 620 $cm^{-1}$ due to S-O stretching mode is well resolved. Furthermore, as the ionic group content in the ionomers increased, the peak intensity increased correspondingly. This correlation, which was nearly linear as estimated by the plot of peak area vs. ionic group content, indicated the successful incorporation of SSBA endcapping reagents in the ionomers. These results also confirmed that FTIR is a reliable quantitative measure of the presence of terminal sulfonate groups.

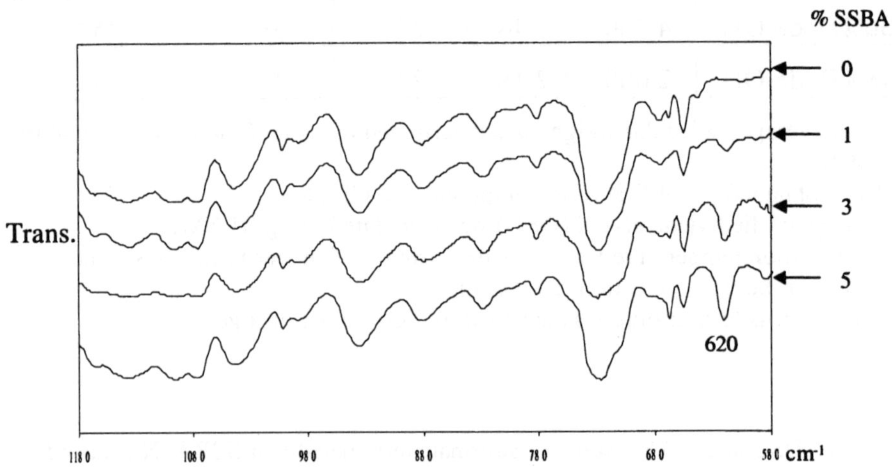


FTIR - MIDAC, M2004 series, under $N_2$, 25°C.

**Figure 1.** FTIR spectra of non-terminated PEI (0% SSBA) and sulfonate terminated PEI-SSBA (1, 3, 5 % SSBA)

### 3.3. DSC and TGA Analysis

DSC and TGA measurements were performed to ascertain the effect of ionic endgroups on thermal properties of the PEI ionomers. DSC indicated that the PEI samples were amorphous and a detectable melting or crystallization processes was not observed. An increase in the level of dodecanol resulted in lower molecular weight polymers and as expected, lower glass transition temperatures (Table 1). Interestingly, the glass transition temperatures of all sulfonate terminated PEI-SSBA polymers were approximately 60 °C, indicating that the values were independent of the ionic group (SSBA) concentration (Table 2) at these ionic levels. The PEI ionomers displayed lower molecular weights as a result of the successful incorporation of the ionic end groups. The formation of ionic aggregates restricts polymer chain end mobility, and results in higher glass transition temperatures.(21,22) Although the molecular weights were significantly lower, the glass transition temperatures remained unchanged (58-60 °C) due to the occurrence of ionic aggregation at the chain ends. It is presumed that the ionic aggregates in the PEI ionomers were multiplets (2-8 ion pairs, < 50 Å), rather than clusters (phase separation, > 50 Å). A single glass transition temperature was observed for the ionomers and ionic concentrations were in the typical range for multiplets.(5) DSC indicated the absence of a crystallization exotherm or melting endotherm, and this further confirmed that the ionic terminated polyesters possessed relatively low crystallinity.

Table 3. Summary of DSC, GPC and NMR results for polyesters prepared using different end capping agents

| PEI with different end cappers | $T_g$ (°C)[a] DSC | $M_n$ (g/mol) GPC[b], NMR[c], or Calculation[d] |
|---|---|---|
| Non-terminated PEI | 57 | 37,800 [b, c] |
| PEI-5%SSBA | 62 | 8,400 [c, d] |
| PEI-12%Dode-OH | 41 | 8,400 [b, c] |
| PEI-5%Benzoic Acid | 57 | 37,300 [b, c] |
| PEI-5%Benzyl Alcohol | 57 | 32,400 [b, c] |

[a] Glass transition temperature was measured using DSC under nitrogen, second scan was used for the data collection, 10 °C/min
[b] Number average molecular weight that was determined using GPC in $CHCl_3$.
[c] Number average molecular weight that was calculated using $^1H$ NMR in $CHCl_3$.
[d] Number average molecular weight that was calculated using the Carothers' equation .

A summary of the thermal properties of the ionic and non-ionic PEI samples is provided in Table 3. Low molecular weight ionic PEI with 5 mol % SSBA (PET-5%SSBA, $M_n$ = 8,400 g / mol), displayed a higher glass transition temperature ($T_g$ = 62 °C) than non-terminated high molecular weight PEI ($M_n$ = 37,800 g / mol, $T_g$ = 57 °C) (Table 3, first two samples). The second and third samples shown in Table 3.3 displayed the same low molecular weight, however, the PEI-12%Dode-OH showed a much lower glass transition temperature. It is proposed that the PEI-5%SSBA ionomer had a higher glass transition temperature due to the telechelic ionic interactions. Benzoic acid and benzyl alcohol were also studied as end capping reagents to control the molecular weight of the PEI, however, higher volatility prevented their utility. TGA indicated that all samples were thermally stable to 400 °C under nitrogen.

## 3.4. Solution Viscometry Study

Solution viscometry was used to study the solution behavior of the PEI samples. For the non-terminated PEI, Huggins ($\eta_{sp}/c$) and Kramer ($\ln\eta_r/c$) equations were employed,(23) and two straight lines were obtained (Figure 2, left). However, telechelic ionomers displayed two curves with upward curvature (PEI-3%SSBA, Figure 2, right). The observed non-linear relationship between solution viscosity and (dilute) polymer concentration is proposed to arise from the ionic interactions in solution. Ionic aggregation varies as a result of different extents of ionic association at various solution concentrations. It is reasonable to expect that the dilute solution viscosity behavior of the ionomers would not resemble the linear behavior of non-ionic polymers. Moreover, these results are consistent with the previous literature for telechelic sulfonated ionic liquid crystalline systems.(24)

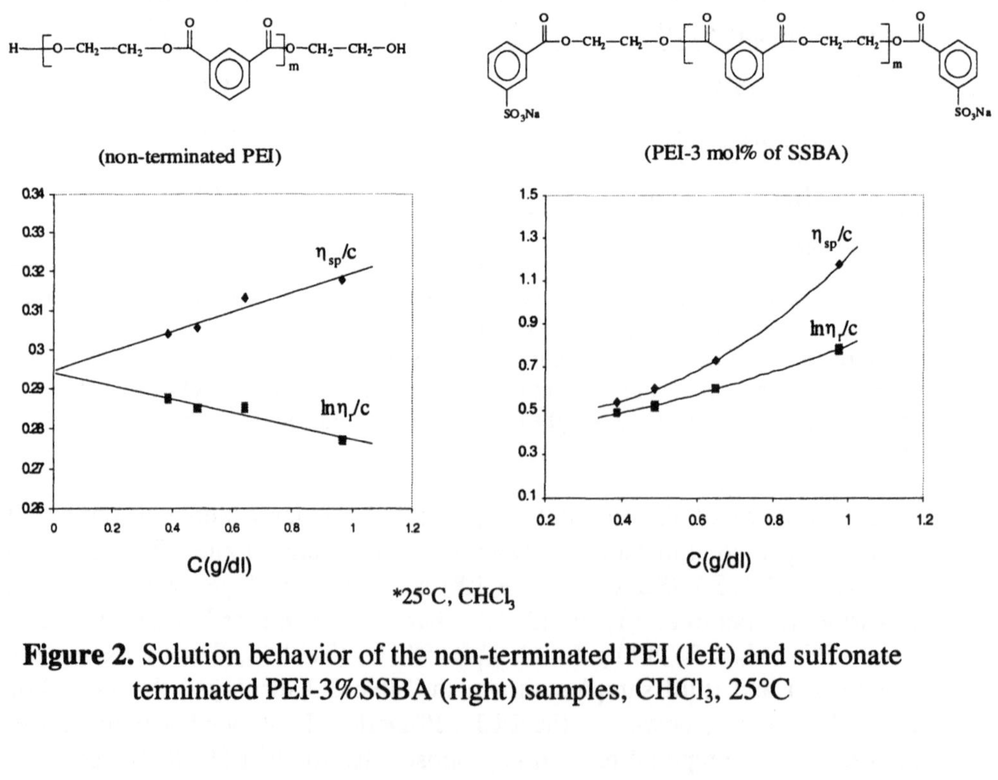


**Figure 2.** Solution behavior of the non-terminated PEI (left) and sulfonate terminated PEI-3%SSBA (right) samples, $CHCl_3$, 25°C

### 3.5. Melt Rheology Study

Although melt rheology is not an efficient technique for the determination of the structure and organization of the ionic aggregates, it is a viable tool for identifying the size and number of aggregates.(25) Thus, it effectively complements direct characterization techniques and provides useful insights.(2-4) Any factor that influences the nature of the ionic aggregation e.g., the anion type, counterion type, degree of neutralization, and flexibility of the polymer backbone, will also influence the melt rheology of the ionomers.(26-31) Figure 3 depicts the results of non-terminated high molecular weight PEI and sulfonate terminated PEI-1%SSBA and PEI-5%SSBA polymers. At lower temperatures (<150 °C), PEI-1%SSBA and PEI-5%SSBA displayed higher viscosity than that of the non-terminated PEI, which behaved similarly to a higher molecular weight polymer. The formation of ionic aggregates at these lower temperatures (<150 °C) is presumed to be responsible for the higher viscosities. However, the ionic clusters dissociated at higher temperatures (>150 °C), and the ionomers behaved as lower molecular weight oligomers. The melt viscosity behavior for both telechelic ionomers was nearly identical. These attractive rheological characteristics may have significant implications in melt processing. At higher or processing temperature, the ionomers will exhibit lower melt viscosities due to the dissociation of ionic aggregates. On the other hand, the materials will maintain sufficient mechanical and thermal properties at application temperatures due to re-association of the ionic aggregates. Thus, the effective molecular weight of an ionomer, is thermo-reversible as a result of the ionic aggregate association/dissociation mechanism.

In summary, the existence of ionic and alkyl groups at polyester chain ends dramatically altered the thermal and rheological properties. The telechelic ionic polyesters displayed relatively higher viscosities and higher glass transition temperatures than non-ionic PEI analogs, primarily due to the formation of ionic aggregations that limited the mobility at the polymer chain end. The higher the ionic group content, the stronger the ionic interaction. This association was confirmed by systematic studies on the various physical properties of the ionomers and dodecanol terminated analogs.

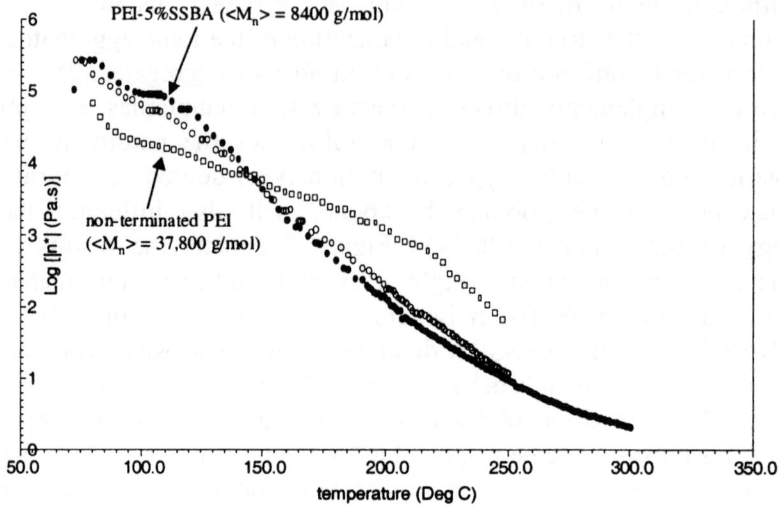


$N_2$, 1 rad/sec, step isothermal

**Figure 3.** Melt rheological behavior of high molecular weight non-terminated PEI and sulfonate terminated PEI-1%SSBA and PEI-5%SSBA, $N_2$, 1 rad/sec, step isothermal

# 4. CONCLUSIONS

The introduction of thermo-reversible bonds to polymeric materials is permits the control of the thermal and rheological properties. Telechelic ionic bonding is an important avenue for the thermo-reversible association. Moreover, telechelic ionic bonding is a good model system due to the well-defined distance between ionic end groups. Aromatic amorphous polyesters, such as PEI, are also a useful model for high performance semicrystalline aromatic polyesters, such as poly(ethylene terephthalate). Novel sulfonate terminated amorphous polyester ionomers (PEI-SSBA) were successfully synthesized using DMI and EG (100 % excess) with varying amounts of sodium sulfonate end capper. Dodecanol terminated non-ionic PEI-Dode-OH polymers were also synthesized for comparison.

It has been shown that melt polymerisation is suitable for the introduction of a sulfonate endgroups forming a novel family of telechelic ionomers. NMR, FTIR and solution viscometry confirmed the presence of the ionic and alkyl end groups. DSC revealed that the low molecular weight ionomers displayed glass transition temperatures comparable to those of high molecular weight non-terminated polyesters. Melt rheology demonstrated that the polyester ionomers displayed higher viscosities than higher molecular weight analogs at temperatures less than 150 °C. The ionic aggregates were significantly weakened above 150 °C and the melt viscosities decreased to levels more typical of non-ionic polymers of similar molecular weight.

## 5. References

1. Eisenberg, A.; King, M. Ion-containing Polymers; Academic Press: New York, 1998; p7.
2. MacKnight, W. J.; Earnest, T. R., Jr. J. Polym. Sci., Macromol. Rev., 16, 41 (1981).
3. Fitzgerald, J. J.; Weiss, R. A. J. Macromol. Sci., Rev. Macromol. Chem. Phys., C28, 99 (1988).
4. Tant, M. R.; Wilkes, G. L. Macromol. Sci., Rev. Macromol. Chem. Phys., C28, 1 (1988).
5. Greener, J.; Gillmor, J. R.; Daly, R. C. Macromolecules, 26, 6420 (1993).
6. Chassenieux, C.; Tassin, J.; Gohy, J; Jerome, R. Macromolecules, 33, 1796 (2000).
7. Sobry, R.; Fontain, F.; Ledent, J.; Foucart, M.; Jerome, R. Macromolecules, 31, 4240 (1998).
8. Broze, G.; Jerome, K.; Teyssie, P.; Macro, G. Polym. Bull., 4, 241 (1981).
9. Broze, G.; Jerome, K.; Teyssie, P. J. Polym. Sci., Polym. Lett. Ed., 28, 8504 (1981).
10. Broze, G.; Jerome, K.; Teyssie, P. J. Polym. Sci., Polym. Lett. Ed., 28, 8504 (1981).
11. Kumar, U.; Kato, T.; Frechet, J. M. J. J. Am. Chem. Soc., 114(17), 6630 (1992).
12. Sijbesma, R. P.; Beijer, F. H.; Brunsveld, L.; Folmer, B. J. B.; Hirschberg, J. H. K. K.; Lange, R. F. M.; Lowe, J. K. L.; Meijer, E. W. Science, *278*, 1601 (1997).
13. Chien, M. L.; Griffin, A. C. Macromol. Symp., 117, 281-290 (1997).
14. Hilger, C.; Drager, M.; Stadler, R. Macromolecules, 25, 2498 (1992).
15. Stadler, R. Macromolecules, 21, 121 (1988).
16 Rao, B. R.; Datye, K. V., Textile Chemist and colorist, 1996, October, 17
17. Toray Industries Inc. JP 8006514, January 1980.
18. Kang, H.; Long, T. E. ACS Polymeric Materials Science and Engineering: American Chemical Society: San Diego, April 2001, Vol. 84, p909.
19. Kang, H.; Long, T. E. Proceedings of the ACS Division of Polymer Chemistry: Biennial meeting: Hawaii, Dec. 2000
20. Nakanishi, K. Infrared Absorption Spectroscopy (Practical); Holden-Day Inc.; San Francisco, 1962; p54.
21. Ogura, K.; Sobue, H.; Nakamura, S. J. Polym. Sci., Polym. Phys., 11, 2079 (1973).
22. Otocka, E. P.; Kwei, T. K. Macromolecules, 1, 401 (1968).
23. Joseph C. Salamone, Polymeric Materials Encyclopedia; Boca Raton: CRC Press, 1996; Vol. 6, p3473.
24. Zhang, B. and Weiss, R. A. Polym. Sci., Part A: Polym. Chem., 30, 989 (1992).

25. Greener, J.; Gillmor, J. R.; Daly, R. C. Macromolecules, 26, 6416 (1993).
26. Weiss, R. A.; Agarwal, P. K. J. Appl. Polym. Sci., 26, 449 (1981).
27. Weiss, R. A.; Agarwal, P. K. J.; Lundberg, R. D. SPE ANTEC Proc. 1984, 468.
28. Weiss, R. A.; Fitzgerald, J. J.; Kim, D. Macromolecules, 24, 1071 (1991).
29. Makowski, H. S.; Lundberg, R. D.; Westerman, L.; Bock, J. In Ions in Polymers; Eisenberg, A. Ed.; ACS Polymer Preprints: American Chemical Sociaty: Washington, DC, 1980; p3.
30. Lundberg, R. D.; Makowski, H. S. In Ions in Polymers; Eisenberg, A. Ed.; ACS Polymer Preprints: American Chemical Society: Washington, DC, 1980; p21.
31. Bagrodia, S.; Pisipati, R.; Wilkes, G. L.; Storey, R. F.; Kennedy, J. P. J. Appl. Polym. Sci., 29, 306 (1984).

Chapter 19

# SYNTHESES, CHARACTERIZATION AND APPLICATION OF FUNCTIONAL CONDENSATION POLYMERS FROM ANHYDRIDE MODIFIED POLYSTYRENE AND THEIR SULFONIC ACID RESINS

*Sakuntala Chatterjee Ganguly*
**SAKCHEM, Consultant, 357A Invermay Road, Mowbray, Tasmania, Australia, button3@hotmail.com*
*Contribution from Department of chemistry, Indian Institute of Technology, Kharagpur 721302, India*

## 1. INTRODUCTION

### 1.1 Functional Condensation Polymer

Recent years, functional condensation polymers are of abundant use due to their functionality. Functional condensation polymers can be broadly classified as follows:

1) Synthesis and chemical modification of a polymer in bulk
2) Surface modification of a polymer by chemical modification and
3) Surface modification of a polymer by Inter penetrating network (IPN) formation

#### 1.1.1 **Synthesis and chemical modification of a polymer in bulk**

In this process, the required functional group can be introduced to the support by two ways:

a)It can either be incorporated during the synthesis of the homopolymer or copolymer

$$X\text{-}Ar(SO_3H)\text{-}X + HO\text{-}Ar'\text{-}OH \rightarrow [\text{-}Ar(SO_3H)\text{-}O\text{-}Ar'\text{-}O]_n$$

b)By chemical modification of a non-functionalized preformed homopolymer or copolymer matrix.

The present study highlights the salient developments in the field of functional condensation polystyrene with monoanhydrides, dianydrides and their potential application as a coating material.

In our previously published work, the author reported the electrophilic substitution of Polystyrene(PS) by phthalic anhydride(PA)(1), trimellitic anhydride(TMA)(2), pyromellitic dianhydride (PMDA)(1), cis-1,2,3,6 tetrahydrophthalic anhydride (THPA)(3). PS was further modified by polycyclic dianhydrides like perylene 3,4,9,10 tetracarboxylic dianhydride(PTDA), naphthalene 1,4,5,8-tetracarboxylic dianhydride(NTDA) and one fluorine containing dianhydride 2,2-bis(3,4-dicarboxyphenyl dianhydride (6FDA) (4) to improve the thermal stability of PS . The substitution reactions involving polystyrene with anhydrides were carried out following the steps illustrated in Figures 1 and 2.

PSPA, PSTMA, PSTHPA, PSPMDA, PSNTDA, PSPTDA and PS6FDA are weak ion exchange resin as containing –COOH group. The morphology of the chemically modified polystyrene by phthalic anhydride in a nonsolvent for modified polymer, i.e., nitrobenzene medium gave a porous surface (5).

During a technology transfer program, we have optimized the best reaction conditions for the synthesis of PSPMDA (6).

It has been observed that Polystyrene-divinyl benzene (PSDVB) based copolymers are characterized by an inhomogeneous pore structure due to different reaction rates of the monomer addition giving a random co-polymer like A-A-B-B-A-A-A type. The inhomogeneous pore structure is believed responsible for their low osmotic ability and insufficient permeability to large ions. This is also an undesirable feature for membrane fabrication because a wide pore size distribution on a microlevel may affect the separation processes. Davankov and Tsyurupa (7,8) have shown that in the case of crosslinking after polymerization process, the crosslinking agent is evenly distributed throughout the high concentrated polymer solution. A crosslinking of some statistical distribution is feasible. The crosslinked products are termed macro net isoporous polymers.

We have developed a novel procedure of membrane fabrication from pyromellitic dianhydride modified polystyrene with controlled pore size on micro- and macrolevels (9) where the easy and controlled reaction procedure of copolymerization crosslinking of the Friedel-Crafts acylation reaction of PSPMDA was exploited.

The gelation time for PSPMDA formation is 15-30 min at 80-100°C temperature and nitrobenzene which is a solvent for the reaction is a non-solvent for the product and upon addition of benzene to the thin film, egression of nitrobenzene by ingression of benzene will take place through the mesh of the network (10).

We have studied the possibility of technology transfer for the synthesis of PSPMDA in large scale in terms of process design, mechanical design of selected equipment and an estimate of cost (11).

The modified PS framework, so-called matrix, consists of three- dimensional network. The aromatic moieties in the network has further facilities to incorporate $-SO_3^-$,$-PO_3^{2-}$ , $-AsO_3^{2}$ groups.

We have, also, reported the syntheses of sulfonic acids resins from PSPA, PSTMA, PSTHPA and PSPMDA (12,13). Both isothermal and non-isothermal stability of the $H^+$ form of PSPAS, PSTMAS, PSTHPAS and PSPMDAS were studied and compared with that of the Na-form of the resins (14). The DTA and TGA studies of the modified polystyrene based resins revealed a better thermal stability compared to sulfonated polystyrene as well as commercial polystyrene divinyl benzene resins like Zeo-carb 225 and Indion 225(available in India) (15).

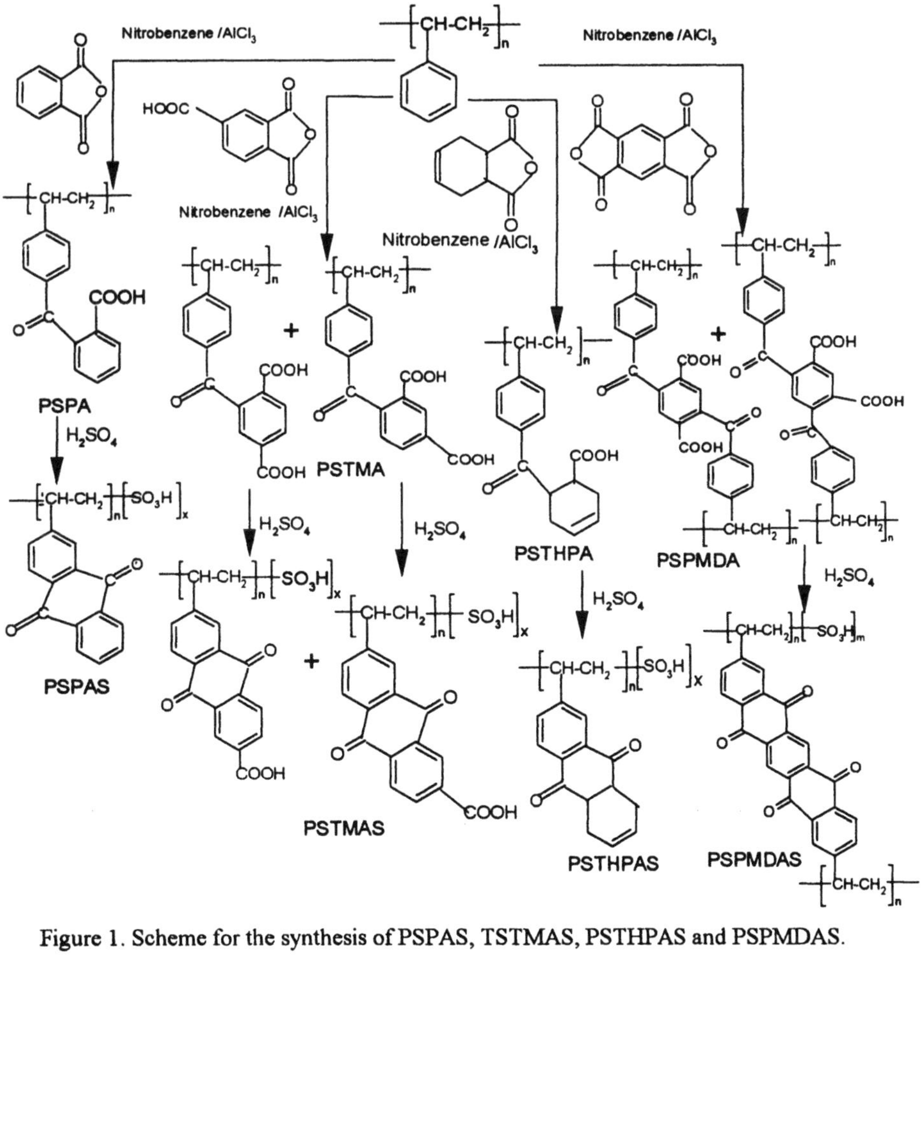


Figure 1. Scheme for the synthesis of PSPAS, TSTMAS, PSTHPAS and PSPMDAS.

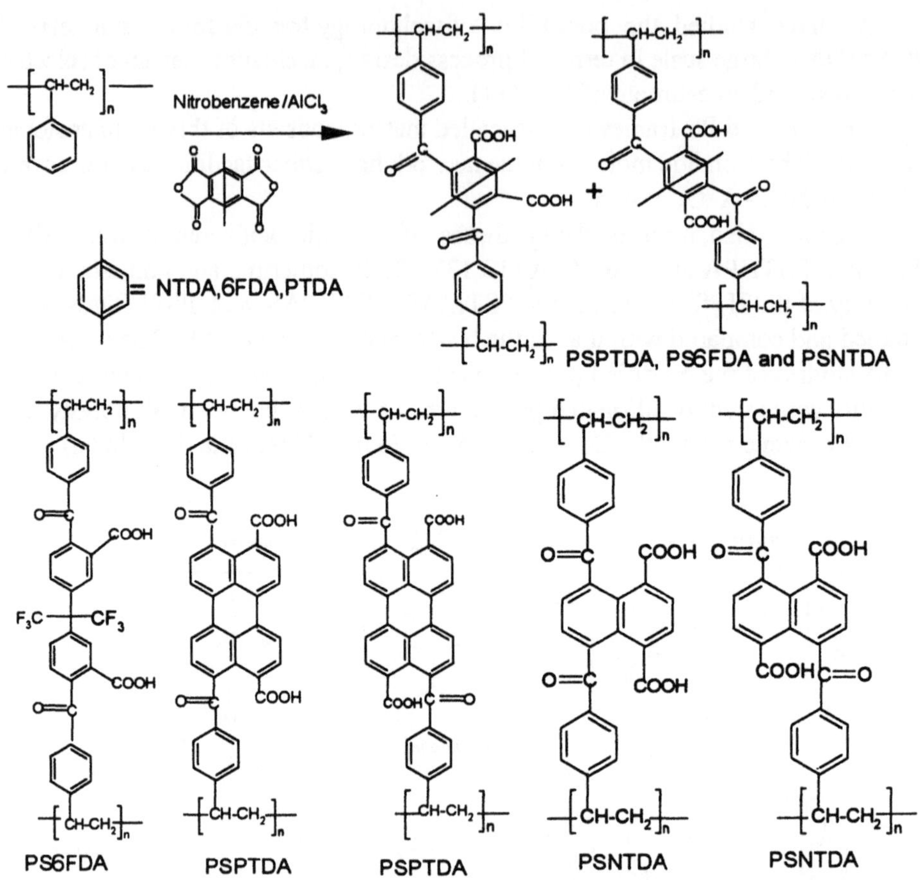


Figure 2. *Scheme for the synthesis of PSPTDA, PSNTDA and PS6FDA*

### 1.1.2 Surface modification of a polymer by chemical modification

Surface modification of a polymer can be done as follows.

**i)By discrete molecule:** Some novel surface modification of polysulfone hollow fibers using the Friedel-Crafts reaction were developed by Higuchi et al. (16-17). A new functional group, $-CH_2CH_2CH_2SO_3^-$, and $-CH(CH_3)CH_2OH$ are introduced on the surface of polysulfone ultrafiltration membranes (Figures 3 and 4).

**ii)By another polymer**: Present work will subsequently describe the surface modification of high strength K-100 Teflon membrane to a fictionalized membrane using chemical modification where competitive Friedel Crafts acylation reaction between pyromellitic dianhydride with polystyrene and Friedel-crafts alkylation reaction between polystyrene over the K-100 Teflon membrane in the presence of

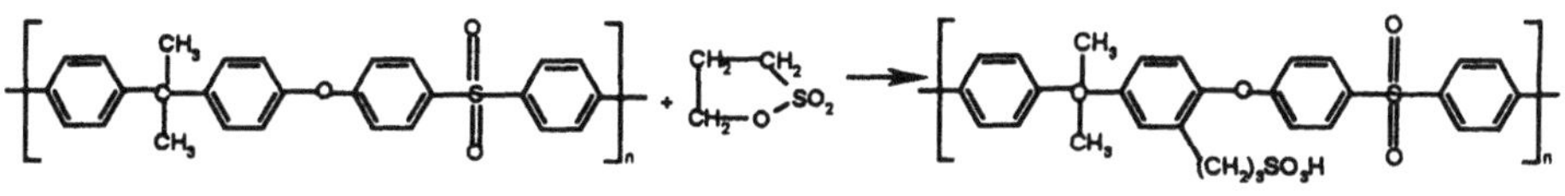

Figure 3. *Scheme for polysulfone modified propane sultone*

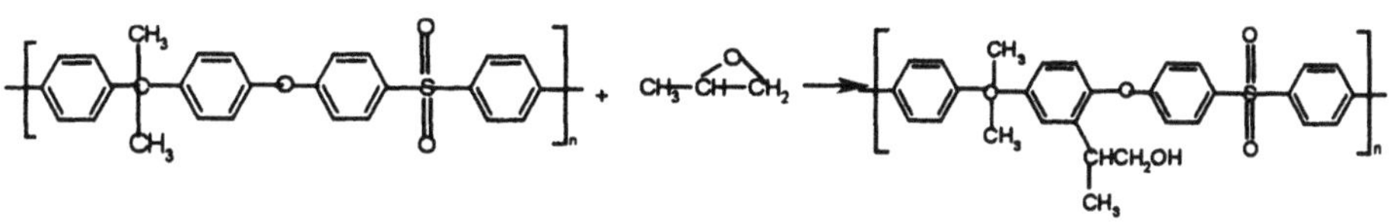

Figure 4. *Scheme for polysulfone modified with propylene oxide*

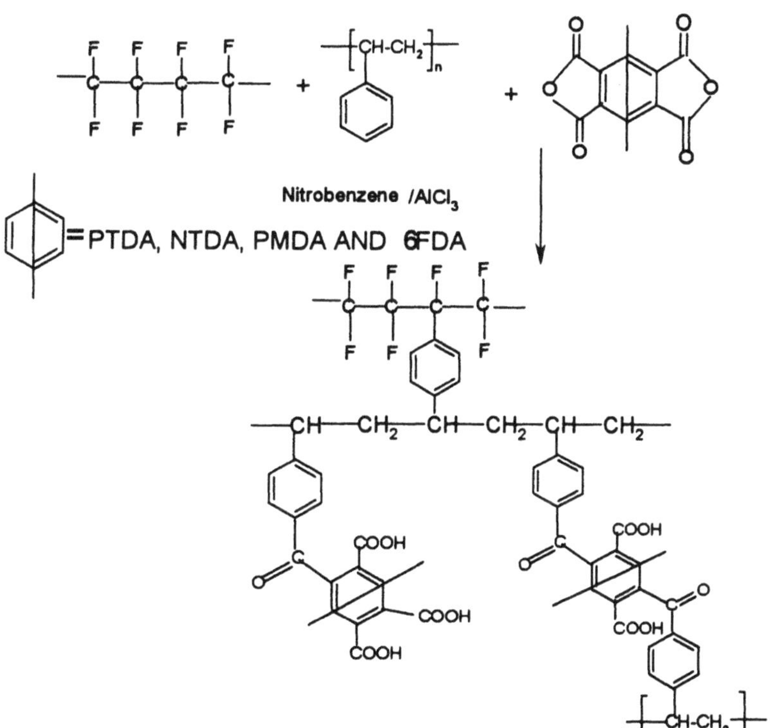


Figure. 5. *Route for the synthesis of surface modified Teflon*

$AlCl_3$ as a catalyst takes place. Using the same way, Nafion 417,perfluoro membrane, was Chemically modified by polystyrene and with PTDA , NTDA and 6FDA respectively. Figure. 5 (18).

### 1.1.3 Surface modification of a polymer by Inter penetrating network (IPN) formation

Interpenetrating networks represents a class of polymer in which crosslinked polymer, specially an ion exchange resin (IER) experiences different physical properties due to the locking of two-polymer network (perfect network or full IPN) or one polymer network and a linear polymer (semi-network or pseudo-IPN) (19). The overall sulfonic acid resins with interpenetrating polymer network, IER-IPN, acts as a functional condensation polymer. The development of the studies on the theory of IER-IPN and its applications were studied by different groups (20-22).

For the preparation of sulfonic acid resin IPN, the method used by Millar (21) is known as the dry method. Another method is known as wet method (23) where the IPN of monomer onto standard polystyrene beads were carried out in suspension.

## 2. EXPERIMENTAL

### 2.1. Material

Polystyrene (DP=480) was supplied by Poly Chem., India, Ltd. Phthalic anhydride(BDH) was purified by sublimation. Trimellitic anhydride, , cis-1,2,3,6 tetrahydrophthalic anhydride, Pyromellitic dianhydride, Napthalene 1,4,5,8 tetracarboxylic dianhydride (NTDA) and Perylene 3,4,9,10 tetracarboxylic dianhydride (PTDA) were obtained from Fluka AG, Buches, Switzerland, and purified at 130°C under vacuum. 2,2-bis (3,4-dicarboxyphenyl)hexafluoropropane dianhydride (6FDA) was received from Hoechst Chemikalien Co which was dried in vacuo at 160°C for 24h prior to use. Anhydrous aluminium chloride RIEDEL, West Germany) certified as "Sublimed for synthesis" was used without further purification. For some reactions fresh anhydrous aluminium chloride was used from Aldrich Chemicals. Fuming sulfuric acid (containing 20% SO3) from International Chemical Industries, India was used. Nitrobenzene (BDH or Aldrich Chemicals) was distilled and a middle fraction was collected. Sulfonated styrene-divinyl benzene copolymers of commercial importance, Indion 225 and Zeo-carb 225, were received from Ion-exchange (India) Ltd. and Iono-Chem (India) Ltd. K-100 Gore of 20 nm was received as a gift. Nafion 417,perfluorinated membrane was obtained from Aldrich Chemicals.

## 2.2. Synthesis

### 2.2.1. Preparation PSPA, PSTMA, PSTHPA and PSPMDA

In a 250 ml round bottomed flask, polystyrene (5g) was dissolved in 50 ml of nitrobenzene followed by slow addition of phthalic anhydride (10 g) and $AlCl_3$ (5g) powder with continuous stirring (magnetic stirrer). The reaction mixture was kept at 0°C for 15-30 min. when color changes to light yellow. Then the reaction mixture was heated at 80°C for 24 h. The dark colored swollen solid mass was separated in nitrobenzene. To the cooled mass, crushed ice was added and the mixture was shaken vigorously and kept overnight to complete the hydrolysis. The insoluble mass was filtered off and washed with concentrated HCl, distilled water, and methanol, respectively. Finally, the condensate was refluxed separately with nitrobenzene and methanol several times removing unreacted PS, PA and nitrobenzene. The product was dried under vacuum at 65°C. Syntheses of PSTMA, PSTHPA and PSPMDA were done under identical conditions.

### 2.2.2. Synthesis of sulfonic acid resins PSPAS, PSTMAS, PSTHPAS, PSPMDAS and their sodium Salts

The sulfonation of PSPA, PSTMA, PSPMDA and PSTHPA was carried out directly by reacting the powder condensate (100 mesh size) with fuming sulfuric acid for 10h in a Pyrex flask under varied conditions such as using different temperatures (35°C, 50°C, 70°C and 100°C) and different quantities of fuming sulfuric acid (25,50,75, and 100g acid/g dry resin).

The contents were cooled to 20°C and were subsequently poured onto crushed ice with constant stirring. The resins, PSPAS, PSTMAS, PSPMDAS and PSTHPAS were filtered, washed several times with deionized water till the washings were free of $SO_4^{=}$ ions and finally dried at 100°C for 8 h.

**The sodium salts of the above resins were prepared** by keeping them in N NaOH solution for 24 h, filtering, and washing the solid resins with deionized water till free of alkali. The resins were finally washed with alcohol and dried.

### 2.2.3. Preparation of PSNTDA, PSPTDA and PS6FDA

For the synthesis of PSNTDA, in a 100 ml flask 2g of polystyrene was dissolved in 12 ml of nitrobenzene and 1g of anhydrous NTDA and 0.5 g of AlC13 powder were added and the reaction mixture was heated at 80°C for three days. Synthesis of PSPTDA was done using 2g of polystyrene in 14 ml of nitrobenzene and 50 mg of PTDA and 1g of AlC13 were added. The reaction mixture was heated on oil bath for three days at 80°C. Processed PSPTDA was a dark red colored solid.

Synthesis of PS6FDA was done employing otherwise identical conditions, except 4g of polystyrene was dissolved in 20cc of nitrobenzene and was treated with 1.6g of 6FDA and 0.8g of $AlCl_3$.

### 2.2.4 Preparation of PSPMDA Film

The condensation of PS with PMDA was carried out after minor modification of our former procedure in which 2g of PS was dissolved in a mixture of 9ml nitrobenzene and 1 ml of 1,2-dichloroethane. To that mixture was added 0.8g of anhydrous PMDA and it was stirred well using a magnetic stirrer. Then 5g of AlC13 was added. The solution was kept at 0°C for 30 min and then filtered through strainer to remove any undissolved lumps of PMDA or $AlCl_3$ particles. The cast solution was spread on a 10cm diameter sealed stainless steel mould of 1 mm depth and kept on a steam bath at 100°C for 4h.. Dark colour film was processed with acidic water, methanol and benzene respectively.

### 2.2.5 Preparation of dianhydride coated Teflon membrane

In a small flask 2 g of PS was dissolved in 9 ml nitrobenzene. To that mixture was added 0.8 g of PMDA, and the mixture was stirred well (magnetic stirrer). Then 5 g of $AlCl_3$ was added. The solution was kept at 0°C for 30min and then filtered through glass wool to remove particles. The cast solution was spread over a piece of membrane in a sealed stainless steel mould and kept at100°C for 4 hr.

## 2.3. Characterization

IR Spectra for PSPA, PSTMA, PSTHPA, PSPMDA and their sulfonic acid resins were taken on a Perkin-Elmer 237B Grating IR Spectrophotometer in KBr pellets. Carbon and hydrogen were estimated in a Thomas CH-Analyzer, 35. Estimation of sulfur in the resins was done by a standard gravimetric method(24). The differential thermal analysis (DTA) and the thermogravimetric analysis (TGA both in air and nitrogen up to 1000°C) of PSPA, PSTMA, PSTHPA, PSPMDA and their sulfonic acid resins were performed with a MOM derivatograph and a Stanton Red Croft TG-750 (U.K) thermo balance, respectively. The DTA and TGA of the PSPMDA membrane were studied with a DuPont 2100 model up to 800°C. The differential scanning calorimetric (DSC) studies of the samples were carried out in a Perkin-Elmer Recorder 56. In all the instruments the heating rate was maintained at 10°C/min.

Thermogravimetric analysis of the PSNTDA, PSPTDA, PS6FDA, Nafion 417 membrane and the coated Nafion 417 membranes were conducted using a SDT 2960 Simultaneous DTA-TGA thermal analyzer (TA Instruments) from room temperature to 1000°C at a heating rate of 10°C/min. The TGA analysis was carried out under an atmosphere of either nitrogen.

The total ion exchange capacity of weak and strong ion-exchange resins was determined by a recommended procedure (25,26).

For the determination of pH metric titration characteristic of PSPA, PSTMA, PSTHPA, PSPMDA and their sulfonic acid resins, incremental quantities of NaOH solution (0.1N) were added to the different mixtures containing a known weight of the resin and 10ml NaCl solution (1.0 N), keeping the total volume at 50 ml by the addition of deionized water. The equilibrium pH and the capacities were measured after 24 h of equilibration (25,26).

Solid state $^{13}$C-NMR, CPMAS, spectrum for PSPMDA film was recorded on a Bruker MSL 300s spectrometer operating at 75.47 MHz. Samples of 300mg of the polymer was packed in a 7 mm o.d. Zirconia rotor and subjected to MAS at 4 kHz. Solid state $^{13}$C-NMR, CPMAS, spectra of PSNTDA, PSPTDA and PS6FDA were recorded on a Varian Inova 300 spectrometer operating at 75.42 MHz. Samples of 300mg of the polymer was packed in a 7.5mm Zirconia rotor and subjected to MAS at 6.5kHz.

Scanning electron micrograph observation for the study of effect of solvents, nitrobenzene and 1,2 dichloroethane ,on PSPA synthesis were made using a ISI-60 model. The fracture surface of PSPA was sputter coated with gold.

SEM of membrane and PSPMDA coated K-100 Teflon membrane samples were imaged in a Hitachi S900 FESEM coated with chromium of thickness 2-4 mm.

# 3. RESULTS AND DISCUSSION

## 3.1. Characterization of PSPA, PSTMA, PSTHPA, PSPMDA

After purification PSPA, PSTMA, PSTHPA and PSPMDA were recovered as white, brown, light yellowish and grayish flaky products, respectively. All of them

Table 1. Elemental analysis data

| *Compound* | | *%C* | *%H* | *%O* |
|---|---|---|---|---|
| *PS* | *Calculated* | *92.30* | *7.69* | ------ |
| | *Found* | *92.00* | *8.13* | ----- |
| *PSPA* | *Calculated* | *76.19* | *4.76* | *19.04* |
| | *Found* | *60.07* | *6.77* | *33.16* |
| *PSTHPA* | *Calculated* | *75.00* | *6.25* | *18.75* |
| | *Found* | *73.40* | *7.05* | *19.69* |
| *PSTMA* | *Calculated* | *83.82* | *6.37* | *9.80* |
| | *Found* | *85.40* | *7.58* | *7.707* |
| *PSPMDA* | *Calculated* | *63.52* | *3.52* | *32.94* |
| | *Found* | *68.75* | *6.35* | *24.90* |

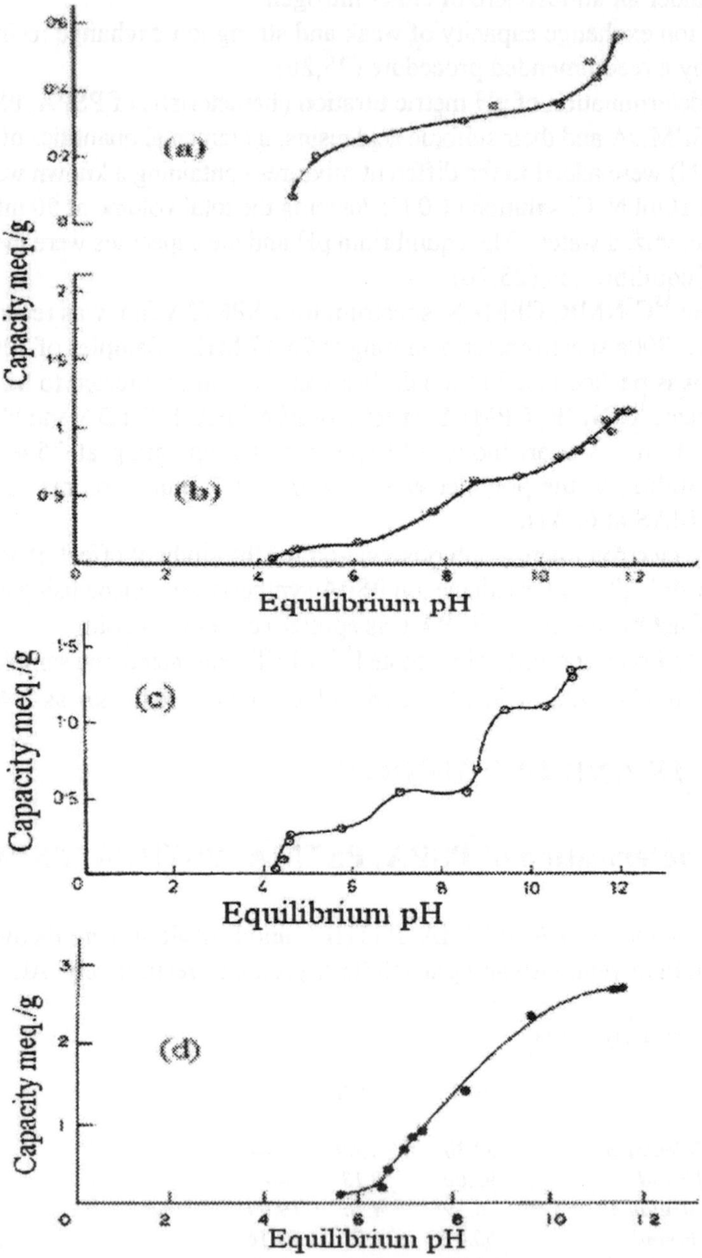


Figure 6. *Capacity vs equilibrium pH curves of a)PSPA, b)PSPMDA, c)PSTMA and d)PSTHPA*

are insoluble in organic solvents including aliphatic and aromatic hydrocarbons in which PS is freely soluble. Table 1. compiles the elemental analysis data based on suggested structures of Figure 1.The IR spectra of PSPA, PSPMDA and PSTMA indicate the stretching bands at 1700 $cm^{-1}$ typical of aromatic carboxylic acids. At 1600$cm^{-1}$ and 1670 $cm^{-1}$ peaks correspond to aromatic C=C stretching and due to diaryl ketone respectively. The IR spectrum of PSTHPA is otherwise same; expect the alicyclic carboxylic acid peak at 1700 $cm^{-1}$.

Table 2. Comparison of apparent pK values of carboxylic acid containing cation exchange resins

| *Resin* | $pK_1$ | $pK_2$ | $pK_3$ |
|---|---|---|---|
| *PSPA* | *6.85* | ------ | ----- |
| *PSPMDA* | *7,05* | *10.56* | ----- |
| *PSTMA* | *4.80* | *8.50* | *8.85* |
| *PSTHPA* | *7.53* | ---- | ------ |

The total capacities of PSPA, PSTMA, PSPMDA and PSTHPA are found to be 0.96, 1.89, 1.60, and 2.72meq/g (dry) polymer respectively. It has been observed from Table 2. that almost all of the four anhydride modified polystyrenes show a higher pK value compared with the literature value for COOH (25). In addition, both PSPA and PSTHPA show one pK value whereas PSPMDA and PSTMA show two and three pK (Figure 6) values respectively which is in accordance with the proposed structure. It is apparent from the proposed structure for PSPMDA and PSTMA that due to ortho, meta and para effect the exchange of Na+ ion in all the –COOH groups are not same. In contrast, considering the case of Matheison and Shet (27), it is not surprising to expect that due to high crosslinking in PSPMDA some –COOH groups will be H-bonded resulting in low degree of dissociation or higher pK values.

Table 3. Comparative studies of DTA peaks of PS, PSPA, PSPMDA, PSTMA AND PSTHPA

| *Polymer* | *DTA peak* |
|---|---|
| *PS* | *Exotherm begins at 240°C sharp endothermic peak at 400 °C* |
| *PSPA* | *Small exothermic peak at 260 °C and 300 °C* |
| | *Sharp exothermic peak at 450 °C and 580 °C* |
| *PSPMDA* | *Broad exothermic peak around 240-320 °C* |
| | *Sharp exothermic peak at 440 °C and 560 °C* |
| *PSTMA* | *Small exothermic effect at 210 °C* |
| | *Sharp exothermic peaks at 430 °C and 575 °C* |
| *PSTHPA* | *Two small exothermic peaks at 340 °C and 390 °C* |
| | *Sharp exothermic peaks at 450 °C and550°C* |

Comparison of TGA analysis of PSPA, PSPMDA, PSTMA and PSTHPA in air and in nitrogen have been replotted in Figure 7.(a and b). The thermal stability of PSPMDA is observed to be more or less the same as that of PSTMA up to approximately 390 °C, though the ultimate stability of the trend in the variation of the thermal stability reveals that PSTMA is thermally more stable. It is possible that additional crosslinks in the network are introduced in a non-facile way in PSTMA in which the –COOH groups are exposed during the course of weak bond formation, while in PSPA only one –COOH group will be available per pendent ring.

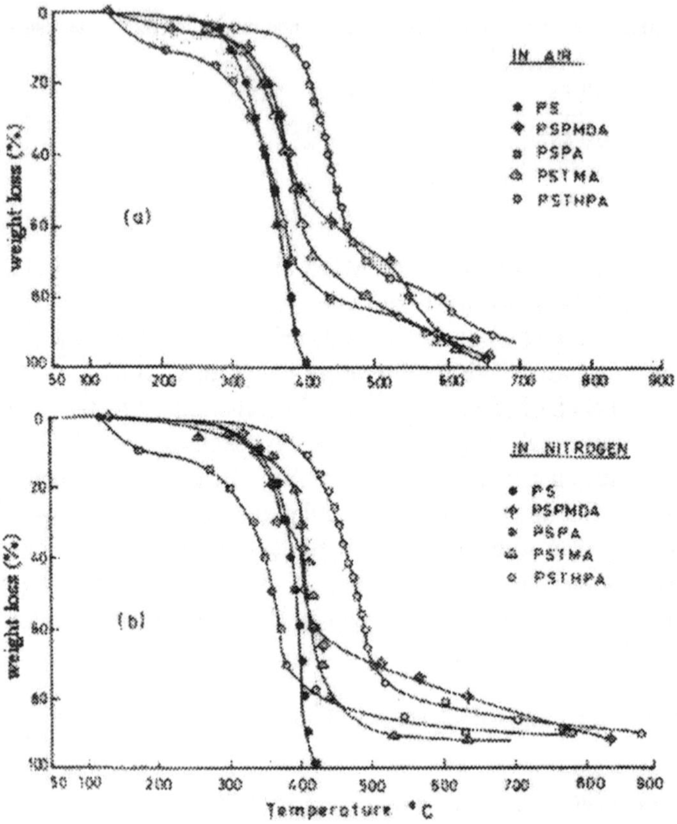


Figure 7. *TGA curves of PS, PSPMDA, PSPA, PSTMA and PSTHPA.*

PSPMDA has greater stability PS, PSPA and PSTMA.The higher thermal stability is apparently consistent with a more rigid network of PSPMDA resulting from interchain crosslinking involving PMDA moieties. About the cross-linked structure of PSPMDA, it has been reported that like the conventional cross-linking process, a small part of PMDA molecules involve at early stage (28). The remaining molecules of pyromellitic dianhydride react at first by only one of their functional groups and wait for favorable conformation of the neighboring polystyrene chain. Ultimately, the other group will react to give least unstrained structure. Hence, macrosyneresis will not stop even when the network becomes sufficiently rigid.

The Tg values as observed from Table 4. appear to be rather consistent with what is to be expected from consideration of the copolymerisation effect coupled with the addition of rigid side groups(29), followed by branching and crosslinking on the polystyrene network and formation of stiff bonds. These effects tend to decrease the flexibility causing hindrance to rotation and hence

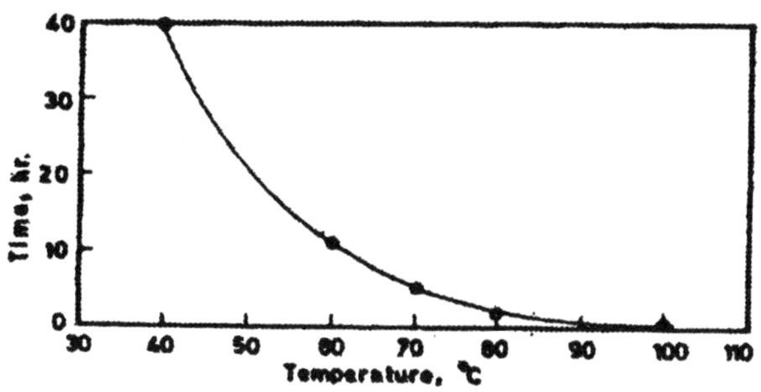


Figure. 8. *Effect of variation of temperature minimum time for complete gelation for PSPMDA.* **Reproduced Author's work from J. Appl. Polym.(6).**

increase Tg. In PSTHPA, the hydrogen bonding is much higher for the aliphatic –COOH group, than that observed for PSPA, PSTMA, whereas in PSPMDA, the crosslinking due to PMDA results in highest Tg. Therefore, the order of Tg of the four anhydride modified polystyrene is justified.

PS<PSPA<PSTMA<PSTHPA<PSPMDA

During the optimization reaction of PSPMDA, we have observed the interesting feature for the reaction is the reduction of gelation time from 40 h to 15-30 min with the change of reaction temperature 40 °C to 100 °C (Figure 8).

Table 4. Comparison of Tg for modified PS

| *Polymer* | *Tg(°C)* |
|---|---|
| *PS* | *101 or100 (2,31)* |
| *PSPA* | *112 (2)* |
| *PSTMA* | *120 (2)* |
| *PSTHPA* | *125 (2)* |
| *PSPMDA* | *142 (2)* |
| *Poly(2-methyl-styrene)* | *118.2 (32)* |
| *Poly(4-methyl-styrene)* | *118.2 (32)* |
| *Poly(2-chlorostyrene)* | *138.6 (32)* |
| *Poly(4-chlorostyrene)* | *138.6 (32)* |

We have controlled the morphology of PSPA using in a (1) nonsolvent for modified polymer, i.e., nitrobenzene medium and (2) in 1,2-dichloroethane medium, which is a solvent as well as a reactant for the reaction(Figures 9&10).

Nitrobenzene medium shows large number of holes, which are created by the removal of nitrobenzene due to the solvent action of benzene. PSPA synthesized in 1,2 dichloroethane medium; rather it shows a corrugated surface because of high substitution on the benzene ring of polystyrene due to competitive Friedel Crafts acylation and alkylation reactions.

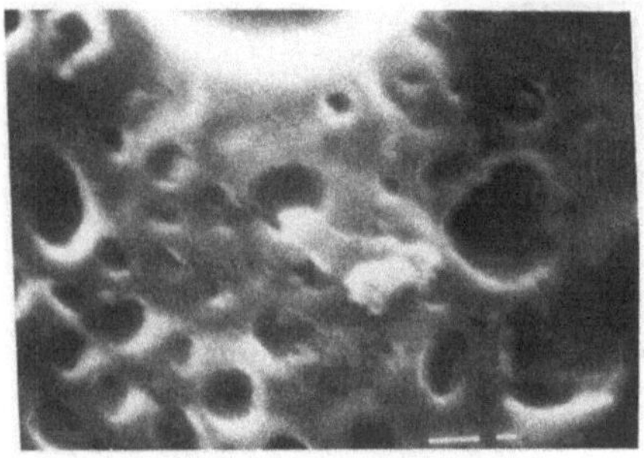

Figure. 9. *SEM of Surface of PSPA prepared from nitrobenzene 5800X*

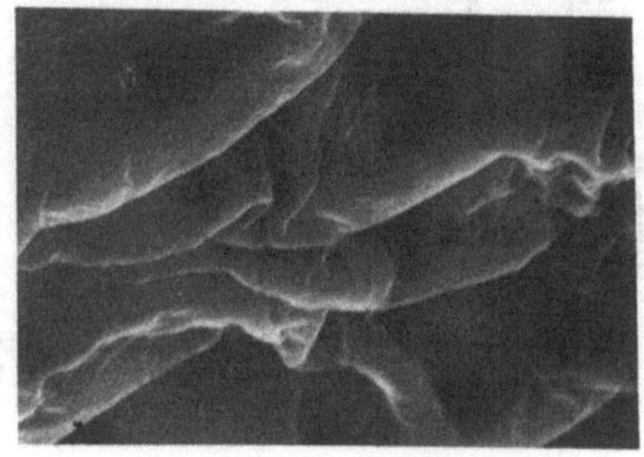

Figure.10. *SEM of Surface of PSPA prepared from dichloro ethane 4700X*

**Reproduced Author's work from J. of Appl. Polym.(5).**

## 3.2 Novel Membrane from Pyromellitic dianhydride Modified Polystyrene with controlled pore size on Micro- and Macrolevels

On the basis of the acquired data, we have fabricated a novel membrane from PSPMDA.The membrane was characterized by TGA, DTA and $^{13}C$ CP/MAS solid state NMR.The cross-section morphology is spectacular. The membrane is generally viewed as a honeycomb structure as shown in Figures 11 and 12. The pore size is approximately 6μm. It appears that the mean diameter of the pores increases from the surface. The magnification in Figure13 shows the interior of the large holes. The walls of those cavities and boundaries are also porous. The average pore size is 0.6μm. Figure 14 shows covered and uncovered holes side by side. Based on our technique of membrane fabrication and the feature of the above morphology, a possible mechanism has been proposed.

## 3.3 Structural Characterization of Sulfonic acids resins PSPAS, PSTMAS, PSTHPAS and PSPMDAS

The sulfonic acid resins, PSPAS, PSTMAS, PSTHPAS and PSPMDAS, are brittle and black in colour and insoluble in organic solvents . However, the resins swell considerably in organic solvents due to gel nature . The sulfonation of PSPMDA results in the presence of a polyquinonic structure in between the polyethylene chain which posses the most rigid cross-links, hence a highly limited conformational mobility. The sodium salt of the resins are brittle and brownish black in color and exhibit the same solubility characteristics of H-form of resins.

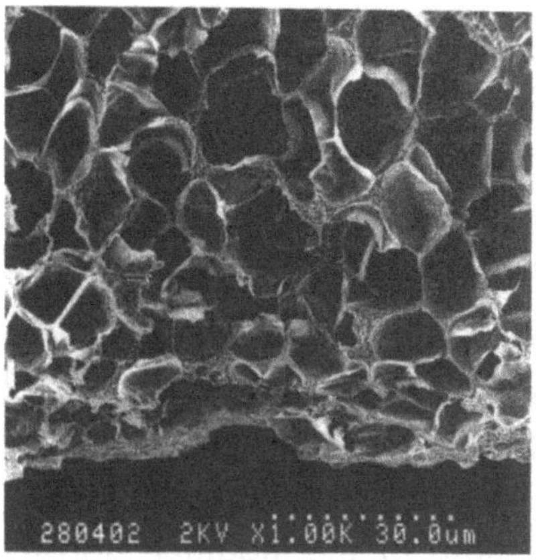


Figure.11. *SEM of side view of PSPMDA at low magnification 500X*

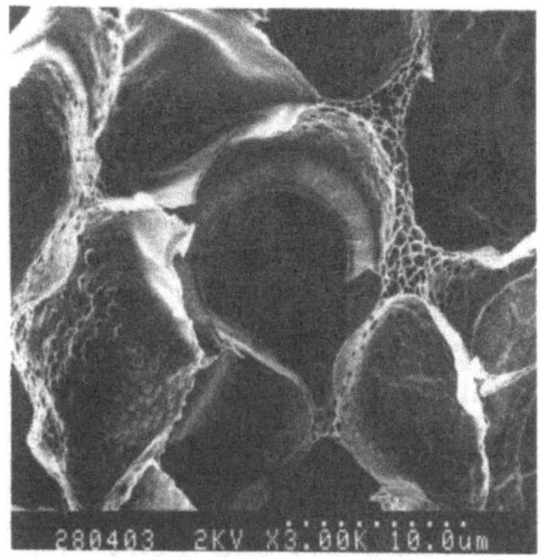


Figure.12. *SEM of side view of PSPMDA at higher magnification 3000X*

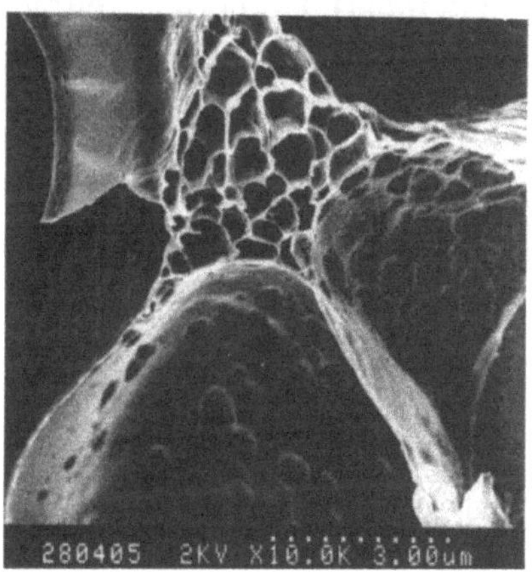


Figure.13. *SEM of side view of PSPMDA at high magnification 10,000X*

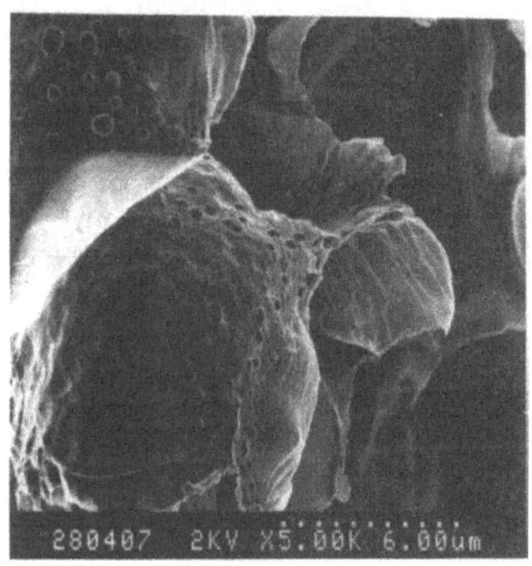


Figure.14. *SEM of side view of PSPMDA at high magnification 5000X*

**Reproduced Author's work from J. of Appl. Polym.(9).**

Table 5. summarizes the pertinent IR absorptions in PSPAS, PSTMAS, PSTHPAS and PSPMDAS, and their assignments. Significantly, no stretching band due to –COOH group appears in the IR spectra.

Incorporation of sulfonic acid group is also endorsed by the results of sulfur analysis. For sulfonation reaction of four resins, effect of variation of sulfuric acid amount, effect of variation of temperature was studied in detail. Table 6 and 7 summarize ion exchange capacities.

Table 5. Characteristic IR absorptions (cm-1)

| *Bands common to PSPAS, PSTMAS, PSTHPAS and PSPMDAS* | *Assignments* |
|---|---|
| *1600* | *C=C aromatic stretching* |
| *1710* | *Strong C=O stretching of quinonic structure* |
| | *One band for SO3H* |
| *1035, 1100-1300* | *Sharp and broad for >$SO_2$ and $SO_3H$* |

pK values for PSPAS, PSTMAS, PSTHPAS and PSPMDAS are 1.64, 1.31, 1.74 and 1.98 respectively which falls in the range of sulfonic acid resins. The pK values follow the trend: PSPMDAS>PSTHPAS>PSPAS>PSTMAS which implies that the dissociation of the resins is somewhat suppressed along the series.

The DTA and TGA (Table 8) studies of the modified polystyrene based resins have a better thermal stability compared to sulfonated polystyrene as well as commercial polystyrene divinyl benzene resins like Zeo-carb 225 and Indion 225 (available in India). Isothermal degradation of the commercial resins was studied at 250 °C +/- 10 °C for 72h in air and the analysis of the degraded product has also been carried out.

Table 6. Compares the maximum capacities and sulfur contents of PSPAS, PSTMAS, PSTHPAS and PSPMDAS

| *Resin* | *Total capacity meq/g* | *Salt splitting capacity meq/g* | *Sulfur content %* |
|---|---|---|---|
| *PSPAS* | *5.80* | *5.42* | *23.35* |
| *PSTMAS* | *5.49* | *4.98* | *19.09* |
| *PSTHPAS* | *5.74* | *5.22* | *20.60* |
| *PSPMDAS* | *5.75* | *4.58* | *16.42* |

Table 7. PSPAS, PSTMAS, PSTHPAS and PSPMDAS behave as a strong exchange resins as compared with the commercial available resins

| *Resin* | *Inorganic group.* | *Capacity(meq/g)* |
|---|---|---|
| *PSPMDAS* | *$SO_3H$* | *5.75* |
| *PSPAS* | *$SO_3H$* | *5.80* |
| *PSTMAS* | *$SO_3H$, COOH* | *5.49* |
| *PDTHPAS* | *$SO_3H$* | *5.74* |
| *Amberlite IR-120(SDV)* | *$SO_3H$, COOH* | *4.20* |
| *Amberlyst (ROHM & HAAS)* | *$SO_3H$, COOH* | *4.8* |
| *SDV copolymer (Bayer)* | *$SO_3H$, COOH* | *4.2-5.1* |

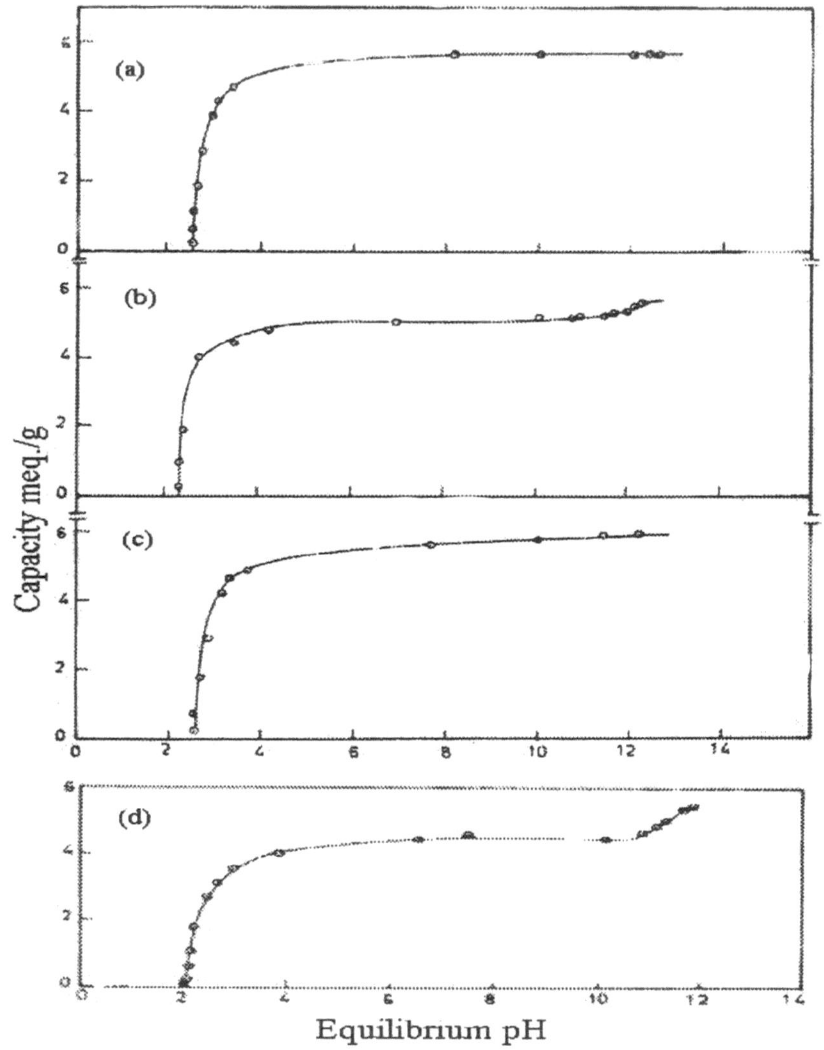


Figure 15. *Capacity vs. equilibrium pH curves of a)PSPAS, b)PSTMAS,c)PSTHPAS and d)PSPMDAS*

Isothermal degradation studies of PSPAS, PSTMAS, PSTHPAS and PSPMDAS at 150 °C+/- 10 °C for 72 h in air and nitrogen reveal significant change in IR, decrease in sulfur content, and increase in ion-exchange capacity values. The pH-metric titration characteristics of PSPMDAS after isothermal heating under the above conditions indicate weak acid behavior, in contrast to the strong acid nature of the original resin. The isothermal decomposition of the resin in water reveals a maximum capacity loss of ~ 15% in all the resins except for the PSPMDAS for which the corresponding loss is ~3.5%. The sodium salts of the resins exhibit, in general, better thermal stabilities.

Table 8. TGA, DTA for sulfonic acid resins

| | *Temperature (°C)for weight loss% in N2* | | | | | |
|---|---|---|---|---|---|---|
| *RESIN* | *10* | *30* | *50* | *70* | *100* | *MAX LOSS* |
| *PSS* | *35* | *87* | *275* | *448* | -------- | *836(78%)* |
| *PSPAS* | *51* | *280* | *366* | *536* | *629 (100%)* | |
| *PSTMAS* | *51* | *306* | *511* | *676* | ------ | *734(75%)* |
| *PSPMDAS* | *92* | *280* | *637* | ----- | ------ | *895(58%)* |
| *ZC225* | *87* | *420* | *653* | *820* | ------ | *820(70%)* |
| *IN225* | *35* | *170* | *264* | *391* | | *812(88%)* |

## 3.4 Structural Characterization of PSPTDA, PS6FDA and PSNTDA

Comparison of DTA (Figure 16) TGA (Figure 17) analyses of PS with that of PSPTDA, PSNTDA and PS6FDA confirms the incorporation of dianhydride moieties into PS . The very high thermal stability of PSPTDA, PSNTDA and PS6FDA compared to that of PS are apparently consistent with a more rigid network due to polycyclic aromatic system and presence of fluoro derivatives combining with interchain crosslinking involving dianhydride moieties.

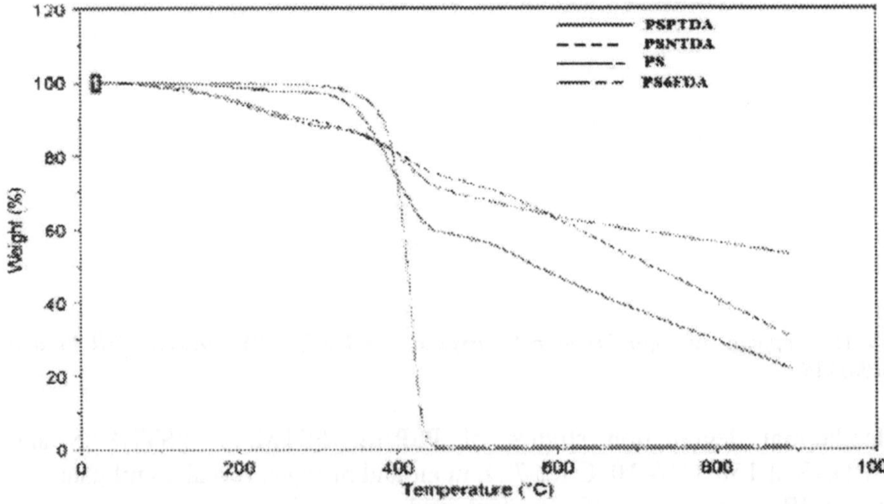


Figure. 16. *TGA Curves in nitrogen for PS; PSPTDA; PSNTDA and PS6FDA*

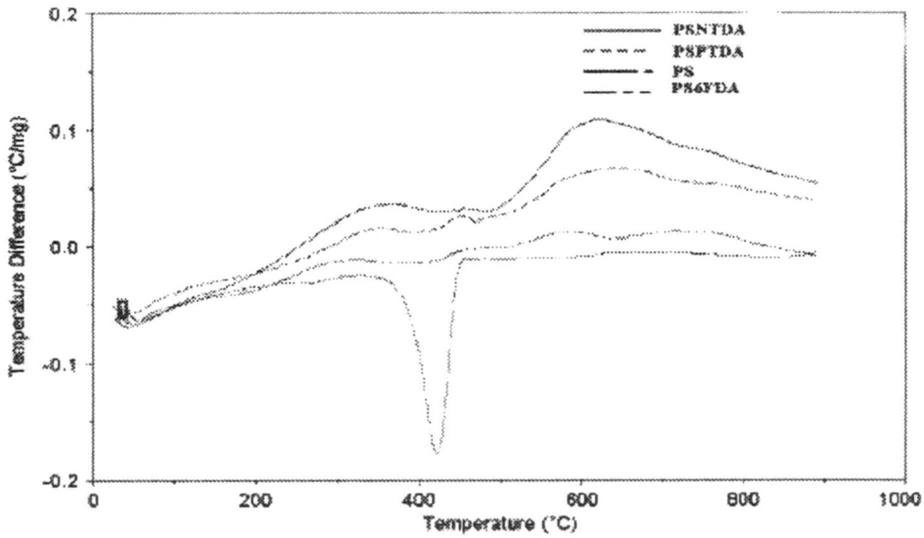


Figure. 17. *DTA Curves in nitrogen for PS; PSPTDA; PSNTDA and PS6FDA*

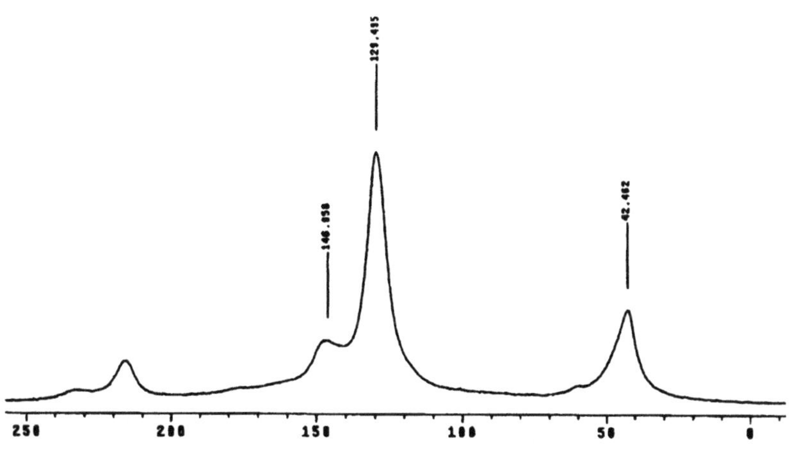

Figure. 18. *$^{13}$C CPMAS NMR of PSNTDA*

PS shows traditional endothermic peak at 400 °C whereas PSPTDA, PSNTDA and PS6FDA show exotherms. The $^{13}$C CPMAS NMR spectra of hypercrosslinked PSNTDA(Figure18), PSPTDA(Figure19) and PS6FDA(Figure 20) show peaks at 129ppm sharp peak for proton bearing aromatic carbons and small peak at 146-

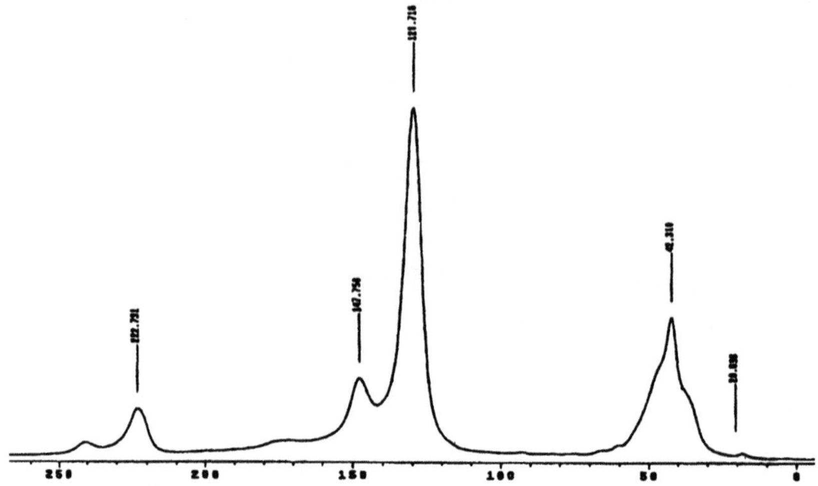


Figure. 19. [13] *C CPMAS NMR of PS6FDA*

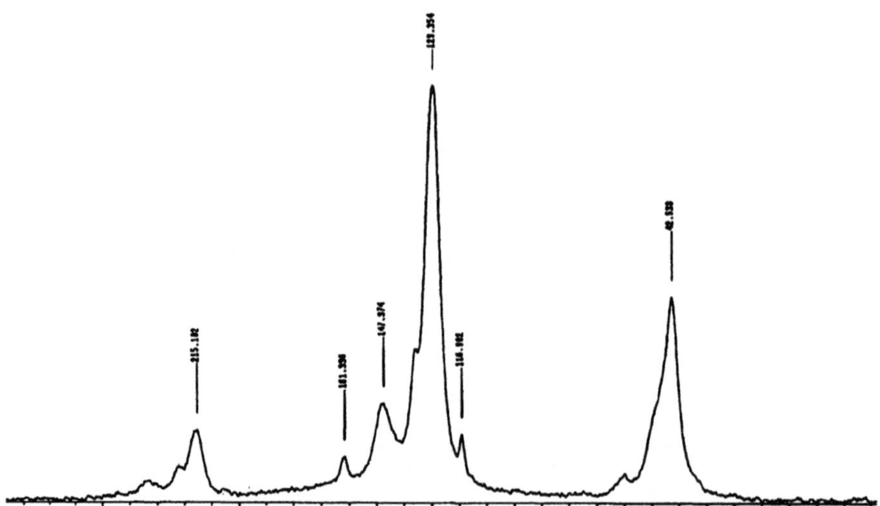


Figure. 20. [13] *C CPMAS NMR of PSPTDA*

147 for non proton bearing carbon atoms (31).

However, PSPTDA has molecular overcrowding due to poycyclic aromatic system as a result it shows three peaks at 161.33, 147.37 and 118.9 ppm for non proton bearing aromatic carbons. Peak at 42 ppm corresponds the overlapped peaks for aliphatic carbons. –COOH and –C=O peaks appeared at 215ppm and 220 ppm respectively.

## 3.5 Low Voltage Scanning Electron Microscopy of a Surface Modified K-100 Teflon Membrane and Thermalanalysis Studies of Several Anhydride Modified Nafion 417 Membrane

Kim et.al (32) have shown that with the change of beam energies on the field emission (FE) SEM images of K-100 TFE surface image from 2, 5 and 20kV, detail of the surface diminishes progressively with increasing beam due to in-

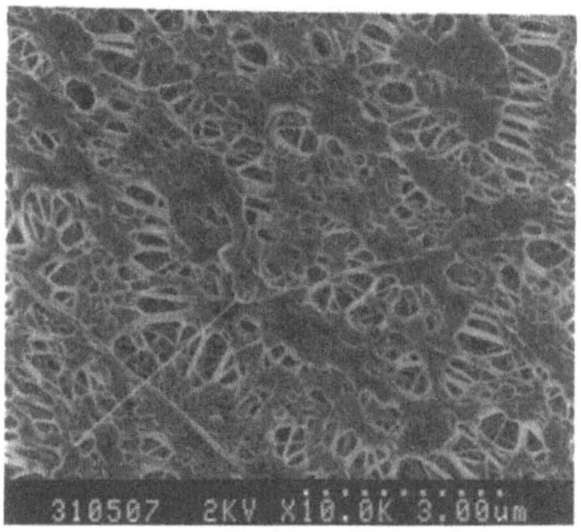


Figure 21. *SEM of top surface of a K-100 teflon membrane without PSPMDA coating*

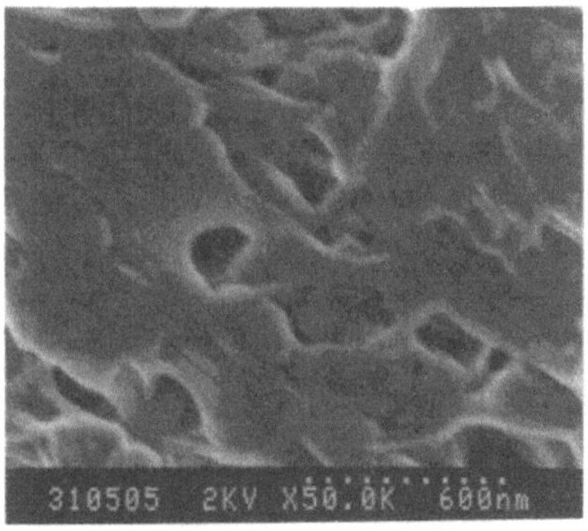


Figure. 22. *SEM of top surface of a K-100 teflon membrane lightly coated with PSPMDA*

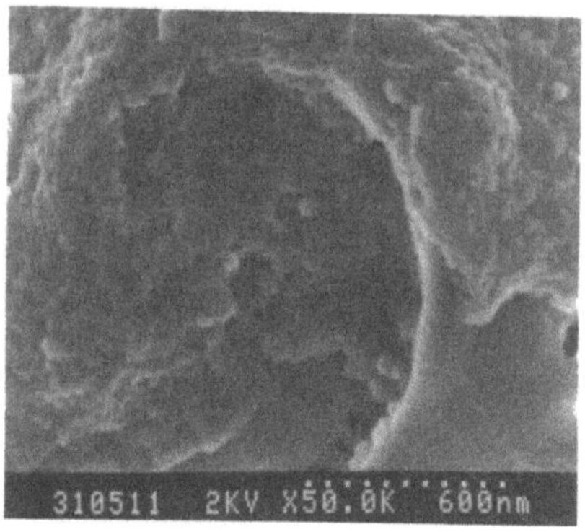


Figure. 23. *SEM of top surface of a K-100 teflon membrane with a layer of PSPMDA*

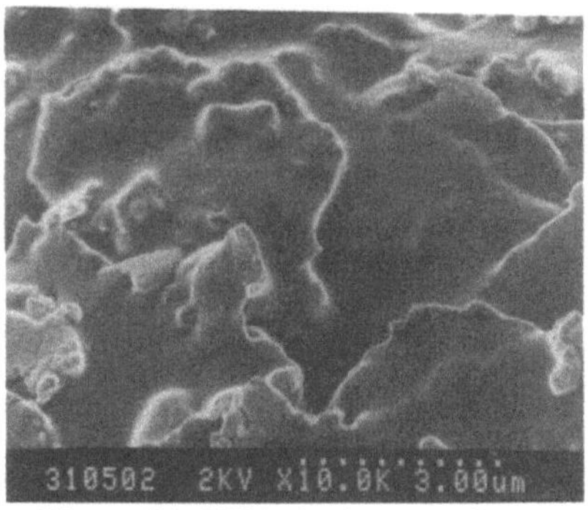


Figure. 24. *SEM of side view of a K-100 teflon membrane coated with PSPMDA*

depth sampling at higher energy. Accordingly, we have studied the modified membrane surface at 2kV beam energy. Gore's expanded PTFE membrane, K-100, is composed of billions of tiny interconnected continuous fibrils.

Figure 21 is the standard web like morphology of K-100 Teflon membrane without any coating. Figure 22 shows the top surface of K-100 Teflon membrane

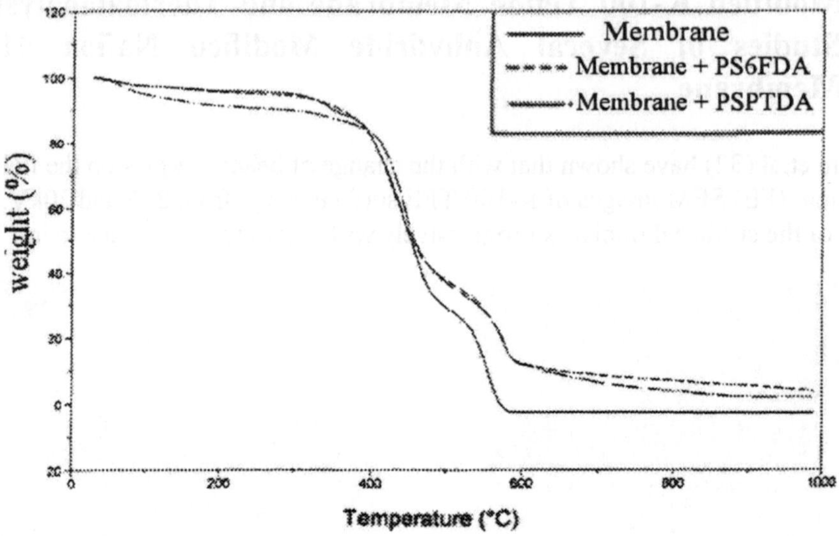


*Figure. 25.* TGA of coated Nafion417

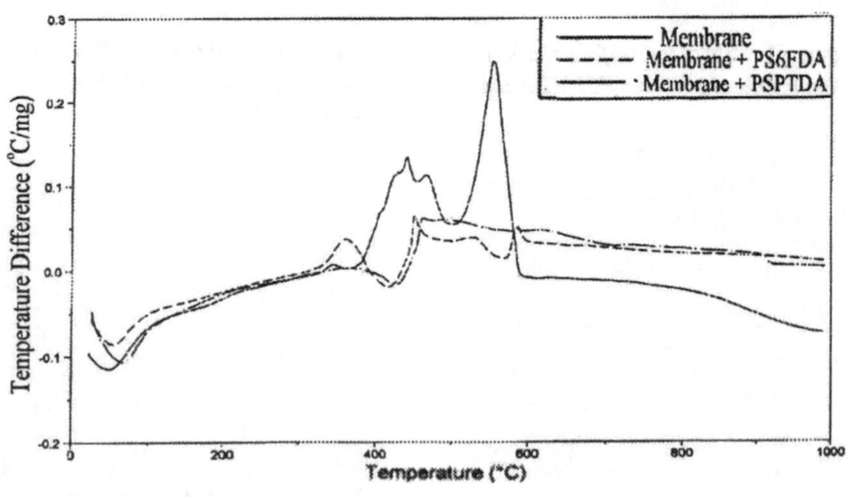


*Figure. 26. DTA of coated Nafion417*

coated with dilute solution of PSPMDA in nitrobenzene and aluminium chloride as a catalyst. Then several flicking from other side of the membrane gave a very light coating and heated as mentioned above. The morphology indicates the presence of foreign material but the pores are intact. The modification of top surface of K-100 teflon membrane with a layer of PSPMDA (Figure 23) shows that the web like surface has been filled up with a layer of polymer with some assembly of nodules and partly as a concave upper surface has shown.

However, some uncovered pores are visible there. The side view of the modified membrane is shown in Figure 24 . TGA (Figure 25) curves of anhydride modified Nafion 417 membranes show some improvement of thermal stability compared to that of unmodified Nafion 417 membrane. Surface modified Nafion 417 show better thermal stability after 420 $^0$C compare to that of unmodified Nafion membrane, which confirms the incorporation of PSPTDA and PS6FDA moiety. DTA data (Figure 26) for Nafion 417 shows sharp exothermic peak where as PSPTDA, PSNTDA and PS6FDA modified membranes show the reduction of those peaks and appearance of a small endothermic hump, which is characteristics of PS Chain.

# 4. CONCLUSION

Recycling of PS foams can be done to several functionally graded new polymeric coating materials for fluoro polymers and polystyrene itself. Ionic groups such as $-SO_3^-$,$-PO_3^{2-}$, $-AsO_3^2$ groups -can be incorporated into chemically modified polystyrene. PSPMDA is compatible with PTFE in terms of its chemical inertness and thermal stability make it ideal for a wide range of applications where exposure to harsh chemical conditions or extremes in temperatures are expected, or where biocompatibility or low chemicals extractable are required. The process has several advantages, including low cost starting material, easy reaction procedure, commercial viability as we have studied the technology transfer for the process as a) Process description and optimization, b) Process Design, c) mechanical design. The cost for the product, PSPMDA as estimated in 1984 was $8.126/kg. Surface modified membrane with pore is useful as a new membrane and surface modification with out pore and ion containing film may be used for Fuel cell, Battery Industry and Laser Technology.

**Acknowledgments:**
The author expresses her sincerest gratitude to Prof. Charles E. Carraher of Florida Atlantic University, USA for his constant encouragement and moral support to publish this work. Also, thanks to my son Amit for all his help in using my computer.

## 5. REFERENCES

1 Biswas, M.; Chatterjee, S. J. appl. Polym. Sci. 27, 3851(1982).
2. Biswas, M.; Chatterjee, S.Angew. Makromolek. Chem. 113, 11(1983).
3. Biswas, M.; Chatterjee, S, Euro. Polym. J. 19, 317(1983).
4. Ganguly, S.C.; Hook, J and Bhattacharyya, B.C.;PMSE preprint. 82, 196(2000).
5. Chatterjee, S.; Biswas, M., J.Appl. Polym. Sci., 44,619(1992).
6. Chatterjee, S.; Bhattacharyya, B.C., J. appl. Polym. Sci.33, 2769(1987).
7. Tsyurupa,M.P.; Davankov, V.A.; Rogozhin, S. V., J. Polym. Sci., Polym.Symp.,47,189(1974).
8. Tsyurupa,M.P.;Davankov,V.A.,J.Polym.Sci.,Polym.Chem.Ed.,18,1399(1980).
9. Ganguly, S.C.; Bhattacharyya, B.C. J. Appl. Polym. Sci. 69, 709(1998)
10. Hariharan, D.;Peppas, N.A. J.Membr. Sci.,78,1(1983).
11. Chakrabarti, M. Thesis, I.I.T, KharagpurIndia, 1986.
12. Biswas, M.; Chatterjee, S., J. appl. Polym. Sci. 27, 4645(1982).
13. Biswas, M.; Chatterjee, S., J. appl. Polym. Sci. 29, 829(1984).
14. Biswas, M.; Chatterjee,S.,J.Macromol.Sci.Chem.A21,1507(1984).
15. Chatterjee,S.;Biswas,M., Polymer 26,1365(1985).
16. Higuchi, A.; Iwata, N.; Nakagawa,T., J.Appl.Polym.Sci. 41,709(1990)
17.Higuchi,A.;Iwata,N.;Tsubaki,M.;Nakagawa,T.,J.Appl.Polym.Sci.,41,1973(1990).
18. Ganguly, S.C.; Matisons, J.G.;PMSE preprint.,84,347(2001).
19.Xu, H.,South China Teacher's College,2,141(1981).
20. Frisch, K.C.., J. Appl.Polym.Sci.,689(1974).
21.Millar, J.R.,J.Chem.Soc.,p1311(1960),p.1789(1962),p218(1963).
22. Kolarz, B.J.Polym.Sci.,Polym.Symp.,47,197(1974).
23. Ding, J., J.South China Teacher's College, 1,44(1981)
24. Vogel, A.I, **A Text Book of Quantitative Inorganic analysis**, The English Language Book Society, Longmans Green, London, 1962.
25. Helfferich, F., **Ion Exchange**, McGraw-Hill, New York,1962.
26. Biswas, M.; John,K.J., Angew.Makromol.Chem.,72,57(1978).
27. Mathieson, A.R.; Shet, R.T., J. Polym.Sci.,Part A-14,2945(1966).
28. Davankov, V.A.; Tsyurupa,M.P., Angew.Macromol.Chem.91,127(1980).
29. Bares, J.; Billmeyers, F.W., **Experiments in Polymer Science**, John Wiley, Inc., New York, 435(1973).
30. Van Krevelen, D.W., **Properties of Polymers, Their Estimation and Correlation with Chemical Structure**, Elsevier Scientific Publishing Company, Amsterdam,p.574(1980).
31.Joseph,R.; Ford,W.T.; Zhang,S.; Tsyurupa,M.P.; Pastukhov,A.V.; Davankov,V.A.; J.Polym.Sci. Polym.Chem.35,695(1997).
32. Kim, K.; Fane, A.G. J. Membrane Sci., 88, 103(1994).

Chapter 20

# CONDENSATION COPOLYMERIZATION VIA Ru-CATALYZED REACTION OF *o*-QUINONES OR α-DIKETONES WITH α,ω-DIHYDRIDO-OLIGODIMETHYLSILOXANES

Joseph M. Mabry and William P. Weber*
*K. B. and D. P. Loker Hydrocarbon Research Institute*
*Department of Chemistry, University of Southern California*
*Los Angeles, CA 90089-1661*

## 1. INTRODUCTION

### 1.1 Poly(silyl ether)s

Condensation polymers are those in which the repeating unit lacks certain atoms present in the monomer(s) from which it is formed (1). By comparison, addition polymers are formed from monomers without loss of small molecules.

Copolymers of poly(dimethylsiloxane) (PDMS) are of interest for a variety of applications. Copolymers, which incorporate oligodimethylsiloxane (ODMS) units may have $T_g$s similar to that of PDMS, but exhibit no detectable $T_m$s (2). They are attractive candidates for low temperature sealants and adhesives. There is also interest in the utility of poly(silyl ether)s as elastomers (3), sensor materials (4), and polymer membranes (5). Symmetrical poly(silyl ether)s have been prepared by the condensation polymerization of dialkoxysilanes and α,ω-diols with loss of alcohol (6-8).

### 1.2 Transition Metal Catalysis

There is considerable interest in transition metal catalyzed polymerizations (9,10). The instability of poly(silyl ether)s and poly(silyl enol ether)s in acidic or basic media makes transition metal catalyzed polymerization under neutral conditions attractive. High molecular weight poly(silyl ether)s have been obtained by Rh or Pd-catalyzed condensation of bis(silane)s with diols (figure 1) (11-14). We have reported the Ru-catalyzed addition polymerization of aromatic diketones and α,ω-dihydrido-oligodimethylsiloxanes to yield high molecular weight materials (15,16). Poly(silyl ether)s were recently produced by Pd-catalyzed condensation copolymerization of dihydridosilanes and *p*-quinones (figure 2) (17).

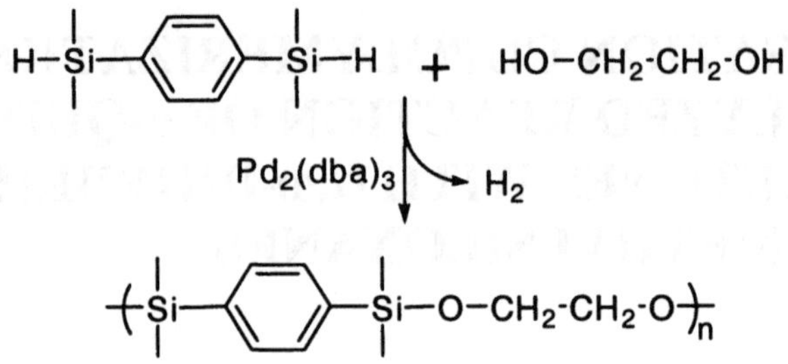


*Figure 1.* Condensation of bis(silane)s with diols.

## 1.3 Poly(silyl enol ether)s

Poly(silyl enol ether)s have been produced by radical ring-opening polymerization of trimethylsilyloxy substituted vinylcyclopropanes (18) or by radical copolymerization of 2-trimethylsiloxy-1,3-butadiene with vinyl monomers (19-21). In these polymers, the C-C double bond is part of the polymer backbone, while the trimethylsiloxy group is pendant.

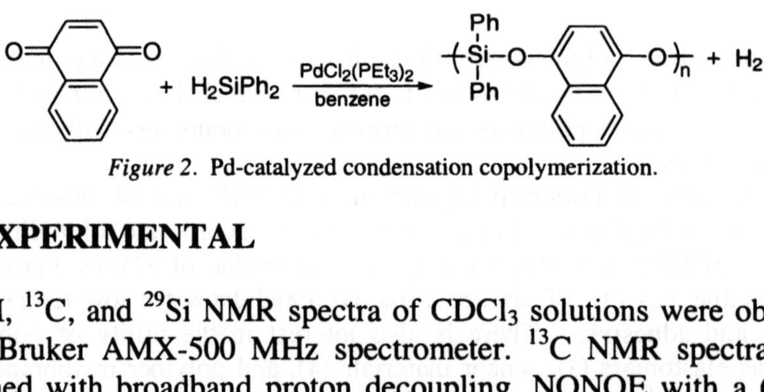


*Figure 2.* Pd-catalyzed condensation copolymerization.

## 2. EXPERIMENTAL

$^{1}H$, $^{13}C$, and $^{29}Si$ NMR spectra of $CDCl_3$ solutions were obtained on a Bruker AMX-500 MHz spectrometer. $^{13}C$ NMR spectra were obtained with broadband proton decoupling. NONOE with a 60 sec delay was used to acquire $^{29}Si$ NMR spectra. Residual $CHCl_3$ was used as an internal standard for $^{1}H$ and $^{13}C$ NMR. $^{29}Si$ NMR spectra were referenced to internal TMS. IR spectra of neat films on NaCl plates were recorded using a Perkin Elmer Spectrum 2000 FT-IR spectrometer. UV spectra of $CHCl_3$ solutions were obtained on a Shimadzu UV-260 spectrometer. Fluorescence spectra of degassed $CHCl_3$ solutions were taken on a PTI fluorimeter.

GPC analysis of the $M_w/M_n$ of the polymers was performed on a Waters system equipped with a 401 RI detector. Two 7.8 mm x 300 mm Styragel columns (HR4 and HR2) in series were used for the analysis. The eluting solvent was toluene at a flow rate of 0.3 mL/min. The retention times were calibrated against known monodisperse PS standards: (929,000; 212,400; 13,700; and 794).

TGA of the polymers was measured on a Shimadzu TGA-50 instrument at a flow rate of 40 cc of $N_2$ per min. The temperature was increased 4 °C/min from 25 to 800 °C. The $T_g$ of the polymers was determined on a Perkin-Elmer DSC-7. The DSC was calibrated from the thermal transition temperature (-87.06 °C) and mp (6.54 °C) of cyclohexane (22). The temperature was increased 10 °C/min from -150 °C to 25 °C.

9,10-Phenanthrenequinone (**I**), 1,2-acenaphthenequinone (**II**), benzil (**III**), 2,3-butanedione (**IV**), toluene, and styrene were obtained from Aldrich. Hexamethylcyclotrisiloxane (**$D_3$**), 1,7–dihydrido-octamethyltetrasiloxane (**VI**), 1,5-dihydridohexamethyltrisiloxane (**VII**), and 1,3-dihydridotetramethyldisiloxane (**VIII**) were purchased from Gelest. **I** and **II** were recrystallized. **IV**, **V**, **VI**, **VII**, and **VIII** were distilled.

All reactions were conducted in flame-dried glassware under argon.

**$RuH_2(CO)(PPh_3)_3$ (Ru)** was prepared from $RuCl_3$ (23) and activated by reaction with styrene in a 20 mL Ace pressure tube at 125 °C for 3 min. The color of the activated catalyst is red (24).

**V** was prepared by the acid-catalyzed reaction of **$D_3$** with **VIII** (16).

***alt*-Copoly(phenanthrene-1,2-dioxy/2,2,4,4,6,6,8,8,10,10-decamethyl-1,11-pentasiloxanylene) (IX).** **I** (1.24 g, 6.0 mmol), **V** (2.13 g, 6.0 mmol), and activated **Ru** catalyst were placed in a 25 mL rb flask, which was heated to 125° C. After 18 h at 125° C, the reaction was stopped. The polymer was dissolved in THF and was precipitated from methanol. In this way, 3.0 g, 88.8 % yield of material, $M_w/M_n$ = 26,100/16,900, $T_g$ = -38 °C was obtained. $^1$H NMR δ: -0.011 (s, 12H), 0.013 (s, 6H), 0.322 (s, 12H), 7.59 (m, 6H), 8.30 (d, 2H, $J$ = 7.5 Hz), 8.64 (d, 2H, $J$ = 7.5 Hz). $^{13}$C NMR δ: 0.09, 0.86, 1.05, 122.24, 123.33, 124.90, 126.20, 127.76, 129.77, 136.68. $^{29}$Si NMR δ: -21.77 (s, 1Si), -21.04 (s, 2Si), -11.22 (s, 2Si). UV $\lambda_{max}$ nm(ε): 309(20,830), 297(19,240), 272(34,650), 258(92,930), 226(24,980). TGA: **IX** is stable in $N_2$ to 200 °C. Between 200 and 650 °C, catastropic decomposition occurs. 90 % of the initial weight is lost. Between 650 and 800 °C, no additional weight is lost. **IX** is stable in air to 150 °C. Between 150 and 550 °C, 69 % of the initial sample weight is lost. To 800 °C, no additional weight is lost.

***alt*-Copoly(acenaphthene-1,2-dioxy/2,2,4,4,6,6,8,8,10,10-decamethyl-1,11-pentasiloxanylene) (X)** was prepared by reaction of **II** (0.49 g, 2.7 mmol) and **V** (0.96 g, 2.7 mmol) as above. After precipitation with methanol, 1.2 g, 82.8 % yield, $M_w/M_n$ = 8600/3800, $T_g$ = -54 °C was obtained. $^1$H NMR δ: 0.11 (s, 18H), 0.320 (s, 12H), 7.41 (d, 2H $J$ = 6.5 Hz),

7.50 (t, 2H, $J$ = 6.5 Hz), 7.61 (d, 2H, $J$ = 6.5 Hz). $^{13}$C NMR δ: -0.14, 1.05, 119.97, 126.02, 127.20, 127.40, 127.55, 129.51, 134.35. $^{29}$Si NMR δ: -21.90 (s, 1Si), -20.60 (s, 2Si), -11.49 (s, 2Si). UV $\lambda_{max}$ nm(ε): 327(17,523), 239(19,958). TGA: **X** is stable in $N_2$ to 200 °C. Between 200 and 225 °C, 15 % of the initial weight is lost. Between 225 and 400 °C, an additional 10 % of the initial weight is lost. Between 400 and 625 °C, an additional 65 % of initial weight is lost. **X** is stable in air to 125 °C. Between 125 and 225 °C, 40 % of the initial sample weight is lost. From 225 to 550 °C, an additional 35 % of the initial weight is lost. To 800 °C, no additional weight is lost.

***alt*-Copoly(phenanthrene-1,2-dioxy/2,2,4,4,6,6,8,8-tetramethyl-1,9-tetrasiloxanylene) (XI)** was prepared by reaction of **I** (1.04 g, 5 mmol) and **VI** (1.41 g, 5 mmol) as above. After precipitation, 2.1 g, 85.6 % yield, $M_w/M_n$ = 12,900/8800, $T_g$ = -35 °C was obtained. $^1$H NMR δ: -0.10 (s, 12H), 0.25 (s, 12H), 7.55 (m, 4H), 8.24 (t, 2H, $J$ = 7 Hz), 8.60 (t, 2H, $J$ = 7 Hz). $^{13}$C NMR δ: 0.05, 0.80, 122.21, 123.27, 124.89, 126.19, 127.68, 129.68, 136.61. $^{29}$Si NMR δ: -21.09 (s, 2Si), -11.22 (s, 2Si). UV $\lambda_{max}$ nm(ε): 310(5473), 258(23,846). TGA: **XI** is stable in $N_2$ to 225 °C. Between 225 and 675 °C, catastrophic decomposition occurs and 89 % of the initial weight is lost. To 800 °C, no additional weight is lost. **XI** is stable in air to 225 °C. Between 225 and 575 °C, 59 % of the initial weight is lost. To 800 °C, no additional weight is lost.

***alt*-Copoly(acenaphthene-1,2-dioxy/2,2,4,4,6,6,8,8-tetramethyl-1,9-tetrasiloxanylene) (XII)** was prepared by reaction of **II** (1.00 g, 5.5 mmol) and **VI** (1.55 g, 5.5 mmol) as above. After precipitation, 2.1 g, 82.3 % yield, $M_w/M_n$ = 20,300/12,600, $T_g$ = -43 °C was obtained. $^1$H NMR δ: 0.08 (s, 12H), 0.30 (s, 12H), 7.38 (t, 2H $J$ = 7.5 Hz), 7.48 (d, 2H, $J$ = 7.5 Hz), 7.59 (d, 2H, $J$ = 7.5 Hz). $^{13}$C NMR δ: -0.14, 1.03, 119.97, 120.98, 126.03, 126.65, 127.20, 134.33, 136.79. $^{29}$Si NMR δ: -20.47 (s, 2Si), -11.42 (s, 2Si). UV $\lambda_{max}$ nm(ε): 315(4142), 241(5705). TGA: **XII** is stable in $N_2$ to 225 °C. Between 225 and 600 °C, catastrophic decomposition occurs and 90 % of the initial sample weight is lost. To 800 °C, an additional 5 % of the initial weight is lost. **XII** is stable in air to 200 °C. Between 200 and 550 °C, 80 % of the initial sample weight is lost. To 800 °C, no additional weight is lost.

**XIII** was prepared by reaction of **III** (1.47 g, 7.0 mmol) and **V** (2.50 g, 7.0 mmol) as above. After precipitation, 3.4 g, 85.6 % yield, $M_w/M_n$ = 22,200/16,600, $T_g$ = -65 °C was obtained. $^1$H NMR: [-0.22, -0.20, -0.18, -0.15, -0.13, -0.11, -0.09, -0.07, -0.04, -0.01, -0.00, 0.04, 0.06, 0.08] (30H), 4.64 (s, 0.35H), 4.82 (s, 0.17H), [7.05, 7.09, 7.10, 7.12, 7.18, 7,19, 7.21, 7.22, 7.24, 7.30, 7.77] (10H). $^{13}$C NMR: -1.31, -0.78, -0.60, -0.36, -0.26, 0.08, 0.67, 0.77, 1.04, 78.98, 126.84, 127.05, 127.18, 127.37, 127.43, 127.55, 127.58, 128.70, 129.35, 129.54, 136.18, 137.98, 141.23, 142.35. $^{29}$Si NMR: [-22.13, -22.10, -22.07, -22.01, -21.98, -21.91, -21.85, -21.76, -21.66, -21.51] (3Si), [-13.21, -13.03, -12.83, -12.41] (2Si). UV $\lambda_{max}$ nm(ε): 301-

(4826), 230(3823). TGA: **XIII** is stable in $N_2$ to 150 °C. Between 150 and 400 °C, 10 % of the initial weight is lost. Between 400 and 625 °C, an additional 80 % of the initial weight is lost. **XIII** is stable in air to 125 °C. Between 125 and 550 °C, 72 % of the initial weight is lost. To 800 °C, no additional weight is lost.

**XIV. IV** (0.60 g, 7.0 mmol), **V** (2.50 g, 7.0 mmol), were reacted as above in an Ace pressure tube. After precipitation, 2.7 g, 87.1 % yield, $M_w/M_n$ = 7300/4900, $T_g$ = -98 °C was obtained. $^1$H NMR: 0.01 (s, 18H), 0.03 (s, 9H), 0.07 (s, 3H), 1.04 (d, 2.3H, $J$ = 4 Hz), 1.10 (d, 2.5H, $J$ = 4 Hz), 1.71 (s, 0.8H), 1.74 (s, 0.4H), 3.61 (m, 0.6H), 3.75 (m, 0.3H). $^{13}$C NMR: -0.55, -0.14, 1.03, 17.68, 17.73, 19.99, 71.45, 72.80, 129.22, 133.52. $^{29}$Si NMR: -22.02 (s, 1Si), -22.01 (s, 1Si), -21.96 (s, 1Si), -14.10 (s, 0.5Si), -14.07 (s, 0.5Si). TGA: **XIV** is stable in $N_2$ to 200 °C. Between 200 and 625 °C, catastrophic decomposition occurs.

**XV** was prepared by reaction of **IV** (0.62 g, 7.2 mmol) and **VI** ( 2.04 g, 7.2 mmol) as above. After precipitation, 2.17 g, 81.6 % yield, $M_w/M_n$ = 4000/3300, $T_g$ = -91 °C was obtained. $^1$H NMR: 0.02 (s, 6H), 0.026 (s, 6H), 0.033 (s, 4.5H), 0.04 (s, 4.5H), 0.08 (s, 3H), 1.05 (d, 2.2H, $J$ = 4.5 Hz), 1.11 (d, 3.1H, $J$ = 4.5 Hz), 1.72 (s, 0.5H), 1.75 (s, 0.2H), 3.62 (m, 0.6H), 3.76 (m, 0.3H). $^{13}$C NMR: -0.61, -0.19, 0.96, 17.57, 17.68, 19.87, 71.39, 72.74, 128.41, 132.02. $^{29}$Si NMR: -22.14 (m, 2Si), -14.23 (m, 2Si). TGA: **XV** is stable in $N_2$ to 175 °C. Between 175 and 625 °C, catastrophic decomposition occurs; 90 % of initial weight is lost.

**XVI** was prepared by reaction of **IV** (0.86 g, 10.0 mmol) and **VII** (2.09 g, 10.0 mmol) as above. In this way, 2.5 g (84.9% yield) of polymer with $M_w/M_n$ = 5300/3700, $T_g$ = –92 °C, was obtained. $^1$H NMR: 0.046 (s, 3H), 0.052 (s, 3H), 0.06 (s, 4.5H), 0.07 (s, 4.5H), 0.11 (s, 3H), 1.08 (d, 2.2H, $J$ = 5 Hz), 1.14 (d, 2.6H, $J$ = 5 Hz), 1.75 (s, 0.8H), 1.78 (s, 0.4H), 3.63 (m, 0.6H), 3.78 (m, 0.3H). $^{13}$C NMR: -0.53, -0.29, -0.11, 1.03, 17.65, 17.73, 19.96, 71.48, 72.84, 128.79, 133.93. $^{29}$Si NMR: -22.00 (m, 1Si), -14.16 (m, 2Si). TGA: **XVI** is stable in $N_2$ to 175 °C. Between 175 and 600 °C, catastrophic decomposition occurs; 90 % of initial weight is lost.

**XVII** was prepared by reaction of **IV** (0.86 g, 10.0 mmol) and **VIII** (1.34 g, 10.0 mmol), as above. In this way, 1.9 g (86.2% yield) of polymer with $M_w/M_n$ = 13,300/6900, $T_g$ = -95° C, was obtained. $^1$H NMR δ: 0.067 (s, 5H), 0.072 (s, 5H), 0.11 (s, 2H), 1.08 (d, 2.4H, $J$ = 5 Hz), 1.13 (d, 3H, $J$ = 3.5 Hz), 1.74 (s, 0.4H), 1.78 (s, 0.2H), 3.64 (m, 0.6H), 3.77 (m, 0.5H). $^{13}$C NMR δ: -0.56, -0.16, 1.02, 17.68, 19.89, 19.97, 71.45, 72.78, 126.55, 135.50. $^{29}$Si NMR δ: -14.10 (s). TGA: **XVII** is stable in $N_2$ to 175 °C. Between 175 and 650 °C, catastrophic decomposition occurs; 90 % of initial weight is lost.

# 3. RESULTS AND DISCUSSION

## 3.1 Results

Herein, we report the activated **Ru**-catalyzed condensation copolymerizations of **I**, **II**, **III**, or **IV** with **V** or **VI**. Condensation copolymerizations of **I** or **II** with **V** or **VI** yield *alt*-copoly(arylene-1,2-dioxy/decamethylmethylpentasiloxanylene)s (figure 3). **III** and **IV** also undergo a **Ru** catalyzed reaction with **V**, but in these reactions, addition (hydrosilylation) polymerization competes with condensation (dehydrogenative silylation) polymerization to yield copoly(silyl ether/silyl enol ether)s. Copolymerizations of **VI**, **VII**, and **VIII** with **IV** also yield copoly(silyl ether/silyl enol ether)s. These copolymers have been characterized spectroscopically by $^{1}$H, $^{13}$C, and $^{29}$Si NMR, as well as by IR, UV, and fluorescence measurements. Their $M_w/M_n$ have been determined by GPC, their thermal stability by TGA, and their $T_g$s measured by DSC.

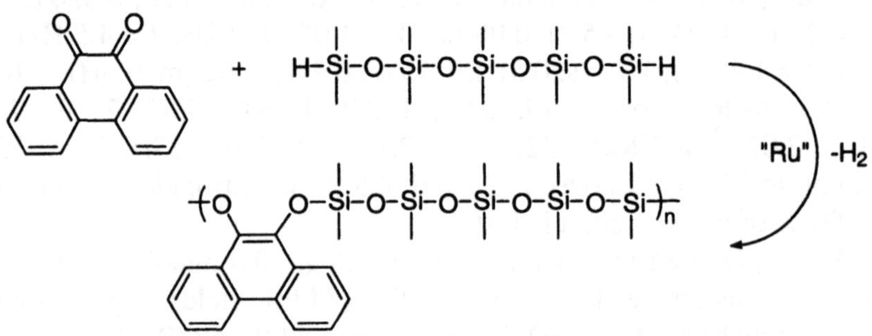


*Figure 3.* Ru-catalyzed condensation copolymerization.

Activated **Ru** catalyzes the condensation copolymerizations of **I** or **II** with **V** to yield *alt*-copoly(arylene-1,2-dioxy/decamethylpentasiloxanylene)s **IX** or **X**, respectively. Similar reaction of **I** or **II** with **VI** yield *alt*-copoly-(arylene-1,2-dioxy/octamethyltetrasiloxanylene)s **XI** or **XII**, respectively. **Ru** also catalyzes the competitive addition/condensation copolymerization of **III** with **V** to yield copoly(silyl ether/silyl enol ether) **XIII** (figure 4). Similar competitive addition/condensation copolymerization of **V**, **VI**, **VII**, or **VIII** with **IV** yields copoly(silyl ether/silyl enol ether)s **XIV**, **XV**, **XVI**, or **XVII**, respectively (figure 5).

**IX**, **X**, **XI**, and **XII** were produced via condensation polymerization with the loss of hydrogen. Addition polymerization competes with condensation in **XIII**, **XIV**, **XV**, **XVI**, and **XVII**. **XIII** is composed of ~75% silyl enol ether, ~25% silyl ether units. **XIV**, which is typical of 2,3-butanedione copolymers, is composed of ~25% silyl enol ether, ~75% silyl ether units.

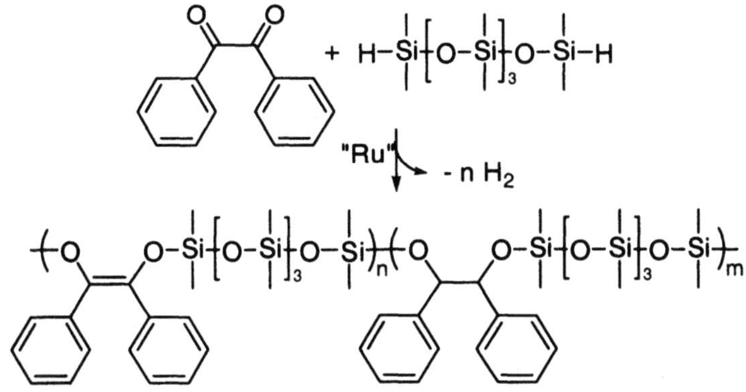


*Figure 4.* Ru-catalyzed addition/condensation copolymerization.

Despite the low molecular weights observed in some cases, no Si-H or carbonyl end groups were detected by IR in any of the polymer products.

Among the advantages of the Ru catalyst is that the reaction does not equilibrate the ODMS units. Equilibration is observed in many siloxane reactions (25). Good yields are obtained. The incorporation of aromatics units into the polymer backbone increases the thermal stability. Several polymers show only a 10% weight loss at 400 °C.

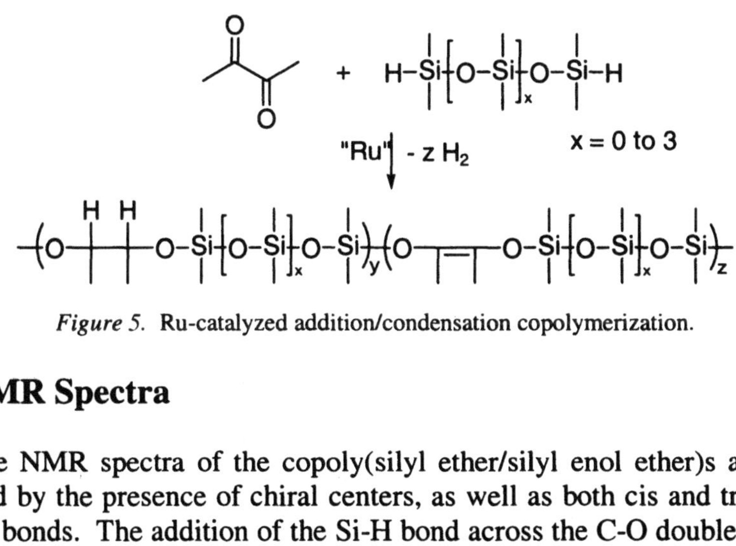


*Figure 5.* Ru-catalyzed addition/condensation copolymerization.

## 3.2 NMR Spectra

The NMR spectra of the copoly(silyl ether/silyl enol ether)s are complicated by the presence of chiral centers, as well as both cis and trans C-C double bonds. The addition of the Si-H bond across the C-O double bond of the ketone results in formation of a single chiral center. Therefore, each saturated polymer unit contains two chiral centers. The stereochemical relationship of these to one another can be RR, SS, or RS (meso). This results in two different stereochemical environments for the methyl groups on the outer silicon atoms. Condensation results in the creation of cis or trans C-C double bonds. Each unsaturated polymer unit contains either a cis or trans environment. The result is four possible chemical environments. The effect of the chiral centers on the Si-$CH_3$ groups in $^1$H NMR spectra has been described (16). The C-C double bonds further complicate the spectra. In fact, every $^{13}$C signal from the aliphatic portion of the copolymer is split into at least two peaks. The outer Si-$CH_3$ signals are also split due to their proximity to the chiral centers. The inner Si-$CH_3$ peaks of **XI** through **XV** are too far removed from the chiral centers for splitting to occur.

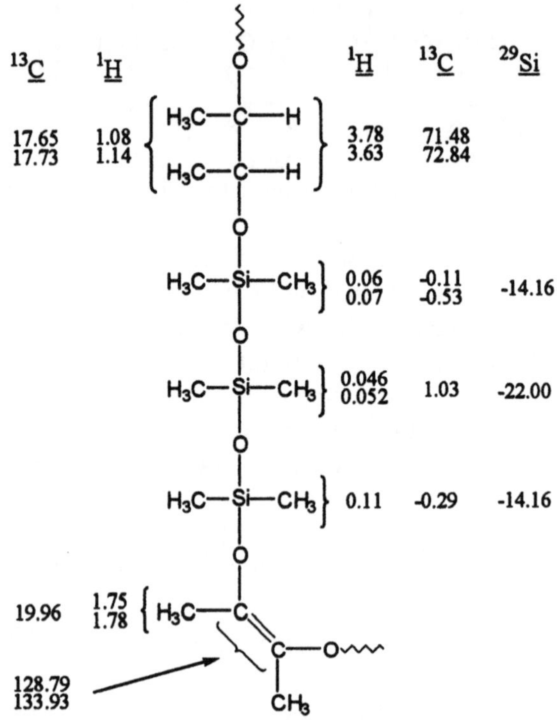

*Figure 6.* NMR chemical shift data for copolymer XV.

**XIII** is most complicated. The $^1$H NMR spectrum shows a large number of peaks in both the Si-$CH_3$ and aromatic regions. This results from the splitting of the thirty Si-methyl protons into the four different diastereotopic environments mentioned above. Similar splitting is seen in the $^{13}$C and $^{29}$Si NMR spectra. Integration is consistent. There are thirty protons in the Si-$CH_3$ region and ten protons in the aromatic region. The integration of the benzyl protons (4.64 and 4.82 ppm) indicates the amount of dehydrogenation that has taken place. If there were complete condensation, the integration would be zero. If there were complete addition, the integration would be 2. The actual integration of the benzyl protons is 0.52 protons, which is consistent with 74% condensation and 26% addition. These chemically equivalent protons are split into two peaks by chiral centers as above. The integration of the saturated methyl groups of **XIV** at 1.04 ppm (2.3H) and 1.10 ppm (2.5H) can be compared to the unsaturated methyl groups at 1.71

ppm (0.8H) and 1.74 ppm (0.4H), indicating a 4 to 1 ratio, or 20% condensation to 80% addition. The Si-$CH_3$ peaks can also be compared with each other. The peak at 0.07 ppm (3H) may result from the condensation of monomer units while the peak at 0.03 ppm (9H) appears to correspond to the outer Si-$CH_3$ groups of addition units. A comparison of the integration gives a 1 to 3 ratio, or 25% condensation to 75% addition. These values are in reasonable agreement with those above. The other copolymers of **IV** show similar values. The chemical shift data is shown for copolymer **XV** (figure 6). Multiple peaks assigned to chemically equivalent atoms result from splitting due to chiral centers. The one exception is the pair of carbon peaks at 128.79 and 133.93 ppm, resulting from cis and trans double bonds. Note that the reactions of the aliphatic diketones were performed in pressure tubes, due to the volatility of **IV** at the reaction temperature of 125 °C. This does not allow the hydrogen gas to escape.

## 3.3 Mechanism

**Ru** reacts with styrene to form the active catalyst (figure 7).

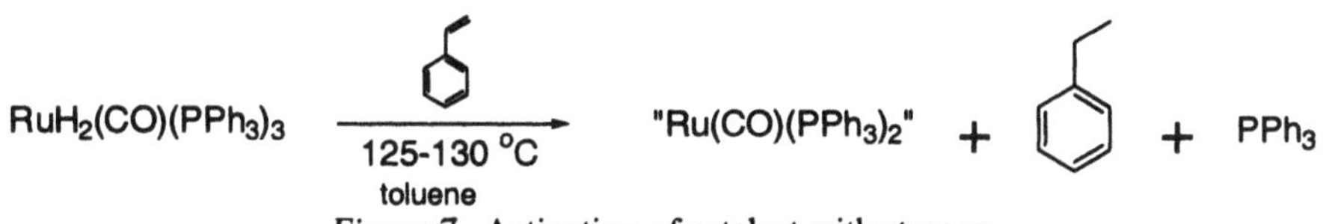


*Figure 7.* Activation of catalyst with styrene.

The initial steps of the mechanism involve a hydrosilylation addition reaction of an Si-H bond across a C-O double bond. An Si-H oxidatively adds to activated **Ru** to yield **1**. Coordination of a ketone yields **2**. 1,2-Insertion of the ketone into the Ru-Si bond yields **3**. Reductive elimination produces an α-keto silyl ether (**4**) and regenerates the catalyst (figure 8).

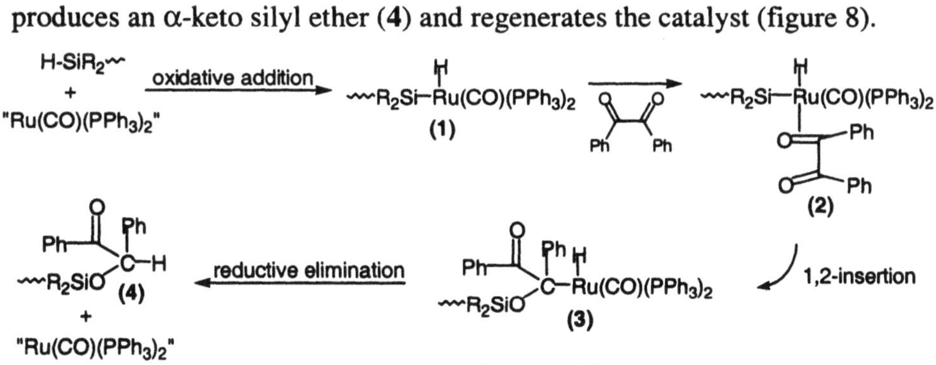


*Figure 8.* Hydrosilylation addition.

The next steps are similar. Oxidative addition of an Si-H produces **1**. Coordination of the second carbonyl group of **4** yields **5**. 1,2-Insertion of the carbonyl of **5** into the Ru-Si bond produces **6,** which has a hydrogen β to the

Ru center (figure 9). Competition occurs between β-hydride elimination, to produce the silyl enol ether, and reductive elimination, which yields the silyl ether (figure 10).

Condensation is favored by the formation of conjugation. **XIII** yields ~75% condensation, whereas **XIV** through **XVII** yield only 15-25% condensation.

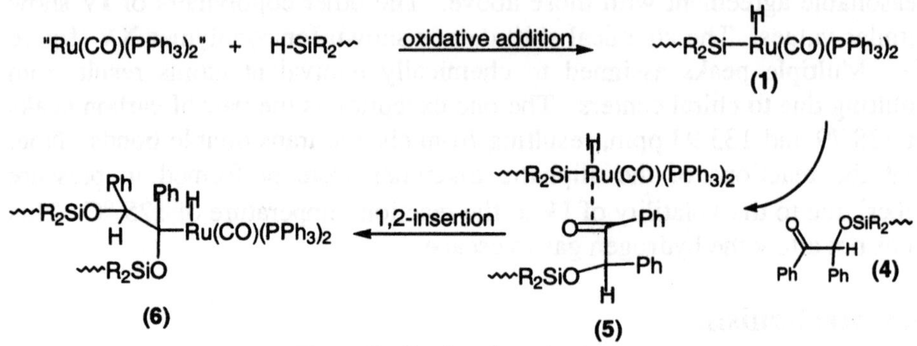


*Figure 9.* Continued mechanism.

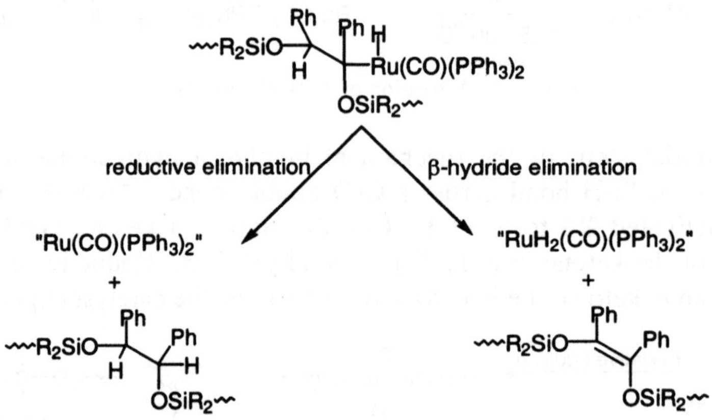


*Figure 10.* Competition between processes.

## 3.4 Luminescence

These reactions allow the incorporation of polycyclic aromatic 1,2-dioxyacenaphthene, 9,10-dioxyphenanthrene, or 1,2-dioxystilbene units into polymer backbones. These units exhibit interesting photochemical properties. Shown are the UV, excitation, and fluorescence spectra of **IX** (figure 11) and **X** (figure 12). **IX** exhibits a large Stokes shift while **X** does not. The reason for this difference is not completely understood at this time and is currently under investigation.

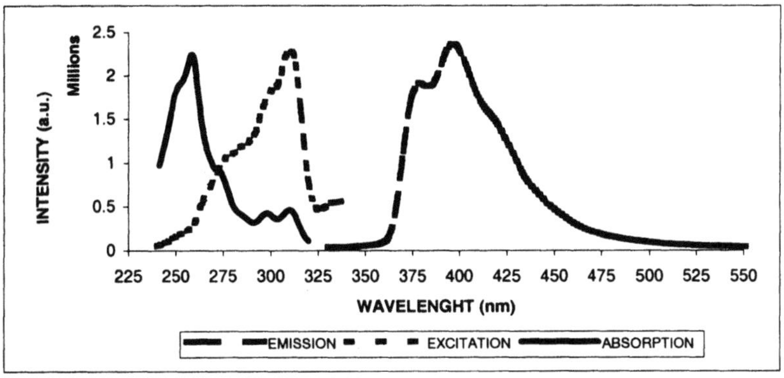


*Figure 11.* Luminescence spectrum of IX.

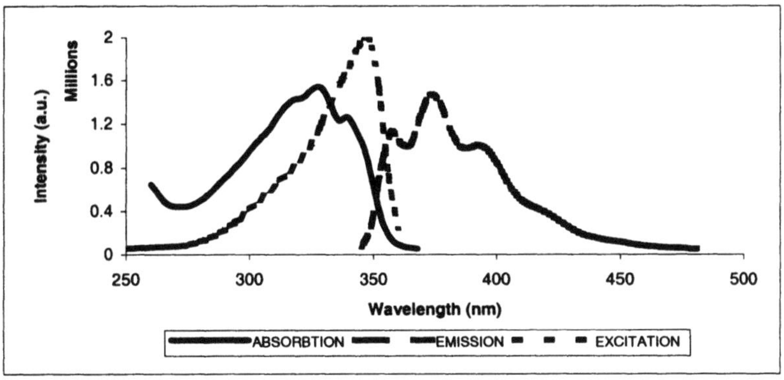


*Figure 12.* Luminescence spectrum of X.

## 4. REFERENCES


1. Odian, G., **Principles of Polymerization**, John Wiley and Sons, 1991.

2. Clarson, S. J.; Semlyen, J. A., **Siloxane Polymers**, Prentice Hall, 1993.

3. Curry, J. E.; Byrd, J. D., Macromolecules, 1, 249 (1968).

4. Kaganove, S. N.; Grate, J. W., Polym. Prepr., 39(1), 556 (1998).

5. Stern, S. A.; Shan, V. M.; Hardy, B. J., J. Polym. Sci., Part B: Polym. Phys., 25, 1263 (1987).

6. Bailey, D. L.; O'Connor, F. M., Brit. Pat. 880022, 5-22-58.

7. Koepnick, H.; Delfs, D.; Simmler, W. German Pat. 1108917, 6-15-61.

8. Bailey, D.L.; O'Connor, F. M., German, Pat. 1012602, 7-25-57.

9. **Transition Metal Catalysis in Macromolecular Design**, (L. S. Boffa, B. M. Novak, Eds.), ACS Symposium Series 760, American Chemical Society, 2000.

10. **Transition Metal Catalyzed Polymerization in Step-Growth Polymers for High-Performance Materials**, (J. L. Hedrick, J. W. Labadie, Eds.), ACS Symposium Series, 624, American Chemical Society, 1996.

11. Li, Y.; Kawakami, Y. Macromolecules, 32, 8768 (1999).

12. Li, Y.; Kawakami, Y. Macromolecules, 32, 6871 (1999).

13. Li, Y.; Kawakami, Y. Polym. Prepr., 41(1), 534 (2000).

14. Li, Y.; Seino, M.; Kawakami, Y. Macromolecules, 33, 5311 (2000).

15. Paulasaari, J. K.; Weber, W. P. Macromolecules, 31, 7105 (1998).

16. Mabry, J. M.; Paulasaari, J. K.; Weber, W. P. Polymer, 41, 4423 (2000).

17. Reddy, P. N.; Chauhan, B. P. S.; Hayashi, T.; Tanaka, M. Chem Lett., 250 (2000).

18. Mizukami, S.; Kihara, N; Endo, T. J. Am. Chem. Soc., 116, 6453 (1994).

19. Penelle, J.; Mayné, V.; Touillaux, R. J. Polym. Sci., Part A: Polym. Chem., 34, 3369 (1996).

20. Mayné, V.; Penelle, J. Macromol. Chem. Phys., 199, 2173 (1998).

21. Penelle, J.; Mayné, V. Tetrahedron, 53, 15429 (1997).

22. Aston, J. G.; Szabz, G. J.; Fink, H. L. J. Am. Chem. Soc., 65, 1135 (1943).

23. Levison, J. J.; Robinson, S. D. J. Chem. Soc. A, 2947 (1970).

24. Guo, H.; Wang, G,; Tapsak, M. A.; Weber, W. P. Macromolecules, 28, 5686 (1995).

25. Noll, W. **The Chemistry and Technology of Silicones**, Academic Press, 1968.

# Chapter 21

# GEL-DRAWN POLY(p-PHENYLENEPYROMELLITIMIDE)

Jiro Sadanobu and Rei Nishio
*Polymer Research Institute, Teijin Ltd., Iwakuni, Yamaguchi 740-8511, Japan*

## 1. INTRODUCTION

### 1.1 History

Poly(p-phenylenepyromellitimide) (PPPI), is the simplest form of aromatic polyimide with completely rigid backbone structure (Figure 1). The fully-extended crystal structure was characterized by Russian researcher in as early as 1970's (1). Later theoretical young's modulus was evaluated as 505 Gpa. (2), that is the highest value for realistic polymer ever reported. From this potential rigidity there has been enormous efforts on fabricating PPPI into fibre and film with ultra-high modulus. However, all trials ended up identifying PPPI as too brittle a material to apply any practical usages.

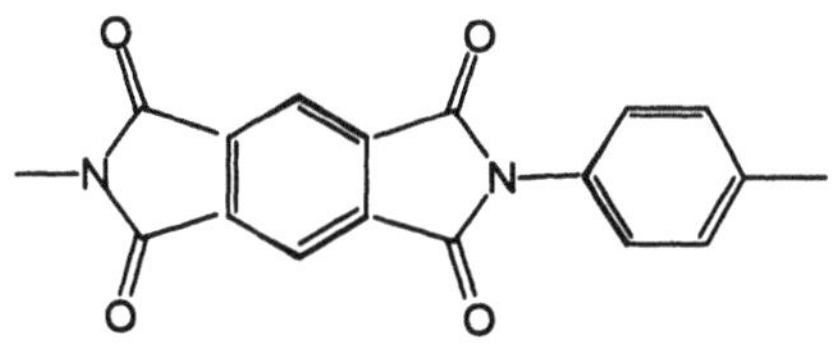


*Figure 1.*Chemical Structure of PPPI

To remedy the situation, the techniques of precursor orientation have been investigated. Kakimoto et al used polyamic acid long alkyl esters as precursors and fabricated fibres by wet spinning (3); Yokota et al devised the drawing method for polyamic acid film in swollen state and obtained uniaxially drawn films (4). Both methods improved Young's modulus, however, did not overcome the brittleness of the resultant fibres nor films.

### 1.2 New procedure

One of the origins of brittleness of PPPI is in hydrolytic instability of precursor polyamic acid that is easily degraded by ambient moisture and water generated on imidization. Another is in a coarse grained structure developed by intrinsic high-crystallinity of PPPI, that may induce stress concentration on applied force.

To break through these problems we propose new concept of fabrication technique: the gel drawing of precursor polyisoimide. The flow chart of the procedure is shown in Figure 2.

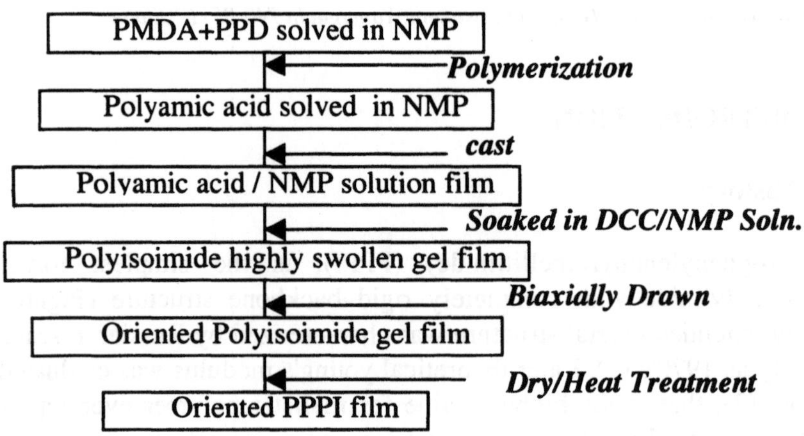


*Figure 2.* Flow Chart of the Novel Gel-drawing Process for PPPI Film

Polyisoimide is an isomeric of polyimide and is thermally converted to polyimide without any generation of water. The choice of polyisoimide is based on stability to moisture and low crystallinity that are well feasible to precede gel-drawing. Since precursor polyisoimide for PPPI is insoluble in any organic solvent, polyamic acid dope is used for film casting and converted to polyisoimide gel film by soaking into the solution of N,N'-dicyclohexylcarbodiimide (DCC) in N-methyl-2-pyrrolidone (NMP). The reaction upon isoimidization is illustrated in Figure 3. Polyisoimide gel film is subjected to biaxially stretching with the highly swollen state kept in. Finally, the heat-treatment at above 350°C provides with the highly oriented PPPI film.

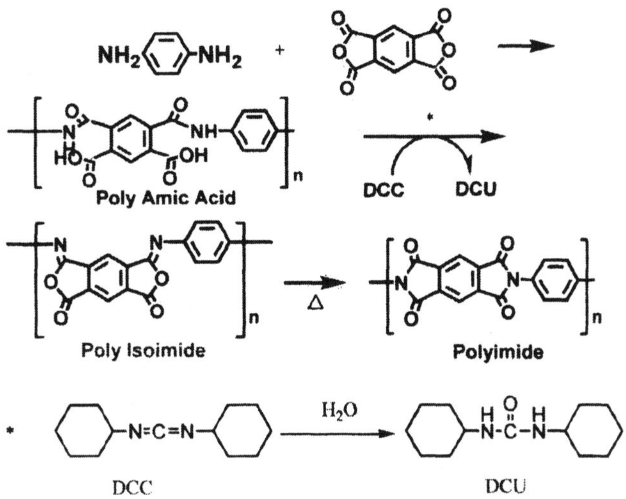


*Figure 3.* Reaction Scheme from Polyamic acid to Polyimide via Polyisoimide

## 2. EXPERIMENTAL

### 2.1 Preparation of Polyamic acid Solution

Pyromellitic anhydride (40.1g) was added to a stirred solution of p-phenylenediamin (19.9g) in NMP (910ml) under N2 at 0°C. After solution was stirred for three hours, further reaction was continued for two hours at room temperature. The resultant solution of polyamic acid was directly used as the dope for film casting. The inherent viscosity of the polyamic acid was 4.12dl/g (0.5g/dl in NMP at 30 °C).

### 2.2 Film Fabrication

The polyamic acid dope was cast on glass substrate using a knife blade, followed by immersing into a gelation bath containing 15wt% solution of DCC in NMP at room temperature. The solution cast film immediately turned reddish and formed highly swollen gel. After soaking in the gelation bath for 8 min., the gel film was stripped from glass substrate and washed by NMP to remove unreacted DCC and generated dicyclohexylurea. The gel film was simultaneously drawn in two perpendicular directions to draw ratio up to 2.3x2.3 in air at room temperature by using a stretching apparatus made by Iwamoto Seisakusho Ltd. In this paper biaxial draw ratio•of •n•x•n• is referred to as $\lambda$ =n. The drawn film was washed by isopropanol to remove NMP prior to drying in air.

The biaxially drawn films were fixed with a metal frame and treated in the air oven at under the prescribed heating condition of 200°C /10 min., 250°C /8 min, 350°C /5 min and over 350 to 450°C /5 min. The thickness of the obtained polyimide film was typically 8μm.

### 2.3 Characterization of Polymer

Tensile properties were determined from 5mm wide strip using Tensilon UTC-1T made by Orientec Corp. An initial gauge length of 25mm and a testing speed of 25mm/min were employed.

Wide angle X-ray diffraction (WAXD) was obtained using Rigaku RAD-B System with Ni-filtered Cu-K radiation, employing both transmission and reflection geometry.

Density was measured by the sink-and-float method using cyclohexane and dichloroethane.

Morphology of the film cross section was studied by mean of scanning electron micrography (SEM) using FE-SEM S-900 made by Hitachi Ltd.

Refractive index was evaluated by using Microwave Molecular Orientation Analyser made by Oji Scientific Instrument.

IR spectra were obtained using Nicolet Maguna-750.

Thermogravimetry was measured by Rigaku 8120.

Viscolasticity was evaluated by RCA-II Solid Analyser.

## 3. RESULTS AND DISCUSSION

### 3.1 Effects of Gel-drawing

Gelation of polyamic acid dope film was induced by the dehydration reaction with DCC upon soaking in DCC-NMP solution as illustrated in Figure 3. The reddish colour of resultant gel film indicates the conversion of polyamic acid to polyisoimide. Figure 4(a) shows FT-IR spectra of gel film, in which isoimidization was evident by the appearance of the bands at 910cm$^{-1}$ and 1803cm$^{-1}$ with only a trace of imide-originated bands.

After gelation was completed, the isoimidized gel film was highly swollen by solvent. The content of solvent in the gel was typically 85wt%. No substantial dissolution of solid content into the DCC solution was observed throughout gelation process. The advantage of using DCC-NMP as a gelation reagent lies in the fact that DCC reacts with polyamic acid in the neutral condition preventing from hydrolysis and that the affinity of the resultant polyisoimide to NMP preserves the swollen state of gel film during the insolubilized process with minimum dimensional contraction.

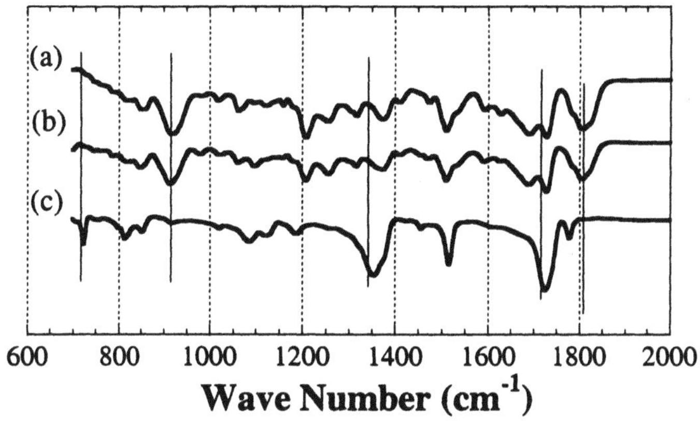


*Figure 4.* FT-IR Spectra of (a) Undrawn Gel Film, (b) Drawn Gel Film ($\lambda$=2.3) and Drawn Dry Film ($\lambda$=2.3) after Heat Treatment at 350°C.

The isoimidized gel film was highly ductile even at room temperature (RT). Maximum draw ratio for biaxial drawing was 2.5 at RT. The degree of swelling was kept almost constant throughout drawing. Upon gel-drawing no significant conversion from isoimide to imide was observed as indicated in Figure 4(b). Figure 5 depicts WAXD through-view pattern for drawn-gel film after extraction of NMP. Most part of drawn gel film consists of amorphous and only a small crystalline portion was identified as polyimide in comparison with the pattern of heat-treated film.

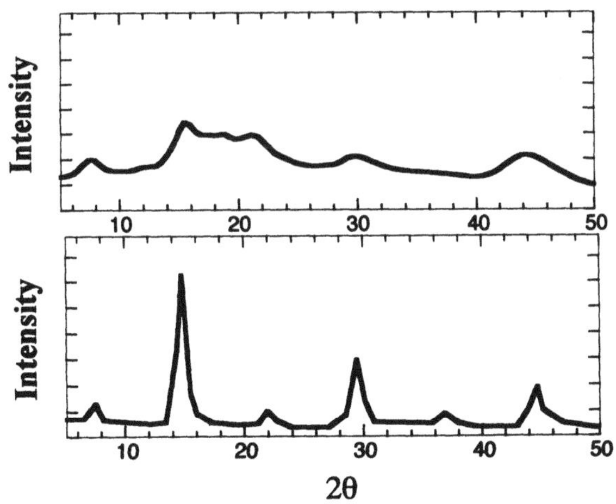


*Figure 5.* WAXD Through View Patterns of As-drawn Gel Film ($\lambda$=2.3), (top) and Imidized Film after Heat Treatment at 350°C (bottom).

To avoid orientation relaxation at elevated temperature, extraction of NMP from drawn gel film was required by using non-solvent, for example isopropanol.

The drawn polyisoimide film was imidized by heat treatment in air. Conversion from polyisoimide to polyimide was completed at around 350°C. Figure 6 show the dependence of in-plane microwave refractive index of heat-treated film upon gel-draw ratio. The systematic increase in refractive index with increasing draw ratio indicates the development of planar orientation induced by gel-drawing.

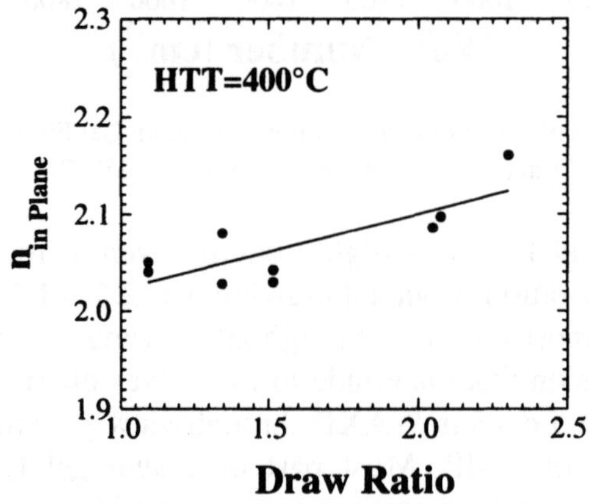


*Figure 6.* Dependence of Microwave In-plane Refractive Index upon Gel-draw Ratio.

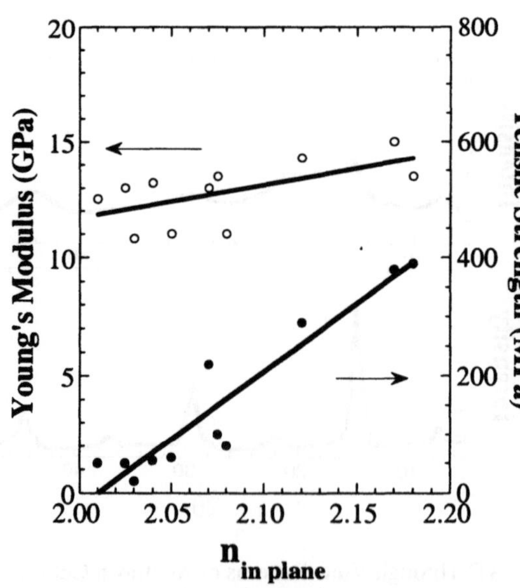


*Figure 7.* Dependence of Young's Modulus and Tensile Strength on Orientation Evaluated by In-plane Microwave Refractive Index

The improvement of mechanical properties for heat-treated PPPI film by orientation is illustrated in Figure 7. We see the Young's modulus increases linearly with increasing orientation. More dramatically, the tensile strength makes remarkable progress by evolution of orientation.

### 3.2 Microstructure developed in imidized film

Figure 8 shows WAX patterns for gel-drawn PPPI film after heat treatment at 450°C.

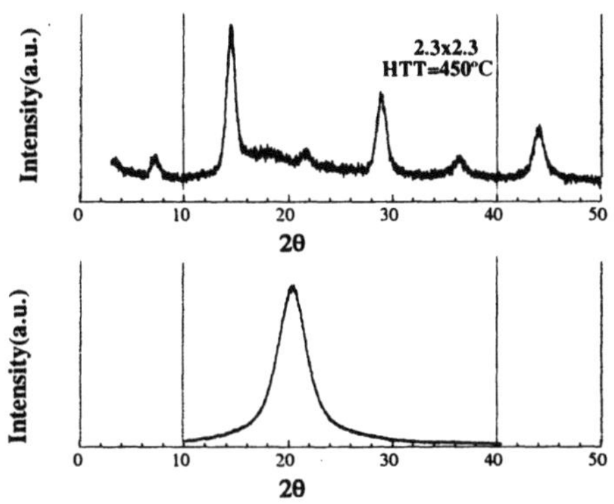


*Figure 8.* WAX patterns for Gel-drawn PPPI Film in Transmission and Reflection Geometry

We see higher order peaks of (00l) type diffraction in transmission and relatively broad single peak in reflection, which suggests high crystallinity and crystal planar-orientation on imidization. The crystalline size was evaluated from peak width in reflection pattern by using Sherrer's equation. The dependence of crystalline size on gel-draw ratio is described in Figure 9. We observe increasing draw ration suppress the growth of crystalline stack.

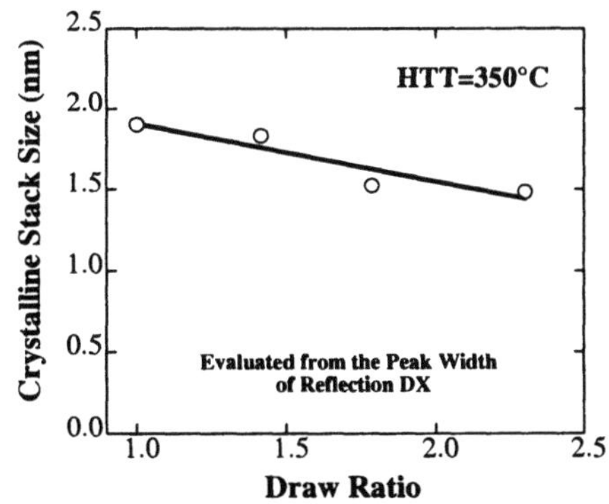


*Figure 9.* Depencence of Crystalline Stack Size upon Gel-draw Ratio

Figure 10 compares SEM Images of the film cross-section for undrawn and drawn ($\lambda$ =2.3) polyimide film. The granular fracture surface was found in the cross section in the undrawn film. On the other hand, the elongated lamella structure that aligned parallel to the film surface was observed for the drawn film. In closer view we can estimate the thickness of thin lamella as 10nm or less.

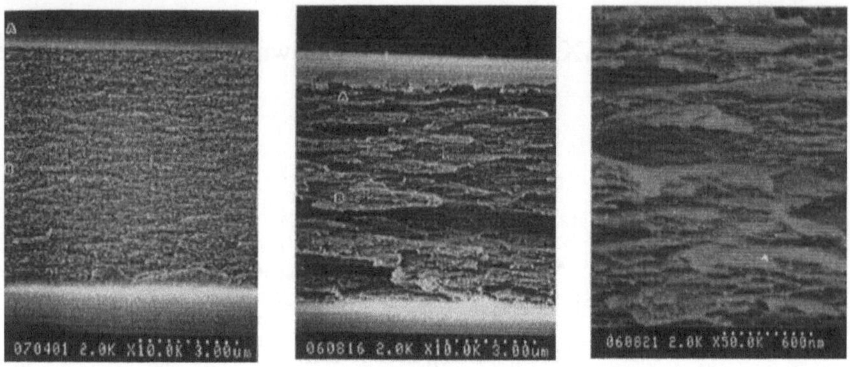


*Figure 10.* SEM Image for Undrawn PPPI Film (Left), Gel-drawn PPPI Film (Center) and Closer View of Gel-drawn PPPI Film (Right)

These observations indicate that the organization of microstructure and morphology is highly affected by gel-drawing. We assume that highly improved tenacity of gel-drawn PPPI film should be attributed to restrained growth of crystalline and fining of microtexture, that result in releasing internal stress and suppressing stress concentration by applied force.

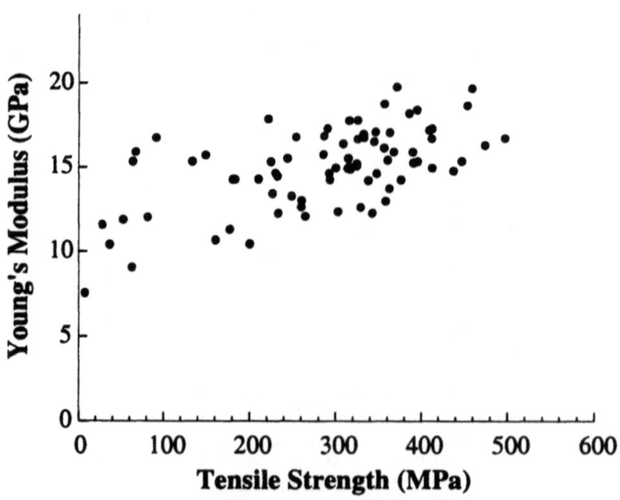


*Figure 11.* Correlation between Young's Modulus and Tensile Strength for Gel-drawn PPPI Film Prepared in Various Conditions

## 4. Properties of PPPI film

Figure 11 illustrates the correlation between Young's modulus and tenacity for gel-drawn PPPI film prepared by various conditions of draw ratio and heat treatment. At optimized condition the biaxially balanced PPPI film with 20 GPa of Young's Modulus and 500 MPa of tenacity was realized.

Figure 12 compares thermogravimetry of gel-drawn PPPI film with Kapton. Gel-drawn PPPI has more improved thermal stability than Kapton both in air and in nitrogen.

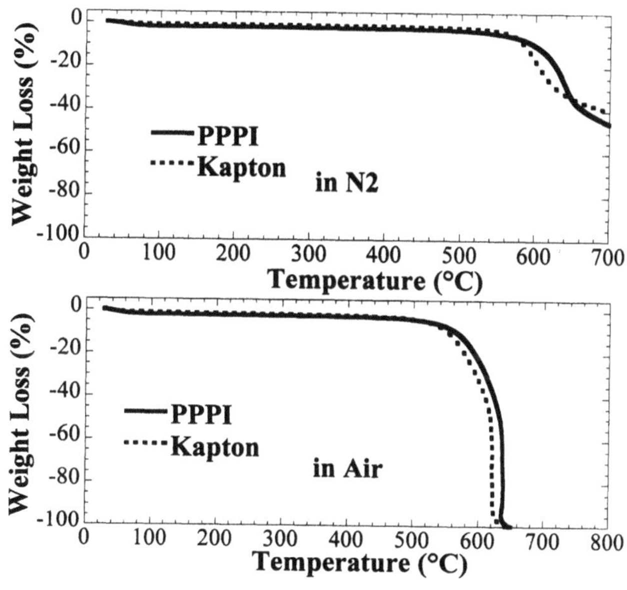


*Figure 12.* Thermogravimetry for Gel-drawn PPPI Film and Kapton in Nitrogen (top) and in Air (bottom)

Figure 13 also compares viscoelasticity of gel-drawn PPPI film with Kapton. Gel-drawn PPPI film indicated no relaxation up to 500°C in contrast to Kapton, which shows significant increase in tan $\delta$ at around 400°C.

Finally we draw a map in the reference frame of Young's modulus and water absorption for commercial heat-resistant films (Figure 14). We clearly understand that gel-drawing PPPI film located unique position with lower water absorption and higher Young's modulus in this map. From this special characteristic features the future applications are prospective as substrate for next generation microelectronics, base film for high-density magnetic recording media and high-performance electric insulator.

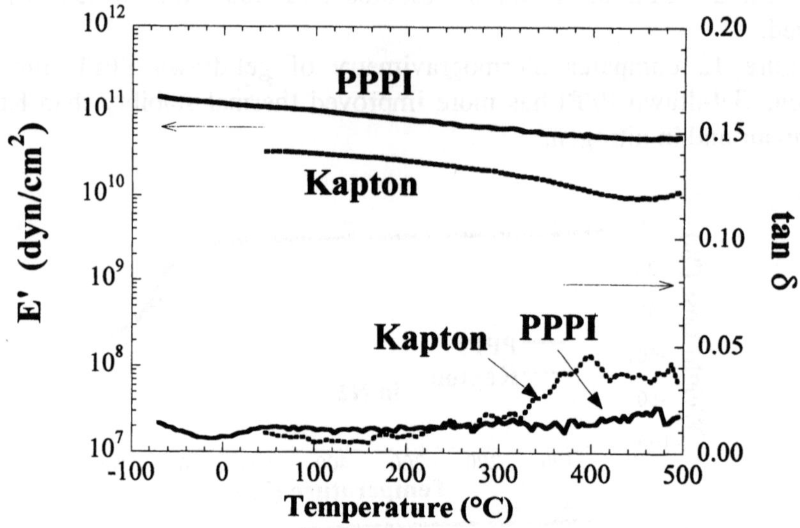


*Figure 13.* Storage Modulus, E' and Mechanical Loss, tan δ for Gel-drawn PPPI Film and Kapton

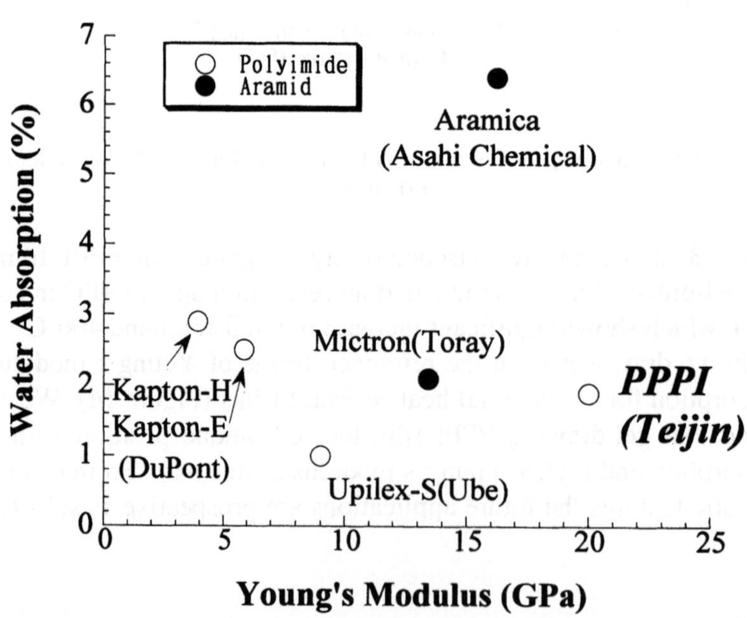


*Figure 14.* Mapping for Commercial Heat-resistant Films Based on Water Absorption and Young s Modulus

## 5. CONCLUSIONS

We have prepared high performance biaxially oriented PPPI film with high Young's modulus. The film was converted from polyisoimde precursor oriented by gel-drawing. The gel-drawing process controls microstructure and morphology and results in break-through in overcoming intrinsic brittleness of long discarded material: PPPI.

PPPI film with high Young's modulus and excellent thermal stability is prospective material for next generation electronics use.

## 6. REFERENCES

1. Korzhavin, L. N.; Prokopchuk, N. R. ; Baklagina, Yu. G.; Florinskii, F. S.; Yefanova, N. V. ; Dubnova, A.M.; Frenkel, S. Ya. and Koton, M. M., Visocomol. Soyed., 1976, A18(6), 1235

2. Tashiro,K.; Kobayashi, M., Sen-i Gakkaishi 1987, 43, 78

3. Kakimoto, M.; Orikabe, H.; Imai, Y. Polymer Preprints 1993, 34 (1), 746

4. Masuda, A.; Kotobuki, S.; Nakamura, S.; Oshida, S.; Kochi, M.; Yokota, R. Kobunshi Ronbunshu 1999, 56 (5), 282.

# Index

MIX
Papier aus verantwortungsvollen Quellen
Paper from responsible sources
FSC® C105338

If you have any concerns about our products,
you can contact us on
**ProductSafety@springernature.com**

In case Publisher is established outside the EU,
the EU authorized representative is:
**Springer Nature Customer Service Center GmbH**
**Europaplatz 3, 69115 Heidelberg, Germany**

Printed by Libri Plureos GmbH
in Hamburg, Germany